T0227509

Financial Mathematics

This book is a study of the mathematical ideas and techniques that are important to the two main arms of the area of financial mathematics: portfolio optimization and derivative valuation. The text is authored for courses taken by advanced undergraduates, MBA, or other students in quantitative finance programs.

The approach will be mathematically correct but informal, sometimes omitting proofs of the more difficult results and stressing practical results and interpretation. The text will not be dependent on any particular technology, but it will be laced with examples requiring the numerical and graphical power of the machine.

The text illustrates simulation techniques to stand in for analytical techniques when the latter are impractical. There will be an electronic version of the text that integrates Mathematica functionality into the development, making full use of the computational and simulation tools that this program provides. Prerequisites are good courses in mathematical probability, acquaintance with statistical estimation, and a grounding in matrix algebra.

The highlights of the text are:

- A thorough presentation of the problem of portfolio optimization, leading in a natural way to the Capital Market Theory
- Dynamic programming and the optimal portfolio selection-consumption problem through time
- An intuitive approach to Brownian motion and stochastic integral models for continuous time problems
- The Black-Scholes equation for simple European option values, derived in several different ways
- A chapter on several types of exotic options
- Material on the management of risk in several contexts

Chapman & Hall/CRC Financial Mathematics Series

For more information about this series please visit: https://www.crcpress.com/Chapman-and-HallCRC-Financial-Mathematics-Series/book series/CHFINANCMTH

Financial Mathematics
From Discrete to
Continuous Time

Kevin J. Hastings

CRC Press
Taylor & Francis Group
Boca Raton London New York

CRC Press is an imprint of the
Taylor & Francis Group, an **informa** business

A CHAPMAN & HALL BOOK

First edition published 2023
by CRC Press
6000 Broken Sound Parkway NW, Suite 300, Boca Raton, FL 33487-2742

and by CRC Press
4 Park Square, Milton Park, Abingdon, Oxon, OX14 4RN

CRC Press is an imprint of Taylor & Francis Group, LLC

Library of Congress Cataloging-in-Publication Data

Names: Hastings, Kevin J., 1955- author.
Title: Financial mathematics : from discrete to continuous time / Kevin J. Hastings.
Description: First edition. | Boca Raton : CRC Press, 2023. | Includes
bibliographical references and index.
Identifiers: LCCN 2022030740 (print) | LCCN 2022030741 (ebook) | ISBN
9781498780407 (hardback) | ISBN 9781032403892 (paperback) | ISBN
9780429113659 (ebook)
Subjects: LCSH: Business mathematics.
Classification: LCC HF5691 .H348 2023 (print) | LCC HF5691 (ebook) | DDC
332/.0151--dc23/eng/20220720
LC record available at https://lccn.loc.gov/2022030740
LC ebook record available at https://lccn.loc.gov/2022030741

ISBN: 978-1-498-78040-7 (hbk)
ISBN: 978-1-032-40389-2 (pbk)
ISBN: 978-0-429-11365-9 (ebk)

DOI: 10.1201/9780429113659

Typeset in Latin Modern
by KnowledgeWorks Global Ltd.

Publisher's note: This book has been prepared from camera-ready copy provided by the authors.

To my students, for their unfailing inspiration

Contents

Preface

This book is envisioned as a sequel to my book: *Introduction to Financial Mathematics*, although it can be used independently. Appropriate for advanced undergraduates, MBA, or other students in quantitative finance programs, it is a study of the mathematical ideas and techniques that are important to the two main arms of the area of Financial Mathematics: portfolio optimization and derivative valuation. The approach is mathematically correct but informal, balancing theory and application, sometimes omitting proofs of the more difficult results and stressing interpretation. The text will not be dependent on any particular technology, but it will be laced with examples requiring the numerical and graphical power of the machine. We will also be illustrating simulation techniques to stand in for analytical techniques when the latter are impractical. As with the prequel, there will be an electronic version of the text that integrates *Mathematica* functionality into the development, making full use of the computational and simulation tools that this program provides. Prerequisites are good courses in mathematical probability, acquaintance with statistical estimation, and a grounding in calculus and matrix algebra.

The highlights of the text are: (1) thorough presentation of the problem of portfolio optimization, leading in a natural way to the Capital Market Theory; (2) dynamic programming and the optimal portfolio selection-consumption problem through time; (3) an intuitive approach to Brownian motion and stochastic integral models for continuous-time problems; (4) the Black-Scholes equation for simple European option values, derived in several different ways; (5) material on several types of exotic options. The organization leads the reader from problems in a single time period, through those in multiple discrete time periods, and finally into continuous-time analogs.

Chapter 1 sets the stage for what will be done later. So that this text can stand on its own, independently of the prequel, the key concepts of asset price models, risk aversion, portfolios and their optimization, arbitrage, derivatives of underlying assets, and single-period valuation of derivatives are reviewed in this chapter.

Then, in Chapter 2, Lagrange multiplier and matrix methods are used to determine the optimal balance of risky assets in general portfolios, leading to the famous separation theorem and the Capital Asset Pricing Model (CAPM). Optimizing by use of utility functions satisfying basic axioms is discussed next. At this point we live in a world of "now" vs. "later", where we invest, a single time period passes, and we reap the benefits of the investment at the end of that period. But we go on to solve problems in which assets change value at discrete time instants and the investor has the choice of how much wealth, if any, to consume from the system and how to reinvest the remaining wealth.

In Chapter 3 we extend the material in the first chapter about derivatives valuation to multiple periods. In order to facilitate the passage to the limit and preview the results in continuous time, and also to motivate the classical Fundamental Theorems of Asset Pricing, we pause and introduce basic concepts from probability involving random variables, measurability, conditional expectation, and martingales in the discrete-time setting. With the Fundamental Theorems in hand, we are able to extend the martingale valuation approach to several kinds of exotic options, including barrier and Asian options. The chapter ends with a study of how simulation techniques may be used to value options in complicated situations.

The purpose of the first portion of Chapter 4 is to review key facts from continuous probability that will be used in the rest of the book, and to introduce measure-theoretical ideas such as measurability and conditional expectation in continuous time. This allows for better understanding of the Brownian motion process and its relatives, used as price models for underlying assets. The chapter ends with a friendly introduction to stochastic differential equations, sufficient to set up the next chapter.

Chapter 5 begins with a statement of the classical Black-Scholes solution of the European call option valuation problem in continuous-time. Three approaches to deriving the solution are taken as we move through the chapter: first, the solution viewed as the limit of the Cox-Ross-Rubinstein discrete solution as both the time period and step size become shorter; second, by the martingale approach, justified by the continuous-time version of the Fundamental Theorems of Asset Pricing; and third, by the differential equation argument originally used by Black and Scholes. The chapter and the book end with an examination of the stochastic assumptions in the Black-Scholes model, showing how to use real data to cast some doubt on those assumptions.

The experience of using this book will be greatly enhanced by obtaining the electronic version from the publisher in the form of *Mathematica* notebooks, one per chapter. These allow for deeper explorations and make it easy to obtain the graphs and computational results that are in the print text. Occasional allusions are made to the *Mathematica* version in the print text that preview what is available. But again, it is not necessary to have *Mathematica* or be particularly proficient with this program in order to understand what is in this book.

At Knox College we have a two-course sequence in Financial Mathematics, the first geared toward a sophomore undergraduate level student for which I have used the earlier text successfully for years. The advanced course is meant for junior-senior level undergraduates, and could conceivably fit quantitatively oriented finance graduate students. I have used the material in the incomplete manuscript several times in this advanced course, and students have reacted very well to it, occasionally being very challenged, but invariably successful.

The fact that my students have been so involved with this project encouraged me to dedicate the book to them, but I cannot in good conscience end without mentioning the consistent support from Taylor & Francis/CRC Press during this and earlier projects. And finally, to my wife Gay Lynn who has had to exercise utmost patience for many years as I wrote, and wrote, and wrote some more. Another one is done, honey. Maybe now we can both rest a little easier.

Kevin J. Hastings
June 13, 2022

Author

Kevin J. Hastings is a professor of mathematics and holds the Rothwell C. Stephens Distinguished Service Chair at Knox College. He has a PhD from Northwestern University. His interests include applications to real-world problems affected by random inputs or disturbances. He is the author of three other books for CRC Press:

Introduction to Financial Mathematics, CRC Press, 2016. CHOICE Highly Recommended Selection and 2017 Top Books for Colleges.

Introduction to Probability with Mathematica, 2nd ed., Chapman & Hall/CRC Press, 2009.

Introduction to the Mathematics of Operations Research with Mathematica, 2nd edition, Taylor & Francis/Marcel Dekker, 2006.

Introduction to Probability with Mathematica. CRC Press/Chapman & Hall, 2000. Also available as an e-book.

1

Review of Preliminaries

This book focuses on two branches of the mathematics of finance: optimization of portfolios and valuation of derivative assets, which have been cornerstones of the field since the early 1960's. Our approach will be to build from results in the single time period case, through the multiple but discrete time period case, and from there to financial problems in continuous time. Basic calculus in one and several variables, some matrix algebra, and calculus-based probability theory will be required. We will review certain key aspects of this mathematical background in context as needed. The level of discourse that we will attempt to maintain can be described as intuitive, yet correct, without being overly formal or rigorous.

There are several good sources that have served me well in bringing together the most interesting material relevant to financial mathematics, including Sharpe [21], Baxter and Rennie [1], Stampfli and Goodman [22], and Roman [17]. My earlier, more elementary book (Hastings [11]) is a good prequel to the current text. In fact, this book is designed to be a continuation of the first book, which will include a recapitulation of the important ideas and results so that this book can be read independently of the prequel.

The organization of Chapter 1 is as follows. In the first section of the chapter, we focus on stochastic process models for the behavior of individual risky assets, in order to set down basic language and assumptions. Then, in Section 1.2, we set up the idea of a portfolio of risky assets and a measure of the degree of tolerance of investors toward risk. We review a few key results about means, variances and covariances in Section 1.3. With these in hand we can proceed in Section 1.4 to basic portfolio optimization using only elementary calculus in the single time period case. The study will be continued in Chapter 2 using more sophisticated tools from multivariate calculus and linear algebra. The rest of Chapter 1 sets up the problem of valuation of assets that are contingent on the values of other basic underlying assets, also in a single time period. Section 1.5 introduces the crucial idea of arbitrage, on which valuation is based. We also look at a few examples of such derivative assets and some of their basic properties. Finally, the problem of valuation of single-period derivatives in solved in Section 1.6.

1.1   Risky Assets

It is not a deep insight that assets in financial markets are risky precisely because their values cannot be predicted ahead of time. Thus, the probabilistic notions of random variable and stochastic process are obvious choices to use as models for the values of risky assets, or the rates of return on those assets.

Definition 1. A **random variable** X is a numerical valued function on the space of possible results of a random phenomenon. A **discrete-time stochastic process** $X_0, X_1, X_2, \ldots$ is a sequence of random variables. A **continuous-time stochastic process** is a family $(X_t)_{t \in I}$ of random variables indexed by a time set I of real numbers (typically an interval $[0, T]$ or the set of non-negative real numbers). ∎

Thinking in the context of financial modeling, the value of the random variable X_t could be the price of a risky asset at time t. We will also use random variables and stochastic processes to model random rates of return on assets. The two main cases, as below, are when the random variable can only take on discrete values, and when it can take on values in continuous intervals. From basic probability theory, we have the following.

Definition 2. A discrete random variable X with possible values $x_1, x_2, x_3, \ldots$ has **probability mass function** $p(x)$ if

$$p(x_i) = P[X = x_i] \tag{1.1}$$

for each x_i. A continuous random variable X with possible values in a set $S \subset \mathbb{R}$ has **probability density function** $f(x)$ if

$$P[X \in B] = \int_B f(x)\,dx \tag{1.2}$$

for subsets $B \subset S$. ∎

Example 1. Suppose that the current value of an asset is \$50, and that it has single period rates of return of either -4%, -2%, 0%, 2%, or 4%, with probabilities .10, .20, .40, .20, and .10, respectively. Since the next price of the asset is the current price multiplied by 1 plus the rate of return, we can algebraically describe the probability mass function of the random variable X which is the next value of the asset as:

$$p(x) = \begin{cases} .10, & \text{if } x = 50(1 - .04) = 48; \\ .20, & \text{if } x = 50(1 - .02) = 49; \\ .40, & \text{if } x = 50(1 + 0) = 50; \\ .20, & \text{if } x = 50(1 + .02) = 51; \\ .10, & \text{if } x = 50(1 + .04) = 52. \end{cases}$$

A probability histogram of this discrete distribution is displayed in Figure 1.1. For instance the probability that this asset at least breaks even is

$$P[X \in \{50, 51, 52\}] = .40 + .20 + .10 = .70. \ \blacksquare$$

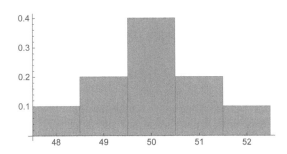

FIGURE 1.1 Discrete asset price distribution.

Example 2. A continuous model similar to the one in Example 1 is the following. Suppose that the single period rate of return R on the asset has the normal probability density with mean 0 and variance .0004, that is, the density has the form:

$$f(r) = \frac{1}{\sqrt{2\pi \cdot .0004}} \cdot e^{\frac{-(r-0)^2}{2 \cdot .0004}}, r \in \mathbb{R}.$$

As in Example 1, the new price X is $X = 50(1 + R)$. The theory of linear transformations of normal random variables implies that the probability density function $g(x)$ of X is then also normal, with mean and variance

$$\mu = 50(1 + E[R]) = 50(1 + 0) = 50; \sigma^2 = 50^2(\text{Var}(R)) = 2500(.0004) = 1.$$

Figure 1.2 has a picture of the density function of X. The area of the shaded region, that is $\int_{51}^{\infty} g(x)dx$, is the probability that the new price is at least 51. By numerical integration, this probability is about $P[X \geq 51] \approx .159. \ \blacksquare$

Example 3. Suppose that an investor has a contract that will pay $5 per share of the risky asset in Example 2 if the value of that asset exceeds $51.50, and otherwise pays nothing. What is the expected value of this contract? (This is a simple example of a derivative called a **binary option**.)

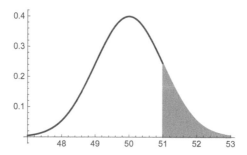

FIGURE 1.2 Continuous asset price distribution.

Solution. On a per-share basis, the profit on the contract can only take on two values: 5 in the case $X > 51.5$ and 0 otherwise. Statistical software yields that the probability that $X > 51.5$ is about .0668. The expected value per share is therefore:

$$0 \cdot P[X \leq 51.5] + 5 \cdot P[X > 51.5] = 5 \cdot (.0668) = .3340. \quad \blacksquare$$

Risky assets can include a variety of investments whose values can change in some random way with time. Some are physical such as real estate and agricultural commodities, while some are financial representations of value, such as stocks, bonds, futures, and options. We will be concerned with the latter financial assets in this book.

1.1.1   Single and Multiple Discrete Time Periods

As illustrated by the examples above, the simplest models of risky assets have only a value "now," which is assumed to be known, and a value "later" after one abstract time period. For these, a single random variable suffices to model the unpredictable behavior of the asset, and the best information possible is to know the probability distribution of that random variable. As we will see in our study of portfolio optimization, it may be sufficient to know only the mean and variance of the distribution. In reality, these usually cannot be known for certain, but past historical data may allow us to estimate those parameters.

For example, if x_0 is the current known value of the asset and X_1 is the random value at the end of one period, then the rate of return on the asset for the period is the change in value divided by the initial value, namely the random variable:

$$R_1 = \frac{X_1 - x_0}{x_0} = \frac{X_1}{x_0} - 1. \tag{1.3}$$

From this we derive the relationship between the next and current asset values:

$$X_1 = (1 + R_1) x_0, \tag{1.4}$$

which formalizes what we observed in Examples 1 and 2. If we are willing to assume that the probability distribution of the single period rate of return remains stable with time, we can treat market rates of return as a random sample from the distribution, and use the sample mean and variance to estimate the rate of return parameters:

$$\mu_r = E[R_1], \sigma_r^2 = \text{Var}(R_1). \tag{1.5}$$

Here is a computational example.

Example 4. Below is a list of 53 end-of-week prices for Walmart stock on the New York Stock Exchange during the calendar year 2016.

prices: 62.32, 60.43, 59.30, 61.54, 63.65, 62.59, 61.44, 65.20, 63.38, 64.60, 65.11, 65.65, 66.12, 65.86, 66.42, 66.10, 66.52, 64.88, 64.53, 67.29, 68.90, 68.99, 69.13, 69.33, 70.11, 71.01, 71.50, 71.67, 71.49, 71.22, 71.28, 71.77, 72.75, 69.73, 71.32, 70.33, 70.89, 70.76, 69.25, 67.91, 66.80, 67.29, 68.37, 68.17, 69.90, 67.74, 69.35, 69.19, 68.87, 70.09, 68.62, 68.30, 68.16.

Suppose that the distribution of the weekly rates of return was stable during that time, and also suppose that these rates are independent random variables. Then we can compute the following random sample of 52 rates of return:

rates: $-0.0303, -0.0187, 0.0378, 0.0343, -0.0166, -0.0184, 0.0611, -0.0279,$ $0.0192, 0.0080, 0.0082, 0.0072, -0.0039, 0.0085, -0.0048, 0.0064, -0.0247,$ $-0.0054, 0.0427, 0.0238, 0.0014, 0.0020, 0.0030, 0.0112, 0.0128, 0.0070,$ $0.0023, -0.0024, -0.0038, 0.0008, 0.0068, 0.0137, -0.0415, 0.0227, -0.0139,$ $0.0079, -0.0018, -0.0213, -0.0194, -0.0163, 0.0073, 0.0160, -0.0029, 0.0253,$ $-0.0308, 0.0237, -0.0023, -0.0047, 0.0178, -0.0210, -0.0047, -0.0020.$

For instance, the first rate of return is:

$$\frac{60.43 - 62.32}{62.32} \approx -0.0303,$$

and the rest are computed similarly. (Some rounding in the reported prices and computed rates of return has been done in these lists.)

The sample average weekly rate of return is

$$\bar{R} = \frac{-0.0303 + -0.0187 + \cdots + -0.0020}{52} \approx .0019,$$

and the sample variance is

$$S_R^2 = \frac{(-0.0303 - .0019)^2 + \cdots + (-0.0020 - .0019)^2}{51} \approx .00039.$$

A histogram of these rates of return indicates a shape that may be consistent with a normal distribution with these parameters, as shown in Figure 1.3. ∎

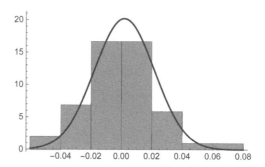

FIGURE 1.3 Empirical distribution of Walmart weekly rates of return.

Easily the more interesting and the more involved cases of financial asset modeling occur when we are able to observe the motion of the value of an asset through a sequence of discrete times. Usually we assume that at the origin of time, the asset value is a known constant $X_0 = x_0$, after which the value is $X_1, X_2, X_3, \ldots$ at times 1, 2, 3, etc.

Later, in Chapter 2 we will discuss thoroughly a particularly important model of this type called the **Cox-Ross-Rubinstein process**, or **binomial branch process**. In this model, at each period of time there is a random rate of return $R_t = \frac{X_t - X_{t-1}}{X_{t-1}}$ which can have just one of two values: b with probability p, or a with probability $1 - p$. Without loss of generality we assume $b > a$. The random variables R_t are assumed to be mutually independent. These "up" and "down" rates of return b and a, as well as the "up" probability p, remain stable throughout time, and the recursive relationship between the current value of the asset and the previous value is:

$$X_t = (1 + R_t) X_{t-1} \tag{1.6}$$

Example 5. A two-step binomial branch process is conceived of as a tree of possible paths in Figure 1.4. Here the initial state is $x_0 = \$50$, the up rate is $b = .10$, the down rate is $a = -.05$, and the up probability is $p = .5$. The root node at level 0 of the tree is the initial price x_0. Nodes at level 1 correspond to the two possible values X_1 of the price process at time 1, and nodes at level 2 correspond to the three possible values X_2 of the price process at time 2. There are four possible paths $X_0 = x_0, X_1, X_2$ from left to right in this tree,

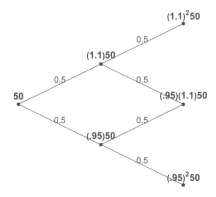

FIGURE 1.4 A two-step binomial branch process.

corresponding to two up moves, one up and then one down, one down and then one up, and two down moves. Each path, by independence, occurs with probability $(.5)(.5) = .25$. We have left the asset values in unevaluated form to enable you to better see how the possible values progress through time: given any tree node, there are two possible values at the next level, obtained by multiplying the current value by $(1 + b) = 1.1$ or by $(1 + a) = .95$. This leads to a recombination of values. For instance, at the second node in level 2 of the tree, the price is $(.95)(1.1)\$50 = \$52.25$, obtained in each of the two cases up-down and down-up. Then for example, the probability that X_2, the price at time 2, is at least $\$52.25$ is therefore the sum of all of the path probabilities of the up-up, up-down, and down-up paths, i.e. $.25 + .25 + .25 = .75$. ∎

1.1.2 Continuous-Time Processes

When neither a single period of time nor a sequence of discrete times is sufficient to describe (at least in ideal terms) the market value of an asset, we must turn to stochastic processes with a continuous time index $(X_t)_{t \geq 0}$. Most often, these processes also are such that the state space of possible values of each individual X_t is continuous as well.

This can lead to some very complicated mathematics, but fortunately the most important models that we will use are variations on a simple process called **standard Brownian motion**. We will cover this process in much more detail in a later chapter, but essentially a standard Brownian motion is a continuous-time version of a random walk, in which in any time interval, the process changes by an increment that has a normal distribution with mean zero and variance equal to the length of the time interval, independently of the past motion. This normality, in view of the independence of the price increments, leads to normality of the asset value X_t. Specifically, the distribution of the value at time t is normal with mean equal to the initial value X_0 and variance equal to t.

Example 6. Suppose that a standard Brownian motion is such that its initial state is $X_0 = 0$ and its value X_t at time t is normally distributed with mean 0 and variance t. Then the value $X_{1.5}$ at time 1.5 is $N(0, 1.5)$-distributed. Hence, by standardization, the probability that $X_{1.5}$ is at most 1.2 is:

$$P[X_{1.5} \leq 1.2] = P\left[\frac{X_{1.5} - 0}{\sqrt{1.5}} \leq \frac{1.2 - 0}{\sqrt{1.5}}\right] = P\left[Z \leq \frac{1.2 - 0}{\sqrt{1.5}} = .980\right] = .836.$$

■

1.1.3   Martingales

When we come to the problem of valuing derivative assets we will find that the idea of a **martingale process** is very useful. Later we will have a more general definition, but for now it will be enough to say the following. A stochastic process $(X_t)_{t \geq 0}$ is a **martingale** if the conditional expectation of any later value given the current value is equal to the current value, that is:

$$E[X_{t+s}|X_t] = X_t \tag{1.7}$$

for all times $t, s \geq 0$. A martingale models a risky asset which is not changing in expected value; i.e. later values of the asset have expectation equal to the current value.

Example 7. The standard Brownian motion process described in Example 6 is an example of a martingale. This is because for any time t, the increment of the process $X_{t+s} - X_t$ during time interval $[t, t + s]$ is independent of the past prior to time t, and it has the normal distribution with mean 0. Hence,

$$\begin{aligned} E[X_{t+s}|X_t] &= E[X_{t+s} - X_t + X_t|X_t] \\ &= E[X_{t+s} - X_t|X_t] + E[X_t|X_t] \\ &= 0 + X_t = X_t. \end{aligned} \tag{1.8}$$

In line 1, we just subtracted and added the same quantity X_t. The expectation is split in line 2. We will justify line three in this derivation more carefully in a later chapter, but intuitively the first expectation is the same as the unconditional expectation $E[X_{t+s} - X_t]$, which is 0 by the assumption that the process is in standard Brownian motion, and the second expectation is just the current value X_t given X_t itself. ■

Example 8. For a particular choice of the one-step transition probability p, the binomial branch process described above is a martingale. Take the parameters of Example 5: initial state $x_0 = \$50$, up rate $b = .10$, and down rate $a = -.05$.

Define:

$$q = \frac{0-a}{b-a} = \frac{-a}{b-a} = \frac{.05}{.15} = \frac{1}{3}. \tag{1.9}$$

(The reason why we include the 0 in the first expression will be made clearer later in this chapter when we have much more to say about the martingale condition relative to the valuation of options. Exercise 11 deals with this issue.) Now whatever state x_t that the process occupies at time t, the expected value of the next state is:

$$\begin{aligned} E\left[X_{t+1} | X_t = x_t\right] &= \tfrac{1}{3}(1+b)x_t + \tfrac{2}{3}(1+a)x_t \\ &= \left(\tfrac{1}{3} \cdot 1.1 + \tfrac{2}{3} \cdot .95\right) x_t \\ &= = 1 \cdot x_t = x_t \end{aligned}$$

In other words, the expected next state is equal to the previous state. This condition carries through inductively to time steps of arbitrary size, so that the martingale condition is satisfied. (See Exercise 12.) ∎

Exercises 1.1

1. Suppose that a risky asset currently has the value $40 per share, and an investor owns 10 shares. The one-step rate of return on the asset has a discrete distribution with four possible values: $-.1$, 0, .05 and .07, occurring with probabilities 1/6, 1/3, 1/3, and 1/6, respectively.
   (a) Find the probability mass function of the total amount the investor gains (or loses).
   (b) What is the probability of a net loss?
   (c) Find the discrete expected value of the investor's profit. (Recall that this is the weighted average of possible profit values, using the respective probabilities as weights.)

2. Suppose that the current (time 0) value of a risky asset is $100, and at time 1 the value follows the normal distribution with mean 102 and variance 9. Find the probability that the stock gains in value by at least 5% during the time interval $[0, 1]$. Can you think of any objections to the normal distribution as a model of price in this case?

3. A random variable has the **lognormal distribution** if its natural logarithm has the normal distribution. The lognormal distribution is then determined by the same parameters as its associated normal distribution. If the price of a risky asset is lognormally distributed with parameters $\mu = 2$, $\sigma = 1$, find the probability that the asset price exceeds 6.

4. Consider the scenario of observing the price of a risky asset at three consecutive times: $t_0 = 0$, $t_1 = 1$, and $t_2 = 2$. Then there will be a rate of return $R_{[0,1]}$ over time interval $[0, 1]$, another one-step rate of return $R_{[1,2]}$

over interval $[1, 2]$, and there will also be a rate of return $R_{[0,2]}$ over the entire time interval. Find a formula to express $R_{[0,2]}$ in terms of $R_{[0,1]}$ and $R_{[1,2]}$.

5. An investor holds a contract called a *future* on a risky asset that will pay the difference between the risky asset's value at time 1 and \$60, which may be negative, resulting in a loss for the investor. If the risky asset currently has value \$55, and its rate of return can be either 0%, 5%, 10%, or 15%, with probabilities $1/2$, $1/4$, $1/8$, and $1/8$, respectively, find the expected profit on the future. (See Exercise 1(c) for a reminder about discrete expected value.)

6. Let a price process follow the binomial branch model with up rate $b = .04$ and down rate $a = -.03$, initial value \$25, and one-step up probability $p = .5$. Construct a three-step tree that displays the possible price values at times 0, 1, 2, and 3 and find the probability mass function of the price at time 3.

7. (Technology required) Below is a price history for 88 consecutive weeks during 2015 and the first half of 2016 for Google Class A stock. Use software to compute the associated list of 87 rates of return as in Example 4. For the rate of return data, find the sample mean and variance, and plot a histogram. Do the rates of return seem to be normally distributed? Plot a time series graph of the rates. Do you observe any patterns to indicate that the rates might not be independent of each other?

529.55, 500.72, 510.46, 541.95, 537.55, 533.88, 551.16, 541.8, 562.63, 572.9, 553., 564.95, 557.55, 541.31, 548.54, 532.74, 573.66, 551.16, 548.95, 546.49, 554.52, 545.32, 549.53, 547.47, 557.52, 553.06, 547.34, 556.11, 699.62, 654.77, 657.5, 664.39, 689.37, 644.03, 659.69, 628.96, 655.3, 660.92, 640.15, 656.99, 671.24, 695.32, 719.33, 737.39, 761.6, 740.07, 777.00, 771.97, 779.21, 750.42, 756.85, 765.84, 778.01, 730.91, 710.49, 745.46, 761.35, 703.76, 706.89, 722.11, 724.86, 730.22, 744.87, 755.41, 754.84, 769.67, 759.47, 780.00, 737.77, 707.88, 725.18, 724.83, 721.71, 747.6, 735.86, 733.19, 704.25, 685.2, 710.25, 717.78, 735.63, 759.28, 791.34, 806.93, 807.05, 799.65, 793.22, 791.92

8. For a stock in standard Brownian motion, i.e., the price of the stock at time t is X_t, find $P[X_2 > 22 | X_0 = 20]$.

9. A generalization of standard Brownian motion that allows for differences in variabilities is a non-standard process in which $\text{Var}(X_t) = \sigma^2 t$, where σ^2 is a constant. For values of σ^2 equal to 10, 20, and 30, find $P[X_2 \geq 60 | X_0 = 50]$. What impact might your computations have on the value of a derivative asset which pays if the original asset price X_2 reaches a value of \$60 or greater at time 2, and otherwise is valueless?

10. We may generalize Brownian motion in a different way than in Exercise 9 by raising e to a Brownian motion, thereby correcting the anomaly that

normally distributed random prices have a non-zero probability of being negative. Suppose that $X_0 = 50$ again, and model X_t as $50e^{10Y_t}$, where (Y_t) is a standard Brownian motion starting at 0. Find $P[X_2 \geq 60 | X_0 = 50]$.

11. Consider a binomial branch process with up and down rates b and a. Find the unique up probability q that guarantees the martingale condition below:

$$E[X_{t+1} | X_t = x_t] = x_t.$$

Furthermore, find the probability q under which the process is expected to behave like a deterministically growing process with growth rate r, i.e. such that:

$$E[X_{t+1} | X_t = x_t] = (1+r)x_t.$$

12. Referring to Example 8, show that $E[X_{t+2} | X_t = x_t] = x_t$.

1.2   Risk Aversion and Portfolios of Assets

The decision of how to allocate wealth among investment opportunities is essentially about balancing the desire for high expected returns with the reluctance to assume the risks that the assets carry. In this section we study a simple way of characterizing investor attitudes toward risk, and then use it to set down a quantitative criterion for choosing optimal collections or **portfolios** of risky assets. The details of the solution of the portfolio selection problem will be shown in Section 1.4 after we review some important concepts involving dependence of random quantities in the next section.

First, recall that the **rate of return** on an investment during an interval of time $[t, t + s]$ is the ratio of the change in value of the investment during the interval to its initial value. Using X to indicate the investment's value, the rate of return would then be:

$$R_{[t,t+s]} = \frac{X_{t+s} - X_t}{X_t}. \tag{1.10}$$

Notice that the rate of return is dependent on the time interval in question.

1.2.1   Risk Aversion Constant

The main idea in our simple characterization of risk aversion is that in order for an investor to tolerate an additional unit of risk, characterized by variance of the rate of return random variable, the investor would demand an increase in the expected return on investment. The higher that required increase, the more **risk averse** the investor is. The model is especially simple because we will

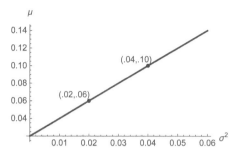

FIGURE 1.5 Linear investor indifference curve.

make the strong assumption that the rate of increase of mean return required per unit increase in variance is constant, regardless of the baseline values.

Figure 1.5 shows what we mean. The linear graph represents, for a particular investor, the collection of all pairs (σ^2, μ) of variances and associated means of financial assets among which the investor is indifferent. We call such a graph an ***indifference curve***. Two points are shown: the first with $\sigma_1^2 = .02$ and $\mu_1 = .06$, and the second with $\sigma_2^2 = .04$ and $\mu_2 = .10$. For this investor, in changing from investment 1 to investment 2, an additional $.04 - .02 = .02$ units of variance is compensated for by an increase of $.10 - .06 = .04$ units of expected return. The slope of the line, $.04/.02 = 2$ is a measure of the degree of risk aversion. Specifically, the investor demands an increase of 2 units of expected return per unit increase in variance. This slope is constant throughout the graph.

Definition 1. The ***risk aversion*** of an investor, denoted by a, is the increase in expected return that the investor demands as compensation for incurring an additional unit of variance. That is,

$$a = \frac{\Delta\mu}{\Delta\sigma^2} = \frac{\mu_2 - \mu_1}{\sigma_2^2 - \sigma_1^2}, \tag{1.11}$$

where (σ_1^2, μ_1) and (σ_2^2, μ_2) are the variance-mean combinations for two assets between which the investor is indifferent. ■

Example 1. Suppose that an investor has a risk aversion of 10. Which of the following would be preferred by this investor: a risk-free investment with mean rate of return $\mu_2 = .02$, or an asset with mean rate of return $\mu_1 = .04$ and variance $\sigma_1^2 = .006$?

Solution. The first investment has variance 0, so that the investor would incur an additional .006 units of risk in moving from this to the risky asset. The increase in mean is $.04 - .02 = .02$. But since the investor's risk aversion is

10, he would demand an additional $10(.006) = .06$ units of mean return in order to choose the risky asset, so this investor would prefer the risk-free asset. Another way to look at this is to compute the slope:

$$\frac{\Delta\mu}{\Delta\sigma^2} = \frac{.04 - .02}{.006 - 0} = 3.33.$$

Since this is less than $a = 10$, the compensation in terms of additional mean return is not sufficient for our investor. ∎

Example 2. If an investor's risk aversion is 8, and one asset has mean rate of return $\mu_1 = .05$ and variance $\sigma_1^2 = .2$, find two other assets that this investor would consider equivalent to the given one.

Solution. By the definition of risk aversion, the following relationship exists between the means and variances of any two equivalent assets:

$$\mu_2 = \mu_1 + a \cdot \left(\sigma_2^2 - \sigma_1^2\right).$$

Using the given information, we have more specifically:

$$\mu_2 = .05 + 8 \cdot \left(\sigma_2^2 - .2\right).$$

Therefore we can substitute in any variance σ_2^2 that we like and find the associated mean rate of return. For instance for $\sigma_2^2 = .22$ we compute:

$$\mu_2 = .05 + 8 \cdot (.22 - .2) = .21,$$

and for $\sigma_2^2 = .23$ the mean is

$$\mu_2 = .05 + 8 \cdot (.23 - .2) = .29. \ \blacksquare$$

Example 3. Suppose that an investment counselor asks a client to rank three investments: one with mean and variance $\mu_1 = .01$ and $\sigma_1^2 = .004$, a second with $\mu_2 = .03$ and $\sigma_2^2 = .009$, and a third with $\mu_3 = .06$ and $\sigma_3^2 = .02$. The client picks the second, then the first, then the third. What does this tell the counselor about the risk aversion of this client?

Solution. Given two assets to compare, the second is better than the first if the risk aversion satisfies

$$a \le \frac{\Delta\mu}{\Delta\sigma^2} = \frac{\mu_2 - \mu_1}{\sigma_2^2 - \sigma_1^2},$$

because in the definition of risk aversion, the second mean return μ_2 is the minimal mean rate of return that would compensate the investor for incurring additional risk. Anything more than that would make the investor prefer investment 2. Since, for this client, asset 2 beats asset 1, we would know that:

$$a \leq \frac{\mu_2 - \mu_1}{\sigma_2^2 - \sigma_1^2} = \frac{.03 - .01}{.009 - .004} = \frac{.02}{.005} = 4.$$

But the client also prefers asset 2 to asset 3, which implies that:

$$a \geq \frac{\mu_3 - \mu_2}{\sigma_3^2 - \sigma_2^2} = \frac{.06 - .03}{.02 - .009} = \frac{.03}{.011} = 2.73,$$

because asset 3 does not have a sufficiently high mean rate of return to make the client indifferent between it and asset 2. So the counselor knows that $2.73 \leq a \leq 4$.

There is one more comparison to make. The client prefers asset 1 to asset 3. This means that:

$$a \geq \frac{\mu_3 - \mu_1}{\sigma_3^2 - \sigma_1^2} = \frac{.06 - .01}{.02 - .004} = \frac{.05}{.016} = 3.125.$$

This gives a somewhat tighter set of bounds: $3.125 \leq a \leq 4$. ■

Remark. While a counselor is likely to be very conversant with the meaning and implications of mean and variance, the typical client may not be. What might the counselor do to indirectly give information about μ and σ to the client for preference analysis? One reasonable option is to prompt the client with a range of likely values of the rate of return, such as $[-.01, .05]$, with the additional information that values in the middle are more likely than those near the extremes. Then, the counselor would translate privately the midpoint to be the mean rate of return μ, and the right endpoint as $\mu + 2\sigma$. For the range $[-.01, .05]$ the midpoint is .02, and the 2σ radius is then $.05 - .02 = .03$, producing a standard deviation $\sigma = .03/2 = .015$. Comparing several such ranges would then allow the counselor to make calculations as in the last example to estimate the client's risk aversion.

1.2.2   The Portfolio Problem

We noted at the beginning of the section that the problem of allocating wealth among possible assets boils down to a return vs. risk tradeoff. It is time to be more precise.

Definition 2. A *portfolio* is a collection of some number n of assets. A *portfolio weight vector* $(w_1, w_2, \ldots, w_n)$ is a list of n real numbers giving the proportions of overall wealth devoted to each of the n assets, i.e., $w_i = W_i/W$, where W_i is the total wealth invested in asset i and W is the total overall wealth invested. ■

Notice that the portfolio weights w_i must sum to 1.

If an investor has risk aversion a, then the investor considers two assets, or portfolios of assets, the same if:

$$a = \frac{\mu_2 - \mu_1}{\sigma_2^2 - \sigma_1^2}, \tag{1.12}$$

where μ_1 and μ_2 are the two mean rates of return, and σ_1^2 and σ_2^2 are the variances of the rates of return. Rearranging formula (1.12), this equivalence of valuation happens if and only if:

$$\mu_1 - a\sigma_1^2 = \mu_2 - a\sigma_2^2. \tag{1.13}$$

A better investment is one for which the difference $\mu_3 - a\sigma_3^2$ is larger than these, since high expected return is beneficial while high variance is not. This gives us a criterion: we can say that an optimal portfolio of assets for an investor with risk aversion a is one that maximizes:

$$\mu_p - a\sigma_p^2, \tag{1.14}$$

where μ_p is the expected rate of return on the portfolio, and σ_p^2 is the variance of the rate of return.

We can go a little farther. Suppose that our portfolio can potentially consist of n random assets, whose rates of return over a single time period are random variables $R_1, R_2, ..., R_n$. How does the random variable R_p that represents the portfolio rate of return depend upon the individual asset rates of return?

Theorem 1. If the rates of return on n individual assets are $R_1, R_2, ..., R_n$, and a portfolio of the assets has weight vector $(w_1, w_2, ..., w_n)$ then the rate of return on the whole portfolio is:

$$R_p = w_1 R_1 + w_2 R_2 + \cdots + w_n R_n \tag{1.15}$$

Proof. The definition of rate of return implies that the final value of wealth is the initial value times 1 plus the rate of return. If W is the total initial wealth invested, then $w_i W$ is the portion that is devoted to asset i, and so for asset i the final wealth is:

$$(1 + R_i) w_i W.$$

Totaling over all assets, the rate of return on the whole portfolio is

$$
\begin{aligned}
R_p &= \frac{\mathrm{finalwealth - initialwealth}}{\mathrm{initialwealth}} \\
&= \frac{(1+R_1)w_1 W + (1+R_2)w_2 W + \cdots + (1+R_n)w_n W - W}{W} \\
&= \frac{(w_1 R_1 + w_2 R_2 + \cdots + w_n R_n)W + (w_1 W + w_2 W + \cdots + w_n W) - W}{W} \\
&= \frac{(w_1 R_1 + w_2 R_2 + \cdots + w_n R_n)W + (w_1 + w_2 + \cdots + w_n)W - W}{W} \\
&= (w_1 R_1 + w_2 R_2 + \cdots + w_n R_n)
\end{aligned}
$$

In line 4, we use the fact that the portfolio weights sum to 1 to subtract away the second and third terms in the numerator, allowing the cancellation of the factor W in the last line. ■

Formulas (1.14) and (1.15) allow us to continue to set up a model for the problem of optimizing a portfolio of assets. By linearity of expectation, the portfolio mean is:

$$\begin{aligned} \mu_p = E\left[R_p\right] &= E\left[w_1 R_1 + w_2 R_2 + \cdots + w_n R_n\right] \\ &= w_1 \mu_1 + w_2 \mu_2 + \cdots + w_n \mu_n. \end{aligned} \tag{1.16}$$

An expression for the portfolio variance is:

$$\sigma_p^2 = \mathrm{Var}\left(R_p\right) = \mathrm{Var}\left(w_1 R_1 + w_2 R_2 + \cdots + w_n R_n\right). \tag{1.17}$$

But here we must pause in the construction of our maximization objective $\mu_p - a\sigma_p^2$ in order to review important results from probability theory about the variance of a linear combination of random variables. We break to a brief section about that, but if you are already familiar with the impact of independence, and the definition and properties of covariance and correlation, you may continue directly to Section 1.4. (See Exercises 11-13 for a sneak preview of where we will be going.)

Exercises 1.2

1. For the investor with indifference curve in Figure 1.5, what mean rate of return is required in order to bear a variance of .05? .07?

2. Sketch a graph and find the equation of the indifference curve of an investor who values two investments equally: the first has mean return .02 and is non-risky, i.e. has variance 0, and the second has mean return .05 and variance .008.

3. Investor 1 considers an asset with mean return .04 and variance .10 to be equal in attractiveness to another asset with mean return .08 and variance .20. Investor 2 is indifferent between one asset with mean .02 and variance .05 and another with mean .06 and variance .055. Is there an asset that the two investors would value equally?

4. If Carla considers an asset with mean rate of return .04 and variance .008 to be superior to a second asset with mean rate of return .02 and variance .004, what can be said about Carla's risk aversion?

5. Fred, an investment counselor, would like to know if the risk aversion of his client Barney is more than 5. Design a question that Fred can ask Barney to determine this.

6. Ophelia presents Hamlet with two possible investments, instructing him that the first is very likely to have a rate of return in the interval $[-.05, .10]$ and the second in the interval $[-.01, .03]$. Being rather bold and perhaps rational, Hamlet selects the first investment. What do you know about Hamlet's risk aversion?

7. Among the following three investments, investor Sid prefers the second one. Find a lower and an upper bound on Sid's risk aversion.

$$\mu_1 = .04, \sigma_1^2 = .002; \quad \mu_2 = .06, \sigma_2^2 = .003; \quad \mu_3 = .07, \sigma_3^2 = .004$$

8. If there are two investments whose rates of return have the continuous uniform distribution, the first on the interval $[-.02, .08]$ and the second on the interval $[-.01, .07]$, is the choice between them clear? That is, would all investors, regardless of risk aversion, make the same choice between these assets? Justify your answer. (Note: For the uniform distribution on interval $[a, b]$, the mean is the midpoint $\frac{b+a}{2}$, and the variance is $\frac{(b-a)^2}{12}$.)

9. Paul begins an investment program by purchasing $1200 of stock A, $2400 of stock B, $1800 of stock C. and $600 of stock D. Find the portfolio weight vector, and compute the mean rate of return on the portfolio if the individual mean rates of return are: A: .04, B: .03, C: .01, D: .10.

10. Three risky assets have rates of return R_1, R_2, R_3 having, respectively, the normal distribution with mean .02 and standard deviation .05, the Poisson distribution with parameter .03, and the exponential distribution with rate parameter 20. Compute the mean rate of return on a portfolio whose weight vector is $(.5, .25, .25)$. (Note that the mean and variance of the Poisson distribution with parameter μ are both equal to μ, the mean of the exponential(λ) distribution is $1/\lambda$, and the variance of the exponential(λ) distribution is $1/\lambda^2$.)

11. To look ahead to Section 1.4, recall that if random variables are independent, then the variance of a linear combination of them is the combination of the variances, with coefficients being squared. For two random variables for example:

$$\mathrm{Var}(cX + dY) = c^2 \mathrm{Var}(X) + d^2 \mathrm{Var}(Y),$$

and similarly for multiple random variables. With this is mind, formulate specifically the objective function for portfolio optimization suggested by formulas (1.14), (1.16), and (1.17) for the three assets in Exercise 10, for a general risk aversion a. Do not attempt to maximize.

12. Refer to Exercise 11 for a reminder about independence and the variance of a linear combination. Consider an investment problem with two risky assets

with independent rates of return. Assume that the mean rates of return are .04 and .06 respectively, and the variances are .001 and .003. An investor will decide on a proportion w of total wealth to place into the first asset, leaving a proportion $1 - w$ in the second. The investor has a risk aversion constant of 4. Use single variable calculus to find the proportion w that maximizes $\mu_p - a\sigma_p^2$.

13. Suppose that there is a non-risky asset with fixed rate of return $r = .02$ and a risky asset with mean rate of return .05 and variance .009. An investor will choose to reserve a proportion w of her wealth to the risk-free asset, and hence invest a proportion $1 - w$ in the risky asset. If her risk aversion is $a = 10$, what value of w maximizes $\mu_p - a\sigma_p^2$?

14. Consider the two assets of Exercise 13. What portfolio would give a mean rate of return of .04, and what is the variance of that portfolio rate of return?

1.3 Expectation, Variance, and Covariance

Before we can continue with the project of finding portfolios that maximize the objective $\mu_p - a\sigma_p^2$, we must remind ourselves how to compute the portfolio variance:

$$\sigma_p^2 = \text{Var}\,(R_p) = \text{Var}\,(w_1 R_1 + w_2 R_2 + \cdots + w_n R_n). \qquad (1.18)$$

The solution depends on some standard definitions and results about expectation from probability theory, which we review here. The reader who is already familiar with the basic definitions and properties of expectation, and the special expectations of variance, covariance, and correlation, may pass on to the next section, although this section provides a good review.

1.3.1 One Variable Expectation

First, we define expected value in the discrete and continuous cases for a single random variable.

Definition 1. If X is a discrete random variable with probability mass function $f(x)$, and g is another function, then the *expected value of g(X)* is:

$$E[g(X)] = \sum_x g(x) \cdot f(x), \qquad (1.19)$$

where the sum is taken over all possible values (or states) x of X. If X is a continuous random variable with probability density function $f(x)$, and g is an integrable function, then the *expected value of g(X)* is:

$$E[g(X)] = \int_{-\infty}^{+\infty} g(x) \cdot f(x)dx. \; \blacksquare \qquad (1.20)$$

In computing the integral above, attention must be paid to the part of the domain where the density function vanishes, so that the effective limits of integration may not be infinite. In both the discrete and continuous cases, the expressions defining the expectation may not converge, in which case we say that the **expectation does not exist**, or the expressions may converge to $\pm\infty$, in which case we will say that the **expected value is infinite**.

The most commonly used expectation is the **mean**, or **expected value of X**, in which g is just the identity function; in the discrete and continuous cases, respectively:

$$\mu = E[X] = \sum_x x \cdot f(x), \text{ or } \mu = E[X] = \int_{-\infty}^{+\infty} x \cdot f(x)dx. \qquad (1.21)$$

The mean represents a probabilistic average value of X where the possible states of X are weighted by their probabilities and summed. Probably the second most important expectation is the **variance of X**:

$$\sigma^2 = E\left[(X-\mu)^2\right] = \sum_x (x-\mu)^2 \cdot f(x), \text{ or } \int_{-\infty}^{+\infty} (x-\mu)^2 \cdot f(x)dx. \quad (1.22)$$

The variance measures spread of the distribution of X, or in a financial context, the riskiness of X. The square root $\sigma = \sqrt{\sigma^2}$ of the variance is called the **standard deviation**.

Here are a few quick examples of computations of expectations. More can be found in the exercise set.

Example 1. (a) If X has the following $\Gamma(\alpha,\beta)$ (that is the gamma) density function, find the mean of X:

$$f(x; \alpha, \beta) = \frac{1}{\beta^\alpha \Gamma(\alpha)} x^{\alpha-1} e^{-x/\beta}, x > 0;$$

(b) If X has the discrete uniform distribution placing equal weights of $1/5$ on the points $1, 2, 3, 4, 5$, find the variance of X;

(c) If X has the Poisson mass function with parameter μ, as below, find both $E[X]$ and $E[X(X-1)]$:

$$f(x) = P[X = x] = \frac{e^{-\mu}\mu^x}{x!}, \quad =0, 1, 2, ...;$$

(d) If X has the $N(0,1)$ (that is, standard normal) density function below, show that the mean of X is 0 and the variance is 1:

$$f(x) = \frac{1}{\sqrt{2\pi}} e^{-x^2/2}, -\infty < x < \infty.$$

Solution. (a) By formula (1.21), the mean is given by the integral:

$$E[X] = \int_0^\infty x \cdot \frac{1}{\beta^\alpha \Gamma(\alpha)} x^{\alpha-1} e^{-x/\beta} \, dx.$$

Note that Γ is the standard mathematical gamma function among whose noteworthy properties is that $\Gamma(\alpha+1) = \alpha\Gamma(\alpha)$ for all non-negative α. We can remove $\Gamma(\alpha)$ from the denominator of the integral, and multiply and divide by β and by $\Gamma(\alpha+1)$ to get:

$$E[X] = \frac{\Gamma(\alpha+1)}{\Gamma(\alpha)} \cdot \beta \cdot \int_0^\infty \frac{1}{\beta^{\alpha+1}\Gamma(\alpha+1)} x^\alpha e^{-x/\beta} \, dt.$$

Since the integrand is now a $\Gamma(\alpha+1, \beta)$ density, by the defining property of density functions the whole integral equals 1. Therefore,

$$E[X] = \frac{\Gamma(\alpha+1)}{\Gamma(\alpha)} \cdot \beta = \frac{\alpha\Gamma(\alpha)}{\Gamma(\alpha)} \cdot \beta = \alpha \cdot \beta.$$

(b) First, the mean of the random variable is:

$$\mu = \frac{1}{5} \cdot 1 + \frac{1}{5} \cdot 2 + \frac{1}{5} \cdot 3 + \frac{1}{5} \cdot 4 + \frac{1}{5} \cdot 5 = \frac{15}{5} = 3.$$

Thus, the variance is:

$$\begin{aligned}
\sigma^2 &= \tfrac{1}{5} \cdot (1-3)^2 + \tfrac{1}{5} \cdot (2-3)^2 + \tfrac{1}{5} \cdot (3-3)^2 + \tfrac{1}{5} \cdot (4-3)^2 + \tfrac{1}{5} \cdot (5-3)^2 \\
&= \tfrac{10}{5} = 2.
\end{aligned}$$

(c) This distribution is discrete, but we have an infinite series to cope with. By the definition of mean we get:

$$\begin{aligned}
E[X] &= \sum_{x=0}^\infty x \cdot \frac{e^{-\mu} \mu^x}{x!} \\
&= e^{-\mu} \cdot \sum_{x=1}^\infty \frac{\mu^x}{(x-1)!}.
\end{aligned}$$

In the second line above, we factored out the constant $e^{-\mu}$, we noted that the initial index $x = 0$ does not contribute to the sum due to the presence of the factor x in the general term, and as a consequence we could divide the factor x away in the $x!$ expression in the denominator. Now if we factor out μ and change indices of summation to $k = x - 1$, we can rewrite the sum as follows:

$$\begin{aligned}
E[X] &= e^{-\mu} \cdot \sum_{x=1}^\infty \frac{\mu^x}{(x-1)!} \\
&= e^{-\mu} \cdot \mu \cdot \sum_{x=1}^\infty \frac{\mu^{x-1}}{(x-1)!} \\
&= e^{-\mu} \cdot \mu \cdot \sum_{k=0}^\infty \frac{\mu^k}{k!}.
\end{aligned}$$

But the summation is now the Taylor series expansion for e^{μ}. Therefore,

$$E[X] = e^{-\mu} \cdot \mu \cdot e^{\mu} = \mu.$$

To compute $E[X(X-1)]$ we work along similar lines. By definition of $E[g(X)]$ we must compute:

$$E[X(X-1)] = \sum_{x=0}^{\infty} x \cdot (x-1) \frac{e^{-\mu}\mu^x}{x!}.$$

The $x = 0$ and $x = 1$ terms of the sum are both zero, so we may start indexing at 2. Factor out $e^{-\mu}$ again, and also take out two factors of μ:

$$
\begin{aligned}
E[X(X-1)] &= \sum_{x=2}^{\infty} x \cdot (x-1) \frac{e^{-\mu}\mu^x}{x!} \\
&= e^{-\mu} \cdot \mu^2 \cdot \sum_{x=2}^{\infty} \frac{\mu^{x-2}}{(x-2)!}.
\end{aligned}
$$

Changing the index of summation to $k = x-2$ yields again that the summation is e^{μ}, hence:

$$E[X(X-1)] = e^{-\mu} \cdot \mu^2 \cdot e^{\mu} = \mu^2.$$

(d) The integral that must be computed to find the mean is:

$$\mu = E[X] = \int_{-\infty}^{\infty} x \cdot \frac{1}{\sqrt{2\pi}} e^{-x^2/2} dx.$$

Since the integrand is an odd function, the integral is zero if indeed it converges. To find out, we would have to evaluate the positive part:

$$\int_{0}^{\infty} x \cdot \frac{1}{\sqrt{2\pi}} e^{-x^2/2} dx$$

You can check that the substitution $u = x^2$ produces an integral of the type $\int_{0}^{\infty} e^{-u} du$, which does converge. Hence $\mu = 0$. For the second part, we must calculate the expectation:

$$\sigma^2 = \text{Var}(X) = E\left[(X - \mu)^2\right] = E\left[X^2\right].$$

By the definition of $E[g(X)]$ in the continuous case, this expectation is:

$$\sigma^2 = \int_{-\infty}^{\infty} x^2 \frac{1}{\sqrt{2\pi}} e^{-x^2/2} dx = 2 \int_{0}^{\infty} x^2 \frac{1}{\sqrt{2\pi}} e^{-x^2/2} dx,$$

where the second expression takes advantage of the symmetry about 0 of the integrand. Integration by parts, with the choices:

$$
\begin{aligned}
u &= x & v &= -\frac{1}{\sqrt{2\pi}} e^{-x^2/2} \\
du &= dx & dv &= x \frac{1}{\sqrt{2\pi}} e^{-x^2/2}
\end{aligned}
$$

successfully completes the computation, showing that the integral itself is $1/2$, so that $\sigma^2 = 1$. ∎

1.3.2   Expectation for Multiple Random Variables

To understand the relationship between two different rates of return, we must extend the definition of expectation to many variables. From probability theory, the **joint probability mass function** of random variables $X_1, X_2, \ldots, X_n$ is the function $f(x_1, x_2, \ldots, x_n)$ such that

$$f(x_1, x_2, \ldots, x_n) = P[X_1 = x_1, X_2 = x_2, \ldots, X_n = x_n]. \qquad (1.23)$$

For continuously distributed variables the **joint probability density function** takes on a similar role. For suitable subsets B of n-dimensional space, the probability that the vector of random variables takes its value in B is the multiple integral:

$$P[(X_1, X_2, \ldots, X_n) \in B] = \underset{B}{\int \int} \cdots \int f(x_1, x_2, \ldots, x_n)\, dx_1 dx_2 \cdots dx_n.$$
$$(1.24)$$

Expectations of functions of many random variables are then defined similarly to the one-dimensional case, as in the next definition.

Definition 2. If $X_1, X_2, \ldots, X_n$ are discrete random variables with joint probability mass function $f(x_1, x_2, \ldots, x_n)$, and g is another function, then the **expected value of $g(X_1, X_2, \ldots, X_n)$** is:

$$E[g(X_1, X_2, \ldots, X_n)] = \sum_{x_1} \sum_{x_2} \cdots \sum_{x_n} g(x_1, x_2, \ldots, x_n) \cdot f(x_1, x_2, \ldots, x_n),$$
$$(1.25)$$

where the sum is taken over all possible joint values x_i of the random variables X_i. If $X_1, X_2, \ldots, X_n$ are continuously distributed random variables with joint probability density function $f(x_1, x_2, \ldots, x_n)$, and g is an integrable function, then the **expected value of $g(X_1, X_2, \ldots, X_n)$** is:

$$E[g(X_1, X_2, \ldots, X_n)] = \begin{array}{l} \int_{-\infty}^{+\infty} \int_{-\infty}^{+\infty} \cdots \int_{-\infty}^{+\infty} g(x_1, x_2, \ldots, x_n) \\ \cdot f(x_1, x_2, \ldots, x_n)\, dx_1 dx_2 \cdots dx_n. \end{array} ∎ \quad (1.26)$$

Fortunately, more often than not we are only interested in two random variables at a time, which limits some of the complication and allows us to live in two-dimensional state spaces.

Example 2. (a) Find $E[X - 2Y]$ if X and Y are discretely distributed with probabilities $1/6$, $1/3$, $1/3$, and $1/6$ on the points $(X, Y) = (0, 0), (0, 1), (1, 0)$, and $(1, 1)$, respectively; (b) If random variables X and Y are continuously distributed with joint p.d.f. $f(x, y) = 1/8$ on the rectangle $[0, 2] \times [0, 4]$, find $E[X + Y]$.

Solution. (a) We can depict the probability distribution that is described in the problem as in Figure 1.6.

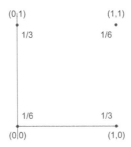

FIGURE 1.6 A discrete, two-variable distribution.

It is helpful to produce a table of values of X and Y individually at the four states, and the resulting values of $X - 2Y$.

point	X	Y	$X - 2Y$	prob
$(0, 0)$	0	0	0	$1/6$
$(0, 1)$	0	1	-2	$1/3$
$(1, 0)$	1	0	1	$1/3$
$(1, 1)$	1	1	-1	$1/6$

Then,

$$E[X - 2Y] = \frac{1}{6}(0) + \frac{1}{3}(-2) + \frac{1}{3}(1) + \frac{1}{6}(-1) = -\frac{1}{2}.$$

(b) The space of possible values (x, y) of the random pair (X, Y) is sketched in Figure 1.7. Because the density function is constant with value $1/8$, we can use formula (1.26) to compute:

$$
\begin{aligned}
E[X + Y] &= \int_0^2 \int_0^4 (x + y) \cdot \frac{1}{8} dy dx \\
&= \frac{1}{8} \cdot \int_0^2 \left(\left(xy + \frac{y^2}{2} \right) \Big|_0^4 \right) dx \\
&= \frac{1}{8} \cdot \int_0^2 (4x + 8) \, dx \\
&= \frac{1}{8} \cdot \left(2x^2 + 8x \right) \Big|_0^2 \\
&= \frac{1}{8} \cdot 24 = 3. \quad \blacksquare
\end{aligned}
$$

Notice that in the last computation in Example 2(b), the mean of the random variable X is easily seen to be 1, and the mean of Y is 2, so $E[X + Y] = 3$ turns out to be the sum of $E[X]$ and $E[Y]$. This is an illustration of one of

$(0,4)$ $(2,4)$

$(0,0)$ $(2,0)$

FIGURE 1.7 State space of a continuous, two-variable uniform distribution.

the most helpful properties of expected value, the **linearity property**, stated in the theorem below. Its proof can be found in all standard texts on probability.

Theorem 1. (Linearity) If X_1 and X_2 are random variables with finite means, and a and b are constants, then:

$$E\left[aX_1 + bX_2\right] = aE\left[X_1\right] + bE\left[X_2\right]. \tag{1.27}$$

Furthermore, the result extends in the natural way to linear combinations of more than two random variables. ■

Linearity allows us to find a useful computational formula for the variance of a random variable. By expanding out $E\left[(X - \mu)^2\right]$ and recombining like terms we find:

$$
\begin{aligned}
E\left[(X - \mu)^2\right] &= E\left[X^2 - 2\mu X + \mu^2\right] \\
&= E\left[X^2\right] - 2\mu E[X] + \mu^2 \\
&= E\left[X^2\right] - \mu^2.
\end{aligned} \tag{1.28}
$$

The linearity property is what allows us to express the mean portfolio rate of return in terms of the individual asset rates of return, because:

$$
\begin{aligned}
E\left[R_p\right] &= E\left[w_1 R_1 + w_2 R_2 + \cdots + w_n R_n\right] \\
&= w_1 E\left[R_1\right] + w_2 E\left[R_2\right] + \cdots + w_n E\left[R_n\right]
\end{aligned} \tag{1.29}
$$

But linearity on its own is not enough to allow us to achieve our main mission, that is to find an expression for $\mathrm{Var}\left(R_p\right) = \mathrm{Var}\left(w_1 R_1 + w_2 R_2 + \cdots + w_n R_n\right)$. For this, we must bring in notions related to independence and dependence of random rates of return.

Remark. Recall the notion of **independence** of random variables. The defining property is that for all sequences of subsets $B_1, B_2, \ldots, B_n$ of the state spaces of $X_1, X_2, \ldots, X_n$,

$$P[X_1 \in B_1, X_2 \in B_1, \ldots, X_n \in B_n] = P[X_1 \in B_1] \cdot P[X_2 \in B_2] \cdots \cdot P[X_n \in B_n].$$

$$(1.30)$$

It can be shown that independence is equivalent to the property that the joint density (or joint p.m.f. in the discrete case) of the random variables factors into the product of the individual densities (or p.m.f.'s) of the n random variables. (See Exercise 11.) ∎

The next theorem helps us handle cases where financial assets do not exhibit dependence on one another.

Theorem 2. If random variables $X_1, X_2, \ldots, X_n$ are independent, then for any functions $g_1, g_2, \ldots, g_n$:

$$E[g_1(X_1) \cdot g_2(X_2) \cdots g_n(X_n)] = E[g_1(X_1)] \cdot E[g_2(X_2)] \cdots \cdot E[g_n(X_n)]. ∎$$

$$(1.31)$$

The proof of Theorem 2 is an easy consequence of the definition of expectation and the Remark above.

Example 3. Recall that the final value of an investment after a period of time elapses is the initial value times one plus the rate of return for that period. In symbols,

$$X_1 = (1 + R_1) X_0,$$

$$(1.32)$$

where X_1 is the value at the end of the period, X_0 is the value at the start, and R_1 is the rate of return during time interval $[0, 1]$. We can think of the investment as changing value over multiple periods as well; adopting similar notations, the value at the end of three periods, for instance, would be:

$$\begin{aligned} X_3 &= (1 + R_3) X_2 \\ &= (1 + R_3)(1 + R_2) X_1 \\ &= (1 + R_3)(1 + R_2)(1 + R_1) X_0. \end{aligned}$$

$$(1.33)$$

If we make the assumption that the rates of return R_1, R_2, and R_3 on the three time periods are independent random variables, then Theorem 2 and linearity together imply that:

$$\begin{aligned} E[X_3] &= E[(1 + R_3)(1 + R_2)(1 + R_1) X_0] \\ &= (1 + E[R_3])(1 + E[R_2])(1 + E[R_1]) X_0. \end{aligned}$$

$$(1.34)$$

In the even more special case that the rates of return have the same mean value r, then the expected value of the investment at time 3 is $E[X_3] = (1+r)^3 X_0$, similarly to ordinary deterministic compound interest. ∎

1.3.3   Variance of a Linear Combination

Now we are in a position to talk about correlations between random variables such as the rates of return on different assets.

Definition 3. Let X and Y be random variables. The **covariance** between X and Y is defined as the following expectation:

$$\sigma_{xy} = \mathrm{Cov}(X,Y) = E\left[(X - \mu_x) \cdot (Y - \mu_y)\right] = E[XY] - \mu_x \mu_y. \qquad (1.35)$$

The **correlation** between X and Y is the covariance between the standardized variables:

$$\rho = \mathrm{Corr}(X,Y) = E\left[\frac{(X - \mu_x)}{\sigma_x} \cdot \frac{(Y - \mu_y)}{\sigma_y}\right] = \frac{\mathrm{Cov}(X,Y)}{\sigma_x \cdot \sigma_y}. \ ∎ \qquad (1.36)$$

(The second representation of σ_{xy} in formula (1.35) follows easily by expanding the product, using linearity of expectation, and simplifying the common terms $\mu_x \mu_y$.)

When random variables X and Y are in an increasing relationship, they tend to exceed their means together for the same outcomes and be less than their means for the complementary outcomes, and they do so with high probability. This makes the product $(X - \mu_x) \cdot (Y - \mu_y)$ positive with high probability, inducing a positive covariance. In a decreasing relationship, when one random variable exceeds its mean the other tends to be less than its mean; hence the product $(X - \mu_x) \cdot (Y - \mu_y)$ is negative with high probability. In this situation the covariance will come out negative. So the covariance measures the degree and direction of the relationship between X and Y. The correlation does the same thing, but by standardizing the random variables first, it corrects for scale differences and gives a truer measure of the dependence of the variables, as the next theorem suggests. Again, its proof can be found in standard probability texts.

Theorem 3. The correlation ρ satisfies the inequality:

$$-1 \le \rho \le 1, i.e., |\rho| \le 1. \qquad (1.37)$$

Moreover, ρ equals its extreme values of ± 1 only when Y is a perfect linear function $Y = a + bX$ of X, with probability 1. ∎

Example 4. Find the covariance and correlation for the random variables in Example 2(a) and (b).

Solution. (a) It helps to reprise the table from the earlier example, adding some helpful columns in order to compute the expectations that we need.

point	X	Y	XY	X^2	Y^2	prob
(0,0)	0	0	0	0	0	1/6
(0,1)	0	1	0	0	1	1/3
(1,0)	1	0	0	1	0	1/3
(1,1)	1	1	1	1	1	1/6

Both X and Y give probability weight of 1/2 to state 0, and also 1/2 to state 1. The individual means are:

$$\mu_x = \mu_y = 0 \cdot \frac{1}{2} + 1 \cdot \frac{1}{2} = \frac{1}{2}.$$

For the variances, we can notice that, since the random variables only take on the values 0 and 1, so do the squares, and the arithmetic comes out the same as for the means, hence $E\left[X^2\right] = E\left[Y^2\right] = 1/2$. Therefore,

$$\sigma_x^2 = \sigma_y^2 = \frac{1}{2} - \left(\frac{1}{2}\right)^2 = \frac{1}{4}.$$

The standard deviations are both equal to 1/2 as a result. The last ingredient in the computation is the expected product:

$$E[XY] = 0 \cdot \frac{1}{6} + 0 \cdot \frac{1}{3} + 0 \cdot \frac{1}{3} + 1 \cdot \frac{1}{6} = \frac{1}{6}.$$

Thus,

$$\sigma_{xy} = \text{Cov}(X, Y) = E[XY] - \mu_x \mu_y = \frac{1}{6} - \frac{1}{2} \cdot \frac{1}{2} = -\frac{1}{12},$$

and:

$$\rho = \frac{\text{Cov}(X, Y)}{\sigma_x \cdot \sigma_y} = \frac{-1/12}{(1/2)(1/2)} = -\frac{1}{3}.$$

(b) As noted in Example 2(b), it is easy to compute that $E[X] = 1$ and $E[Y] = 2$. The expected squares would be:

$$
\begin{aligned}
E\left[X^2\right] &= \int_0^2 \int_0^4 x^2 \cdot \tfrac{1}{8} dy dx \\
&= \tfrac{1}{8} \cdot \int_0^2 4x^2 dx \\
&= \tfrac{1}{8} \cdot \left(\tfrac{4}{3}x^3\right) \big|_0^2 \\
&= \tfrac{4}{3};
\end{aligned}
$$

$$
\begin{aligned}
E\left[Y^2\right] &= \int_0^4 \int_0^2 y^2 \cdot \tfrac{1}{8} dx dy \\
&= \tfrac{1}{8} \cdot \int_0^4 2y^2 dx \\
&= \tfrac{1}{8} \cdot \left(\tfrac{2}{3}y^3\right) \big|_0^4 \\
&= \tfrac{16}{3}.
\end{aligned}
$$

So we have that the variances are:

$$\sigma_x^2 = E\left[X^2\right] - \mu_x^2 = \frac{4}{3} - (1)^2 = \frac{1}{3};$$

$$\sigma_y^2 = E\left[Y^2\right] - \mu_y^2 = \frac{16}{3} - (2)^2 = \frac{4}{3}.$$

The expected value of the product of the random variables is computed as the following double integral:

$$
\begin{aligned}
E[XY] &= \int_0^2 \int_0^4 xy \cdot \tfrac{1}{8} dy dx \\
&= \tfrac{1}{8} \cdot \int_0^2 x \left(\tfrac{1}{2} y^2 |_0^4\right) dx \\
&= \tfrac{1}{8} \cdot \int_0^2 8x dx \\
&= \tfrac{1}{8} \cdot \left(4x^2\right) |_0^2 \\
&= 2.
\end{aligned}
$$

This implies that the covariance is:

$$\text{Cov}(X, Y) = E[XY] - \mu_x \mu_y = 2 - 1 \cdot 2 = 0,$$

and in turn $\rho = 0$. We call such random variables **uncorrelated**. ■

Remark. For the jointly distributed random variables in Example 4(b), the joint density $f(x, y) = \tfrac{1}{8}$ factors into the product of the individual densities $f(x) = \tfrac{1}{2}$ and $g(y) = \tfrac{1}{4}$ of the variables, so that X and Y are independent. It will always happen that when random variables are independent, their correlation equals 0 (see Exercise 14), although the converse is not necessarily true. ■

We are ready now for the theorem for which we have been waiting.

Theorem 4. Let $X_1, X_2, ..., X_n$ be random variables and let $a_1, a_2, ..., a_n$ be constants. Then, if the X_i's are mutually independent,

$$
\begin{aligned}
\text{Var}\left(a_1 X_1 + a_2 X_2 + \cdots + a_n X_n\right) = \; & a_1^2 \text{Var}\left(X_1\right) + a_2^2 \text{Var}\left(X_2\right) \\
& + \cdots + a_n^2 \text{Var}\left(X_n\right).
\end{aligned}
\qquad (1.38)
$$

In the general case, we have

$$\text{Var}\left(\sum_{i=1}^n a_i X_i\right) = \sum_{i=1}^n a_i^2 \text{Var}\left(X_i\right) + \sum_{\substack{j=1 \\ j \neq k}}^n \sum_{k=1}^n a_j a_k \text{Cov}\left(X_j, X_k\right). \quad \blacksquare \qquad (1.39)$$

We will not try to prove the theorem in general, but an inspection of the case $n = 2$ conveys the ideas sufficiently. Consider the random variable

$Y = a_1 X_1 + a_2 X_2$. Then $E[Y] = a_1\mu_1 + a_2\mu_2$ by linearity, where μ_1 and μ_2 are the individual means of X_1 and X_2, respectively. This means that the variance of the linear combination can be expanded as:

$$
\begin{aligned}
\mathrm{Var}(Y) &= E\left[(Y - E[Y])^2\right] \\
&= E\left[(a_1 X_1 + a_2 X_2 - (a_1\mu_1 + a_2\mu_2))^2\right] \\
&= E\left[((a_1 X_1 - a_1\mu_1) + (a_2 X_2 - a_2\mu_2))^2\right] \\
&= E\left[(a_1(X_1 - \mu_1))^2\right] + E\left[(a_2(X_2 - \mu_2))^2\right] \\
&\quad + 2E\left[(a_1(X_1 - \mu_1))(a_2(X_2 - \mu_2))\right] \\
&= a_1^2\mathrm{Var}(X_1) + a_2^2\mathrm{Var}(X_2) + 2a_1 a_2 \mathrm{Cov}(X_1, X_2).
\end{aligned}
$$

This is formula (1.39) in the case $n = 2$, because in the double sum there will be two copies of the term $a_1 a_2 \mathrm{Cov}(X_1, X_2)$. The special case in formula (1.38), where the two random variables are independent, results from the fact that under this assumption,

$$
\begin{aligned}
\mathrm{Cov}(X_1, X_2) &= E\left[(X_1 - \mu_1)(X_2 - \mu_2)\right] \\
&= E[X_1 - \mu_1] \cdot E[X_2 - \mu_2] \\
&= 0 \cdot 0 = 0.
\end{aligned}
$$

Our original goal was to be able to write a formula for the portfolio variance:

$$
\mathrm{Var}(R_p) = \mathrm{Var}(w_1 R_1 + w_2 R_2 + \cdots + w_n R_n).
$$

We have now accomplished that goal. Simply translating notation in Theorem 4, with $X_i = R_i$, $\sigma_i^2 = \mathrm{Var}(R_i)$, and $a_i = w_i$,

$$
\mathrm{Var}(R_p) = \mathrm{Var}\left(\sum_{i=1}^n w_i R_i\right) = \sum_{i=1}^n w_i^2 \sigma_i^2 + \sum_{j=1}^n \sum_{\substack{k=1 \\ j \neq k}}^n w_j w_k \mathrm{Cov}(R_j, R_k). \quad (1.40)
$$

This formula, and formula (1.38) will be used regularly in the development of portfolio theory.

The exercise set for this section reminds you of several important probability distributions, concepts like the moment-generating function, and a few results about expectation, so it is well worth your time to read through all of the problems and practice as many as you can.

Exercises 1.3

1. Find both $E[X]$ and $E\left[X^2\right]$ for the random variable X whose probability mass function is $f(0) = \frac{1}{16}, f(1) = \frac{4}{16}, f(2) = \frac{6}{16}, f(3) = \frac{4}{16}, f(4) = \frac{1}{16}$.

2. Derive the mean of the distribution characterized by the following **binomial probability mass function** (abbr. $b(n, p)$) with parameters n and p.

$$f(k) = \binom{n}{k} p^k (1 - p)^{n-k}, k = 0, 1, 2, ..., n.$$

3. Consider a continuous random variable X with the **exponential distribution** with parameter λ, (abbr. $\exp(\lambda)$) whose density function is below. Find both $E[X]$ and $E[X^2]$. Then, argue that this distribution is a special case of the $\Gamma(\alpha,\beta)$ distribution of Example 1(a), and check that your result for $E[X]$ is consistent with the formula for the mean that was derived in that example.

$$f(x) = \lambda e^{-\lambda x}, x \geq 0.$$

4. If X has the Poisson distribution of Example 1(c), then compute the expectation $M(t) = E[e^{tX}]$, called the **moment-generating function** of X. If you differentiate this function with respect to t, and then set $t = 0$, what do you get?

5. Use the results of Example 1(c) to show that the variance of the Poisson distribution is μ.

6. Use the result of Example 1(d) to show that the mean and variance of the general **normal distribution with parameters μ and σ^2** (abbr. $N(\mu, \sigma^2)$), whose density function is below, are μ and σ^2 respectively.

$$f(x) = \frac{1}{\sqrt{2\pi\sigma^2}} e^{-(x-\mu)^2/2\sigma^2}, -\infty < x < \infty.$$

(Hint: consider the random variable $Z = \frac{X-\mu}{\sigma}$.)

7. Consider two discrete random variables X and Y for which the joint probability $P[X = x, Y = y]$ is the constant $1/6$ at each of the points $(x, y) = (0, 0), (1, 0), (0, 1), (2, 0), (1, 1),$ and $(0, 2)$.
    (a) Find, for each possible x, $P[X = x]$
    (b) Find, for each possible y, $P[Y = y]$
    (c) Compute $P[X + Y \leq 1]$.
    (d) Compute $E[X]$, $E[Y]$, and $E[X + Y]$

8. Suppose that continuous random variables X and Y have joint density $f(x, y) = x + y, x, y \in [0, 1]$. Find $P[X - Y \leq .5]$.

9. Compute $E[2XY]$ for random variables X and Y whose joint probability mass function puts equal weight onto the points in the set:

$$E = \{(1,0), (2,0), (3,0), (4,0), (1,1), (2,1), (3,1),$$
$$(4,1), (1,2), (2,2), (3,2), (4,2)\}$$

10. For the random variables in Exercise 8, compute $E[XY]$.

11. Show that if the joint density $f(x,y)$ of two continuous random variables X and Y factors into the product of a function of x only times a function of y only, then X and Y are independent.

12. Two continuous random variables X_1 and X_2 have joint density $f(x_1, x_2) = cx_1 x_2$ on the set $0 \le x_1, x_2 \le 2$. Find the value of c that ensures that f is a valid joint density, and compute $E[X_1 X_2]$.

13. Prove Theorem 1 in the case of discrete random variables.

14. Show that if random variables X and Y are independent, then $\rho = 0$.

15. Compute the correlation ρ for the random variables of Exercise 9.

16. If random variables X and Y have joint density $f(x,y) = 8xy$ on the set $0 \le y \le x \le 1$, compute $\text{Cov}(X, Y)$ and ρ.

17. Verify formulas (1.38) and (1.39) in the case of three random variables.

18. Suppose that there are two assets with mean rates of return .04 and .05, and variances .001 and .002. Find the portfolio mean rate of return and variance of rate of return for a portfolio that divides wealth evenly between the two assets, if (a) the asset rates of return are independent; (b) the asset rates of return are correlated with $\rho = -.4$.

19. Three risky assets each have normally distributed rates of return (see Exercise 6). The first is $N(.01, .004)$ (this means that the variance is .004), the second is $N(.03, .006)$, and the third is $N(.05, .009)$. Write a general expression for the portfolio mean and variance for a portfolio with weights w_1, w_2, w_3 if: (a) the assets are independent; (b) the assets are correlated, with pairwise correlations $\rho_{12} = .2, \rho_{13} = .1, \rho_{23} = .4$.

20. Two asset rates of return R_1 and R_2 have the joint p.m.f. below. Find the mean and variance of the portfolio rate of return for a portfolio that devotes $1/3$ of its wealth to the first asset and $2/3$ to the second.

$$f(r_1, r_2) = \begin{cases} 1/6 & \text{if } (r_1, r_2) = (.02, .04) \\ 1/3 & \text{if } (r_1, r_2) = (.03, .05) \\ 1/3 & \text{if } (r_1, r_2) = (.04, .08) \\ 1/6 & \text{if } (r_1, r_2) = (.06, .10). \end{cases}$$

21. If one asset has a risk-free rate of return $R_1 = .02$, and another has a risky rate of return R_2 with mean .05 and standard deviation .04, find the portfolio

mean and variance for a portfolio with $1/4$ of its weight on the risk-free asset and the rest on the risky asset. (Hint: what is the covariance of R_1 and R_2?)

1.4   Simple Portfolio Optimization

In the last section we showed two formulas:

$$
\begin{aligned}
E\left[R_p\right] &= E\left[w_1R_1 + w_2R_2 + \cdots + w_nR_n\right] \\
&= w_1E\left[R_1\right] + w_2E\left[R_2\right] + \cdots + w_nE\left[R_n\right],
\end{aligned}
\tag{1.41}
$$

and

$$
\operatorname{Var}\left(R_p\right) = \operatorname{Var}\left(\sum_{i=1}^{n} w_iR_i\right) = \sum_{i=1}^{n} w_i^2\sigma_i^2 + \sum_{j=1}^{n}\sum_{\substack{k=1 \\ j\neq k}}^{n} w_jw_k\operatorname{Cov}\left(R_j, R_k\right), \tag{1.42}
$$

for the mean and variance of the rate of return R_p on a portfolio with n assets and weight vector $(w_1, w_2, ..., w_n)$. We can write $\mu_i = E\left[R_i\right]$ for short. In Section 1.2, we argued that an investor should look to maximize:

$$
\mu_p - a\sigma_p^2,
$$

where a is the risk aversion constant for that investor. Putting this together, we have that the objective for this investor is:

$$
\max_{(w_1, w_2, ..., w_n)} \sum_{i=1}^{n} w_i\mu_i - a\left(\sum_{i=1}^{n} w_i^2\sigma_i^2 + \sum_{j=1}^{n}\sum_{\substack{k=1 \\ j\neq k}}^{n} w_jw_k\operatorname{Cov}\left(R_j, R_k\right)\right), \tag{1.43}
$$

with the constraint that $w_1 + w_2 + \cdots + w_n = 1$.

We will save the multiple asset case until the next chapter. In this short section we just carry out the maximization using single variable calculus for three simple two-asset cases: a risk-free asset and a single risky asset, two independent risky assets, and two correlated risky assets.

Example 1. Suppose that a risk-free asset whose rate of return is .03 and a risky asset with mean rate of return .05 and standard deviation .04 are available for investment. Find the optimal portfolio for an investor with risk aversion 9. What risk aversion would be necessary in order for the portfolio that splits wealth evenly between these two assets to be optimal?

Solution. Let $w_2 = w$ be the proportion of wealth devoted to the risky asset; then $w_1 = 1 - w$ is the proportion that goes to the risk-free asset. By the given information, the objective function is:

$$f(w) = (1 - w)(.03) + w(.05) - a \cdot w^2(.04)^2 = .03 + .02w - .0016aw^2.$$

We leave the risk aversion a general for the moment, anticipating the second question. The derivative is:

$$f'(w) = .02 - .0032aw$$

Setting this equal to 0 gives

$$.02 = .0032aw \Longrightarrow w = \frac{.02}{.0032a} = \frac{.02}{.0032 \cdot 9} = .694.$$

So about 69.4% of the wealth goes to the risky asset and 30.6% to the non-risky asset for an investor with this risk aversion. Now for the second question, we can set the general formula for w above equal to .5 and solve for a:

$$w = \frac{.02}{.0032a} = .5 \Longrightarrow a = \frac{.02}{.5 \cdot .0032} = 12.5.$$

Investors whose risk aversion is 12.5 or greater will prefer a proportion of .5 or more in the non-risky asset, and those with risk aversion less than 12.5 will choose a proportion of less than .5. ∎

Exercise 6 asks you to do the same optimization computation in general for one non-risky and one risky asset.

Notice that having solved for the portfolio weights in an optimal portfolio, we can substitute them into the formulas (1.41) and (1.42) for $E[R_p]$ and $\text{Var}(R_p)$ to find the mean and variance of the return on the optimal portfolio. For example, in the problem of Example 1, the mean portfolio return is:

$$\mu_p = (1 - .694)(.03) + .694(.05) = .0439,$$

and the variance is:

$$\sigma_p^2 = w^2(.04)^2 = (.694)^2(.04)^2 = .00111 \Longrightarrow \sigma_p = \sqrt{\sigma_p^2} = .0333.$$

So the mean return is a blend of the means for the two assets, and the standard deviation is less than that of the risky asset by itself, which is the purpose of diversifying the investment between the two.

Example 2. Brian is considering investing in two assets: A and B. His historical data indicates that the rates of return on these two are approximately uncorrelated. Over the last year the weekly rates of return for A averaged .0013 with a standard deviation of .0009, and for B the rates averaged .0016 with a standard deviation of .0020. If Brian's risk aversion is 6, find his optimal portfolio. By how much does that portfolio change if updated data indicates that the mean rate of return for A is .0015 (all other parameters being equal)?

Solution. Let w_A and w_B be the portfolio weights for the two assets; note that $w_B = 1 - w_A$, which allows us to model the objective in terms of one variable. Because there is no covariance term to consider, the portfolio objective function is:

$$f(w_A) = .0013 w_A + .0016 (1 - w_A) - 6 \left(.0009^2 w_A^2 + .0020^2 (1 - w_A)^2\right)$$

Expansion of this expression gives the quadratic polynomial:

$$f(w_A) = .001576 - .000252 w_A - .00002886 w_A^2.$$

The derivative is:

$$f'(w_A) = -.000252 - .00005772 w_A.$$

The equation $f'(w_A) = 0$ has solution

$$w_A = \frac{-.000252}{.00005772} = -4.3659.$$

The result of this computation is very unusual. The proportion $w_A = -4.3659 = -436.59\%$, hence $w_B = 1 - w_A = 5.3659 = 536.59\%$. This indicates massive short selling of asset A in order to devote extra wealth to asset B. For example, if Brian has \$10,000 to invest, he would short sell a value of \$43,659 in asset A to buy a total of \$53,659 of asset B. The more risky asset B with higher mean return is much more desirable for Brian than asset A, at least with his relatively low risk aversion of 6.

Redoing the calculation with the new mean rate of return for asset A, we get:

$$f(w_A) = .0015 w_A + .0016 (1 - w_A) - 6 \left(.0009^2 w_A^2 + .0020^2 (1 - w_A)^2\right)$$

$$f(w_A) = .001576 - .000052 w_A - .00002886 w_A^2.$$

$$f'(w_A) = -.000052 - .00005772 w_A.$$

$$w_A = \frac{-.000052}{.00005772} = -.9009.$$

Brian still short sells asset A, but now w_A is about $-.9 = -90\%$, and so $w_B = 1 - .(-.9) = 1.9$. With \$10,000 to invest for example, he would short sell \$9,000 worth of asset A, and invest \$19,000 in asset B. This was to be expected, because the increase in mean rate of return for asset A makes it relatively more valuable than it was before. ∎

Remark. In Exercise 11 you will derive general formulas for the proportion of wealth devoted to each of two independent risky assets in terms of the risk aversion parameter, and the means and variances. These formulas are:

$$w_1 = \frac{(\mu_1 - \mu_2) + 2a\sigma_2^2}{2a\left(\sigma_1^2 + \sigma_2^2\right)}; w_2 = 1 - w_1 = \frac{(\mu_2 - \mu_1) + 2a\sigma_1^2}{2a\left(\sigma_1^2 + \sigma_2^2\right)} \tag{1.44}$$

They allow us to look at the collection of possible variance-mean $\left(\sigma_p^2, \mu_p\right)$ combinations among all investors (i.e. risk aversions). We will be looking more deeply at this in Chapter 2 when we study the Capital Asset Pricing Model. With given mean and variance parameters for the rates of return on two independent assets, the weights w_1 and w_2 then become functions of the risk aversion constant a. In turn, the mean portfolio return

$$\mu_p = \mu_p(a) = w_1(a) \cdot \mu_1 + w_2(a) \cdot \mu_2 \tag{1.45}$$

and the variance

$$\sigma_p^2 = \sigma_p^2(a) = (w_1(a))^2 \sigma_1^2 + (w_2(a))^2 \sigma_2^2 \tag{1.46}$$

also are functions of a. Then a can be considered as a parameter, using which we can plot μ_p against σ_p^2. Figure 1.8 shows a case where the underlying asset parameters are $\mu_1 = .02, \sigma_1^2 = .0007, \mu_2 = .04,$ and $\sigma_2^2 = .002$, and the risk aversions range from 2 at the northeast end of the curve through 18 at the southwest end. As the risk aversion increases, more weight is given in the optimal portfolio to the first, safer asset, bringing the mean and variance down. For low risk aversions, short selling of the first asset is done to achieve a mean return that is even higher than that of asset 2, at the cost of a higher variance. ∎

Example 3. The final and most complicated of the two-asset portfolio optimization cases occurs when both assets are risky and they are not independent. Suppose that the correlation between the rates of return on asset 1 and asset 2 is $\rho = -.5$. The mean and variance parameters are $\mu_1 = .04$ and $\mu_2 = .06, \sigma_1^2 = .0008$, and $\sigma_2^2 = .001$. Find the optimal portfolio for an investor with risk aversion $a = 8$.

Solution. As usual, let $w = w_1$ and w_2 be the portfolio weights for assets 1 and 2. The covariance between asset rates is $\text{Cov}\,(R_1, R_2) = \rho \cdot \sigma_1 \cdot \sigma_2 = -.5\sqrt{.0008}\sqrt{.001}$. The maximization objective is:

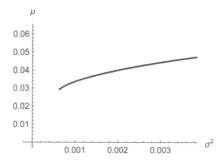

FIGURE 1.8 Optimal portfolio mean vs. variance, $\mu_1 = .02, \sigma_1^2 = .0007, \mu_2 = .04, \sigma_2^2 = .002$.

$$f(w) = (w)(.04) + (1 - w)(.06)$$
$$-8\left(w^2(.0008) + (1 - w)^2(.001) + 2w(1 - w)\left(-.5\sqrt{.0008}\sqrt{.001}\right)\right).$$

Simplifying the expression gives:

$$f(w) = .052 + .003155w - .021556w^2.$$

The derivative of the objective is:

$$f'(w) = .003155 - 2(.021556)w.$$

It is easy to check that the solution of the equation $f'(w) = 0$ is $w = .0732$, and the structure of the objective function shows that this must be a maximum, because of the negative coefficient on the square term. Thus, an investor with risk aversion 8 would devote 7.32% of total wealth to asset 1 and 92.68% to asset 2. ■

Exercises 1.4

1. One risky asset has mean rate of return $\mu_1 = .05$ and variance $\sigma_1^2 = .02$. A second asset, independent of the first, has mean $\mu_2 = .08$ and variance $\sigma_2^2 = .04$. Find the mean and variance of the portfolio rate of return for a portfolio that puts weight $3/4$ on asset 1 and $1/4$ on asset 2.

2. Assume that there are two risky assets with two possible outcomes as below. Find the mean and variance of the rate of return on a portfolio giving equal weight to the two assets.

(R_1, R_2)	$(.07, .04)$	$(.03, .05)$
probability	$2/5$	$3/5$

3. A risky asset has mean rate of return .05 and variance .2. There is a non-risky asset with deterministic rate of return .02. Derive the optimal portfolio of these two assets for an investor of risk aversion (a) 6; (b) 10; (c) 15.

4. (Technology required) This exercise studies the question: by how much does the optimal portfolio change as risky rate sinks, other parameters being equal? Suppose that a non-risky asset has rate of return .03, and a risky asset has mean rate of return $.05 - \epsilon$ and variance .02. Find the optimal weight on the risky asset in terms of ϵ for an investor with risk aversion 4. Plot and interpret a graph of this weight as a function of ϵ for values of ϵ from 0 to .02.

5. Consider a problem in which there is a non-risky asset with rate of return r and a risky asset with mean rate of return .08 and variance .005. Suppose that the investor has risk aversion 10. If the optimal portfolio gives weight $1/4$ to the non-risky asset and $3/4$ to the risky asset, what must have been the risk-free rate r? What is the expected rate of return on the optimal portfolio?

6. Derive the general formula below for the optimal portfolio weights w_1, w_2 in the one risky and one non-risky asset model. Note that μ_1 is the deterministic rate of return on the non-risky asset, μ_2 is the mean rate of return on the risky asset, σ_2^2 is its variance, and a is the risk aversion.

$$w_2 = \frac{\mu_2 - \mu_1}{2a \cdot \sigma_2^2}; \quad w_1 = \frac{\mu_1 - \mu_2 + 2a \cdot \sigma_2^2}{2a \cdot \sigma_2^2}$$

7. (Technology required) Use the formulas in Exercise 6 to plot a graph of optimal portfolio mean vs. variance similar to Figure 1.8, parametrically as a function of risk aversion. Use the asset parameters of Exercise 3, namely $\mu_1 = .02, \mu_2 = .05, \sigma_2^2 = .2$, and let a range from 2 through 15. Comment on what you see.

8. In Example 2, recalculate the optimal portfolio if Brian's risk aversion is (a) 12; (b) 20. (Use the revised mean rate for asset A of .0015 instead of the original rate.) Explain the relation between the change in risk aversions and the changes to the optimal portfolio.

9. Solve the problem of optimizing a two-asset portfolio of independent risky assets with means $\mu_1 = .03, \sigma_1^2 = .2, \mu_2 = .05, \sigma_1^2 = .4$ in three cases: (a) risk aversion 4; (b) risk aversion 8; (c) risk aversion 12.

10. Consider two independent assets with means .03 and .06 and variances .2 and .4 respectively. What range of risk aversions is such that there is no short selling required for the optimal portfolio?

11. Derive the formulas below for the optimal portfolio weights in the problem of two independent assets.

$$w_1 = \frac{(\mu_1 - \mu_2) + 2a\sigma_2^2}{2a(\sigma_1^2 + \sigma_2^2)}, w_2 = \frac{(\mu_2 - \mu_1) + 2a\sigma_1^2}{2a(\sigma_1^2 + \sigma_2^2)}.$$

12. Two risky assets have mean rates of return .04 and .06 and variances .03 and .07. The correlation between them is $-.5$. Find the optimal portfolio of an investor with risk aversion 5.

13. Find the optimal portfolio of the assets in Exercise 2 for an investor with risk aversion 3. Compute also the optimal portfolio mean and variance.

14. Suppose that two risky assets are available for investment, whose mean rates of return are $\mu_1 = .05, \mu_2 = .08$, whose standard deviations are $\sigma_1 = .04, \sigma_2 = .06$, and whose correlation is $\rho = .2$. Derive the optimal portfolio for an investor of risk aversion 6. Compute the mean and variance of the rate of return on this portfolio.

15. (Technology required) Using the means, standard deviations, and risk aversion of Exercise 14, solve for the proportion w of wealth in asset 1 in the optimal portfolio in terms of general correlation coefficient ρ. Plot w as a function of ρ for values of ρ between -1 and 1. Comment on what you see.

1.5   Derivative Assets and Arbitrage

In recent years, a great deal of emphasis in the subject of financial mathematics has been placed on the valuation of so-called **derivative assets**. These are financial instruments that are based on more fundamental underlying assets such as stocks or commodities. The purpose of this section is to introduce a few of the simplest examples of derivative assets, namely **futures** and **options**. We will also utilize an idea that has become a cornerstone of financial reasoning: in a market in equilibrium there should be no opportunities to make profit without initial investment that have no chance of loss. Such opportunities are examples of what is called **arbitrage**, and it turns out that the assumption of no arbitrage can enable us to give correct economic value to derivative assets. We will see how this works on futures contracts in this section. Section 1.6 extends the reasoning to simple options.

1.5.1   Futures

A **futures contract** involves two parties, the buyer and the seller. The contract specifies that at a given date, the buyer will purchase an underlying asset from the seller at a fixed price. No money changes hands when the contract is issued, and the transaction must take place, either by the actual transfer of the asset from the seller to the buyer, or by a cash settlement.

Example 1. Suppose that a futures contract on 10 ounces of gold is entered into by Pam, as the buyer, and Paula, as the seller. They agree that Pam will buy the gold at $1250/oz. in one year. The current market price of gold is $1200/oz. Now suppose that a year passes and the market price of gold moves to $1225/oz. Then Paula wins, because she turns over her gold, worth only $1225/oz., but she will be paid $1250/oz. by Pam. In fact, Paula may not have even owned the gold, but may at the last moment buy it in the market for a total of $12,250, and give it to Pam for $12,500, hence earning a profit of $250. If on the other hand the price of gold rises to $1275 per ounce, then the buyer, Pam, benefits. Pam pays Paula a total of $12,500 for the 10 ounces of gold, and can sell it in the market for $12,750, earning a total profit of $250, while Paula, on the other hand, loses $250. It is clear that the gold never had to change hands. In the first scenario, the contract could be settled by Pam paying Paula $250, and in the second scenario, Paula pays Pam $250. So futures contracts can be used to speculate in this way. A buyer is gambling that the market price of the underlying asset will rise beyond the futures contract price, and a seller is gambling that it will stay below that price. ∎

Besides speculation, futures contracts can be used to help reduce risk, as in the next example.

Example 2. (a) Suppose that a food manufacturer knows that next year it will need 10,000 bushels of wheat to support its operation. The current price of wheat is $4.25/bushel. Fearing that the price will rise drastically, the company enters into a futures contract to buy 10,000 bushels of wheat at a total price of $45,000 in one year. If the price of wheat in the market stays under $4.50/bushel, then the contract is a loser for the company, but if the market price rises above $4.50, then the manufacturer has the security of knowing that the future will be executed and the price to be paid for this supply is controlled at $45,000.

(b) A producer of zinc is concerned that the price of that metal will fall soon. It is currently $1.05 per pound. So the zinc producer enters into a futures contract as the seller of 15,000 pounds of zinc in six months at $1/lb. At the cost of accepting a price that is below the current market price, the producer has the comfort of knowing that its product will be sold at $15,000. If the price of zinc fell to $.90/lb., the savings to the producer is $.10/lb ×15,000 lbs = $1500. ∎

Let us introduce the following notations to describe a futures contract model:

S_0 = current market price of underlying asset;

T = time at which futures contract is to be executed;

S_T = market price of underlying asset at time T;

F = futures contract transaction price.

In Example 2(b) for instance, the underlying asset is 15,000 lbs. of zinc, the initial price is $S_0 = 15,000 \cdot \$1.05 = \$15,750$, the time is $T = 6$ months, the final price is $S_T = 15,000 \cdot \$.90 = \$13,500$, and the futures price is $F = 15,000 \cdot \$1 = \$15,000$. Notice that the zinc producer has benefited by:

$$F - S_T = \$15,000 - \$13,500 = \$1,500.$$

From the preceding examples, it should now be clear that the buyer in a futures contract benefits if the price F is less than the market price, and the seller benefits if it is greater. Specifically:

$$\begin{aligned} \text{Final value of futures contract to buyer} \ &= \ S_T - F; \\ \text{Final value of futures contract to seller} \ &= \ F - S_T. \end{aligned} \tag{1.47}$$

To bring these formulas together, it is customary to say that the buyer is long, or owns one unit of the future, and the seller is short, or owns negative one unit of the future (i.e. owes one unit). Then,

$$\text{Final value of 1 futures contract} = S_T - F. \tag{1.48}$$

Then the buyer's value is $+1 (S_T - F) = S_T - F$ and the seller's value is $(-1)(S_T - F) = F - S_T$.

Example 3. Suppose that the current market price of hogs is \$80/100 lbs. At the end of the season in 4 months, a breeder of hogs believes that the price per hundredweight will either be \$75, \$80, or \$85 with probabilities 1/2, 1/4, and 1/4 respectively. He has approximately 50,000 lbs of hogs to take to market then. What proportion of his value should he put up in a 4 month futures contract at \$80/100 lbs in order that his expected revenue is at least \$39,750?

Solution. Notice that the price per pound will either be \$.75, \$.80, or \$.85. In the three possible scenarios, he will earn:

$50,000 \cdot \$.75 = \$37,500; \ 50,000 \cdot \$.80 = \$40,000; \ 50,000 \cdot \$.85 = \$42,500.$

If he trusts to chance and does not enter into a futures contract, then his expected revenue is:

$$.5(\$37,500) + .25(\$40,000) + .25(\$42,500) = \$39,375.$$

So this does not quite meet the requirement of \$39,750. Now if he took out a futures contract at \$80/100 lbs, or \$.80/lb, then his 50,000 lbs of hogs guarantees him exactly \$40,000, but also limits his possibility to earn more if the price goes up to \$.85/lb. Now suppose that he contracts for a proportion p of his weight in hogs to be sold at \$.80/lb. Then the proportion $1-p$ is sold at whichever of the three market prices prevails in four months. His expected revenue is then:

$$p \cdot 50,000 \cdot \$.80 + (1-p) \cdot 50,000 \cdot ((.5)(\$.75) + (.25)(\$.80) + (.25)(\$.85)).$$

Notice that in the second term of the sum, 50,000 times the parenthesized quantity is just the expected revenue of \$39,375 computed above, and also in the first term, 50,000 times \$.80 is \$40,000. We may equate this expression to the target revenue of \$39,750 and solve for p to get:

$$\$40,000p + \$39,375(1-p) = \$39,750 \implies p = .6.$$

Hence, 60% of the total weight in hogs, or 30,000 pounds should be contracted in the future at \$.80/lb. ∎

1.5.2 Arbitrage and Futures

We mentioned the idea of risk-free, investment-free profit above. More formally:

Definition 1. An **arbitrage** opportunity for an investor is a financial situation in which the investor can have no net monetary outlay at the beginning, and yet have a positive probability of profit after the investment and zero probability of loss. ∎

Notice that we do not require certainty of profit, just a positive probability of profit, although in some examples such certainty does prove to be the case. The point of view that we will take in valuing derivative assets such as futures and options is that the market has no arbitrage opportunities for any investor. If it did, then speaking roughly, astute investors would quickly move in and try to buy undervalued assets and short sell overvalued assets, driving up the price of the former and driving down the price of the latter, until arbitrage-free equilibrium is reached.

Example 4. Suppose that there is a risky asset whose rate of return has the following discrete probability distribution.

rate of return	.035	.04	.045
probability	1/3	1/3	1/3

If there is also a non-risky asset available in the market whose rate of return is 3.5%, then an investor can construct arbitrage by borrowing on the non-risky asset and using the money to purchase the risky asset. For example, suppose that the investor borrows $10,000 at rate 3.5%, and invests that money in this risky asset. No net initial investment by the investor has taken place. After one time period, the investor must repay the loan, which has accrued to a value of $10,000(1.035) = $10,350$. But the risky asset is worth either:

$$\$10,000(1.035) = \$10,350, \ \$10,000(1.04) = \$10,400,$$

$$\text{or} \ \$10,000(1.045) = \$10,450.$$

This means that the investor can sell off the risky asset, pay back the loan, and break even in the first case, profit by $50 in the second case, or profit by $100 in the third case. Thus, there is no possibility of loss, and with probability 2/3, the investor earns positive profit, all without any initial out-of-pocket expense. This situation satisfies the definition of arbitrage. ∎

Example 5. Now suppose that there are two risky assets that move in lock step, in which the four possible pairs of final values are as below.

asset 1 final value	$4000	$5000	$6000	$7500
asset 2 final value	$2000	$2500	$3000	$4000
probability	1/8	1/2	1/8	1/4

Asset 1 has initial value $5000 and asset 2 has initial value $2500. Construct an arbitrage opportunity.

Solution. Inspecting the pairs of possible final values, we notice that in each case but one, just as initially, asset 1 has twice the value of asset 2. The exception is the last case, where asset 1 has less than twice the value of asset 2. This suggests that we may be able to arbitrage by short-selling asset 1 in order to be able to buy asset 2. Suppose that we are able to short sell 10 units of asset 1 at time 0, yielding $50,000, which we use to purchase 20 units of asset 2 at its initial price of $2500. If we do this, there is no net initial investment. At the final time, we must repay the 10 units of asset 1 at whichever of the four prices prevails, and we can cash in the value of the 20 units of asset 2. Thus, our net position in each of the four cases is:

$$
\begin{aligned}
\text{case1} &: \quad 20(\$2000) - 10(\$4000) = \$40,000 - \$40,000 = 0 \\
\text{case2} &: \quad 20(\$2500) - 10(\$5000) = \$50,000 - \$50,000 = 0 \\
\text{case3} &: \quad 20(\$3000) - 10(\$6000) = \$60,000 - \$60,000 = 0 \\
\text{case4} &: \quad 20(\$4000) - 10(\$7500) = \$80,000 - \$75,000 = \$5000
\end{aligned}
$$

So, with probability $1/8 + 1/2 + 1/8 = 3/4$ we have a net profit of 0, but with probability 1/4 the net profit is $5000. Since there was no initial net investment, this is an arbitrage strategy. ∎

We would now like to use the assumption that the market has no arbitrage in order to see that the transaction price F in a futures contract can only be one value, given the initial price S_0 of the underlying asset and the risk-free rate r. Without loss of generality, we suppose that $T = 1$.

For concreteness, suppose that the underlying asset is currently priced at $1000, the risk-free rate is 2%, and a buyer and seller in a futures contract on this asset are both interested in safeguarding their interests. Taking the buyer's point of view first, the concern is that the market price at the transaction time $T = 1$ is less than the contract price F; then the buyer is overpaying for the asset. The buyer can short sell the asset now and receive $1000, then invest this money at the risk-free rate in order to have $1020 at time $T = 1$. With that money the buyer hopes to be able to purchase the asset from the seller, and turn it over to the short-sales partner to eliminate that obligation. Hence the buyer wants the futures price to be no more than $1020. Now consider the seller's point of view. If the final price of the asset S_T exceeds the contract price, then the seller is making the sale for a lower price than would be available in the market, and in fact, if the seller does not currently hold the asset, he would have to purchase it in the market for a higher price S_T, only to give it to the buyer for the lower price F. To guard against this possibility, the seller can purchase the asset now for $1000, borrowing the money at the risk-free rate to do so. At time $T = 1$ the seller will owe $1020 on the loan, so the seller will want the contract price F to be at least this value in order to avoid losing money. So on the one hand, the buyer wants the price F to be no more than $1020, and on the other hand, the seller wants F to be at least $1020, hence $1020 = 1000(1.02)$ is the price that both can agree on to allay the fears of both that the market will go in the wrong direction for them.

We can generalize the ideas of the previous discussion as follows:

Theorem 1. To avoid arbitrage, the transaction price on a futures contract that is executed at time $T = 1$ must be:

$$F = S_0 \cdot (1 + r), \qquad (1.49)$$

where S_0 is the current market price of the underlying asset, and r is the risk-free rate.

Proof. We use a proof by contradiction to establish that neither can $F < S_0(1 + r)$, nor can $F > S_0(1 + r)$, hence F must be identically equal to $S_0(1 + r)$.

If it was true that $F < S_0(1 + r)$, then an arbitrage opportunity can be constructed for the buyer of the contract as follows: the buyer sells the asset short now, and invests the proceeds in a risk-free asset. No net investment occurs at the beginning. When the contract is due to be executed, the risk-free investment has grown to a value of $S_0(1 + r)$, and since this is more than the contract price F, the buyer profits by $S_0(1 + r) - F$, and also has the asset to return to the short sales partner. This is arbitrage, which contradicts the assumption that arbitrage does not exist.

If, on the other hand, it was true that $F > S_0(1+r)$, then the seller has an arbitrage opportunity. Suppose that the seller buys the asset now at a price S_0, and borrows the money at the risk-free rate r in order to do it. The seller has expended no money at the beginning. When the futures contract is executed, the seller receives F and pays back the loan, which has accrued to $S_0(1+r)$. Since we are supposing that $F > S_0(1+r)$, the seller has a certain profit with no initial investment, which is arbitrage. This contradiction indicates that F cannot exceed $S_0(1+r)$. Together, the two parts of the argument establish that $F = S_0(1+r)$. ■

Remark. If the transaction time is a general time T instead of 1, the same argument shows that the futures price is:

$$F = S_0 \cdot (1+r)^T. \tag{1.50}$$

Example 6. A futures contract is available on an asset whose current price is $20. The contract price is $22, and the risk-free rate of return is 5%. Explain how an arbitrageur can create riskless profits with no initial net investment.

Solution. The arbitrage-free price of the future would be:

$$S_0(1+r) = \$20(1.05) = \$21.$$

The contract price is high, so the seller in the contract has an arbitrage opportunity. That person enters into a futures contract, borrowing $S_0 = \$20$ now at the risk-free rate in order to pre-purchase the asset. No money is laid out initially. When the contract is due to be executed, this seller gives the share to the buyer, who pays $22, but since the seller only owes $20(1.05) = \$21$ on the loan, $1 is guaranteed profit, with no expenditure and no risk. ■

Example 7. Suppose that the market price of copper is $2.60/lb. Money can be borrowed at the rate of 2% per semiannum. A futures contract for 1000 lbs. of copper in six months is available with price $F = \$2630$. Is there an arbitrage opportunity? If so, how would it work?

Solution. The arbitrage-free price of this futures contract would be

$$S_0(1+r) = \$2.60 \cdot 1000(1.02) = \$2652.$$

Since the contract price F is just $2630, this creates an arbitrage situation for the buyer. He will short-sell 1000 lbs of copper now, earning $2600, and invest the proceeds at the risk-free rate of 2%, which will give him $2652 at contract execution time. He purchases the copper as per the futures contract, returns this asset to the short-sales partner to eliminate the obligation, and has a net of $2652 - $2630 = 22 in the end, with no risk and no initial net investment. ■

1.5.3 Options

A slightly more complicated type of financial derivative is the following.

Definition 2. An *option* is a derivative in which the holder of the option has the choice of whether to execute it or not. A *call option* gives its holder the right, but not the obligation, to buy an asset at a fixed price at a future time. Similarly, a *put option* gives its holder the right to sell an asset at a fixed price at a future time. The fixed price of transaction is called the *strike price* or *exercise price*. ■

As with futures, there are two parties in an option contract, the party who issues it and the party who owns it, but they are not financially symmetric. The issuer of an option charges the holder of the option a premium at the beginning of the contract, while the purchaser of this option receives in return the right to execute it or not depending on whether it benefits her.

Options further subdivide into two main cases:

Definition 3. A *European option* can only be executed at one final time T. An *American option* can be executed at any time up through T. ■

There are other hybrids, for instance, options that have a small set of times prior to T at which the holder can choose to execute the contract. We will look at some of these later.

What is the final payoff at execution time T of European options? We use the same notations S_0 and S_T for the beginning and ending prices of the underlying asset as we did with futures, and we also denote by K the strike price in the option contract. Consider a holder, Elaine, of a European call. She benefits if the market price S_T at time T exceeds the strike price K, because she is then able to execute the call, purchase the asset from the issuer at K, then sell it in the market for a profit of $S_T - K$. If the market price falls below K, she need not execute the option; if Elaine really wants the asset she can buy it in the market for less than K. This reasoning implies that the payoff of the European call at time T is:

$$V_T^{\text{call}} = \max\left\{S_T - K, 0\right\}. \tag{1.51}$$

Now consider the case of Jim, a holder of a European put option with strike price K and execution time T. Jim hopes that the market price S_T goes down below K, because then he can purchase the asset in the market at S_T, sell it to the option issuer for K, and profit by $K - S_T$. If the market price stays at least as high as K, there is no benefit to Jim to execute the put option. Thus,

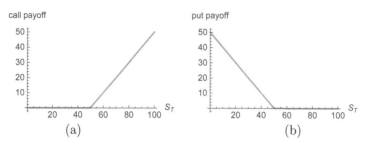

FIGURE 1.9 (a) Payoff function for European call option, $K = \$50$; (b) payoff function for European put option $K = \$50$.

the payoff of a European put at time T is:

$$V_T^{\text{put}} = \max\{K - S_T, 0\}. \qquad (1.52)$$

Figure 1.9 shows a graph of each of the two payoffs as functions of the variable S_T, for contracts for which the strike prices are $K = \$50$. Notice that the call payoff has no upper bound, but the highest that the put payoff can be is $K = \$50$, when the market price falls all the way to zero.

So the final values of European options are easy enough; the trick is to find arbitrage-free initial values.

Example 8. Suppose that an underlying asset has current price $1000 and has five possible prices at time 1, with the probabilities given in the table below. Find each of the possible final payoffs of: (a) a call option with strike price $1020; (b) a put option with strike price $980.

price	$950	$975	$1000	$1030	$1040
probability	.1	.2	.3	.3	.1

Solution. (a) We do not need the probabilities for this example, but they will be used in Example 9. The possible payoffs of a call of strike price $K = \$1020$ are given by the possible values of $\max\{S_1 - \$1020, 0\}$:

$$S_1 = \$950 : 0; S_1 = \$975 : 0; S_1 = \$1000 : 0;$$
$$S_1 = \$1030 : \$1030 - \$1020 = \$10; S_1 = \$1040 : \$1040 - \$1020 = \$20.$$

(b) For the put option with strike price $K = \$980$, the payoffs are the values of $\max\{\$980 - S_1, 0\}$:

$$S_1 = \$950 : \$980 - \$950 = \$30; S_1 = \$975 : \$980 - \$975 = \$5;$$
$$S_1 = \$1000 : 0; S_1 = \$1030 : 0; S_1 = \$1040 : 0. \blacksquare$$

Example 9. For the underlying asset of Example 8, if the one-period risk-free rate is .01, find the expected present values of the two options.

Solution. (a) The present values are found by multiplying the expected payoff by 1.01^{-1}. For the call option, we get:

$$1.01^{-1}(.1(0) + .2(0) + .3(0) + .3(\$10) + .1(\$20)) = \$4.95.$$

(b) The expected present value of the put option is:

$$1.01^{-1}(.1(\$30) + .2(\$5) + .3(0) + .3(0) + .1(0)) = \$3.96. \ \blacksquare$$

The computation of Example 9 might lead you to think that all that is necessary to find initial values of simple European options is to compute the expected present values of their final payoffs. The next section shows that this is not the case. Anti-arbitrage considerations force something different to take place.

For American options, if indeed an American option has not been executed prior to time T, then the time T payoff is the same as that of the corresponding European option on the same asset with the same strike price. But the additional control that the option holder has to choose when, if ever, to execute the option suggests that the value of an American option should be at least as high as that of a European option.

Theorem 2. Let C_A be the initial value of an American call option on an underlying asset with strike price K and expiration time T, and let C_E be the initial value of a European option with the same parameters on the same asset. If there are no arbitrage opportunities in the market, then,

$$C_A \geq C_E. \tag{1.53}$$

Proof. Suppose, on the contrary, that $C_A < C_E$. Then we can construct arbitrage as follows. Issue a European call option with the given parameters, use the proceeds C_E to purchase an analogous American option at price C_A, investing the difference $C_E - C_A$ at the risk-free rate. Hold the American option until T. If the market price S_T of the underlying asset exceeds K, then the European option partner will exercise that call, but simultaneously we can exercise our American call, purchasing the asset at K with the money given to us by the European call partner, and giving the asset to that person. There is no net profit or loss from this, and we have a certain accumulated value of $(C_E - C_A)(1 + r)^T$ from the risk-free investment, without having invested anything at the start. In the other case where $S_T \leq K$, both options expire worthless, and we still have the value of the risk-free investment. This contradicts the assumption of no arbitrage, and so it must be that $C_A \geq C_E. \ \blacksquare$

Exercise 18 asks you to show a similar theorem for put options.

In the next section we will put arbitrage reasoning to use in finding initial values for European call and put options, and in fact, for arbitrary single period financial derivatives. We will value American options in a later chapter, and we will also see that, surprisingly, American calls and European calls have equal initial values, but that is not the case for puts.

Exercises 1.5

1. Dan is the seller in a futures contract on corn, which specifies that in six months time he will sell 1000 bushels of corn to Marlene at a price of $3.70 per bushel. Find the profit or loss to Dan in each of the cases where the market price after six months is: (a) $3.50 per bushel; (b) $3.75 per bushel; (c) $3.80 per bushel. Find Dan's expected profit if all of these three scenarios are equally likely.

2. A food manufacturer needs 20 metric tons of tin for the coating on its cans next year. The current price of tin is $19,000/metric ton. Fearing a price rise, the manufacturer takes the buyer's role in a futures contract for half of the necessary tin at the current price. Find the price paid for all of the 20 tons of tin in each of the cases where the market price of tin next year is: (a) $18,500/metric ton; (b) $18,750/metric ton; (c) $19,250/metric ton; (d) $19,500/metric ton. Find the expected price paid for all of the tin if we assume all of: cases (a) and (b) are equally likely, cases (c) and (d) are equally likely, and case (c) is twice as likely as case (a).

3. Steve owns $10,000 of an asset currently priced at $100 per unit. He plans to sell the asset off in four months, and will safeguard his value by becoming the seller in a futures contract with a transaction price of $98 per unit. If the market price of this asset per unit in four months has the probability distribution below, how many units of the asset should Steve subject to the futures contract so that his expected loss (i.e. current value minus sellout value) is no more than $200?

price per unit	$92	$94	$96	$98	$100	$102	$104	$106
probability	.2	.2	.2	.1	.1	.1	.05	.05

4. In Example 3, what are the variance and standard deviation of the hog breeder's revenue, both in the case where he does not engage in a futures contract and in the case where he places 60% of his hog weight into the contract?

5. Suppose that a risk-free asset exists in the market with constant rate of return $r = .03$, and a risky asset has random rate of return with the distribution in the table below. Show that an arbitrage opportunity exists.

rate of return	.02	.022	.025	.03
probability	1/6	1/6	1/6	1/2

6. Suppose that two risky assets, A1 and A2, begin at values $1000 and $1500 respectively. There are five equally likely possibilities for the pairs of final values, indicated below. Is this an arbitrage situation?

| A1 final value | $900 | $1000 | $1100 | $1200 | $1300 |
| A2 final value | $1350 | $1500 | $1650 | $1800 | $1900 |

7. Suppose that in the country of Potsylvania the currency exchange rate is 1.5 units of their currency, the pot, per dollar, and in the country of Pansylvania, exchange rates are .8 pans (their currency unit) per dollar and 1.3 pots per pan. Create an arbitrage opportunity assuming that monetary exchanges can be done freely and instantaneously.

8. Suppose that Mickey and Minnie are the buyer and seller, respectively, in a futures contract on an asset currently priced at $10,000. The contract specifies the transaction price at the end of one period to be $10,500, and the risk-free rate is $r = .04$ per period. Does an arbitrage opportunity exist for one of them, and if so, how do they carry it out?

9. While not involved in plots to take over Middle Earth, Sauron keeps an Eye on the Middle Earth Board of Trade (MEBOT) for arbitrage opportunities to fund his wars. He knows that the Bank of the Dwarves both loans and gives interest on gold deposits at a rate of 5% per year. The current market price of truesilver (*mithril*) is 60 gold pieces per ounce, and Sauron notes that one-year *mithril* futures on MEBOT are priced at 62.4 gold pieces per ounce. How can Sauron design an arbitrage strategy that will profit him by at least 1000 gold pieces?

10. The table below gives five possible final prices of an underlying asset and their probabilities of occurrence. Find each of the possible final payoffs of: (a) a call option with strike price $95; (b) a put option with strike price $85. In both (a) and (b), find the expected payoff of the option.

| price | $80 | $90 | $100 | $110 | $120 |
| probability | .2 | .1 | .05 | .05 | .1 |

11. An underlying asset is such that in four quarters, its eight equally likely possible final prices per unit are as in the table below. The risk-free rate per quarter is $r = .01$. Find the expected present value of 1000 units of a call option with strike price $72 per unit of asset.

| price | $50 | $55 | $60 | $65 | $70 | $75 | $80 | $85 |
| probability | .125 | .125 | .125 | .125 | .125 | .125 | .125 | .125 |

12. For this problem we use the terminology "long" and "short" in reference to options according to whether the investor has purchased the option or issued it, respectively. A risky asset has current price $S_0 = 25$ and will have price

S_T, a random variable, at time T. A call option with exercise time T and exercise price $E = 26$ is available, and also a put option with exercise time T and exercise price $E = 24$ is available. Find:

    (a) the value of a long call at T if $S_T = 28.5$;
    (b) the value of a short call at T if $S_T = 27.2$;
    (c) the value of a long put at T if $S_T = 26.2$;
    (d) the value of a short put at T if $S_T = 22.8$.

13. A dealer is contracted to sell 10 futures based on the NYSE Dow Jones index. The futures price is \$15,000, with a cash settlement as the culmination of his gamble. Fearing that the index may go up, the dealer purchases 10 call options on the index with strike price \$15,000, due to expire at the same time as the futures. Each option costs \$40. What is the profit or loss for this dealer if the market value of the index at the common time of expiration of the two derivatives is \$16,000? \$14,000? What is the dealer's profit or loss if he had not purchased the call options?

14. Options can be combined to produce more complex derivatives. In one such example, suppose that $K_1 < K_2 < K_3$ are three strike prices on a particular underlying asset. In a **call butterfly spread** the investor holds one call on the asset at strike price K_1, another call on the same asset at strike price K_3, and issues two calls on the asset at strike price K_2. Assume that the option execution dates match and are equal to T. Write a general piecewise formula for the final payoff value of the butterfly spread in terms of K_1, K_2, and K_3. Graph it as a function of the final price S_T of the underlying asset if $K_1 = 8$, $K_2 = 10$, $K_3 = 12$. In light of the graph, what do you suppose that an investor who is interested in a call butterfly spread is expecting the underlying asset to do? What are the largest and smallest possible payoffs for these parameters?

15. Find an expression for, and graph, the final claim value of a portfolio containing a call option with exercise price 110 on an asset and a put option with exercise price 90 on the same asset. Under what market conditions might an investor want to hold such a portfolio?

16. Suppose that you own 100 shares of a stock that is now valued at \$50 per share. You fear a downturn in the market. Put options with a strike price of \$48 are available at \$1.25 per share. What is the smallest number of options to buy or sell in order to limit your worst-case losses in this dismal market to no more than \$500, including what you pay for the options?

17. What is the final claim value of a portfolio containing a share of stock, a put on the same stock of exercise price E, and short a call on the stock with the same exercise price? What financial implication follows from your answer?

18. Give an argument along the lines of Theorem 2 that the value of an American put option is at least as great as the value of a European put option with the same expiration time and strike price.

19. A **binary option** is a derivative in which the holder is paid either a fixed amount or nothing, depending on the behavior of an underlying asset. So these differ from European calls and puts, which give a payoff whose magnitude depends on the final price of the asset. Using the asset of Exercise 10, whose values and probabilities are reproduced below, find the expected payoffs of binary options such that: (a) the payoff is $10 if the value of the asset is greater than $100, and 0 otherwise; (b) the payoff is $5 if the asset value is less than $95 and 0 otherwise.

price	$80	$90	$100	$110	$120
probability	.2	.1	.05	.05	.1

20. As we will see in the next section, this question turns out to have a great deal to do with the valuation of derivatives. Suppose that in a single time period, a risky asset has initial value S_0 and just two possible rates of return, $b > a$, and there is a risk-free asset with rate of return r that lies between a and b. We say that the asset goes up if b is its rate of return; otherwise it goes down. A derivative on that asset has value V_u if the rate of return is b and V_d if the rate of return is a. In general we denote the random value of the derivative at time 1 by V_1.

   (a) Find a probability q for the event that the asset goes up, under which the expected present value of the asset at time 1 is the same as the initial value, that is, $E_q\left[(1+r)^{-1}S_1\right] = S_0$;

   (b) For the value q from part (a), compute the expected present value of the derivative's time 1 value: $E_q\left[(1+r)^{-1}V_1\right]$.

1.6 Valuation of Derivatives in a Single Time Period

Recall from the last section that a derivative is a financial object or contract whose value is based on the value of another, so-called underlying asset. We were able to use the assumption of no arbitrage to value futures contracts in Section 1.5; now we would like to value other derivatives, including European options, in the case of a single time period. We will also be simplifying assumptions to what will seem like an absurd degree. In the single time period in question, we will suppose that the underlying asset can only move to one of two possible values from its current value. As it does so, the derivative has two corresponding possible final claim values. But despite this apparently extreme reduction, we will be able to treat later the case of many time periods with very short time steps, in each of which the binary structure holds.

The methods of this section apply microscopically to each single step, leading to a macroscopic procedure for valuation of derivatives that contains suffi-cient richness and also provides a smooth transition to the continuous-time, continuous-state case.

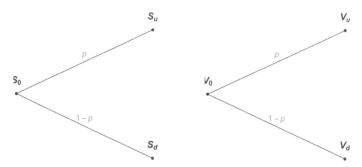

FIGURE 1.10 Single period binomial branch model for underlying asset and derivative.

1.6.1   Replicating Portfolios

The model described above in which the underlying asset has two possible values at time 1 is called a **one-step binomial branch process**. Specifically, as on the left of Figure 1.10, we suppose that at time 0, the underlying risky asset has a known value S_0, and at time 1 its value S_1 will be one of two possible values $S_u > S_d$, the "up" and "down" cases. The asset moves up with probability p and down with probability $1 - p$. Correspondingly, the derivative asset has two possible final claim values V_u and V_d according to whether the underlying asset moves up or down. We assume that these values are known in terms of S_u and S_d. The game that is afoot is to find the present value V_0 of the derivative at time 0.

Remark. When it is convenient, we will reparameterize the risky asset model by defining up and down rates of return b and a such that:

$$S_u = (1 + b)S_0; \quad S_d = (1 + a)S_0. \tag{1.54}$$

This turns out to be a better notational scheme when we progress later to the multiple time period case. Notice the inverse relationships:

$$b = \frac{S_u - S_0}{S_0}; a = \frac{S_d - S_0}{S_0}, \tag{1.55}$$

which show that we are correct in thinking about b and a as the two possible single-period rates of return on the risky asset. ■

In this binomial model we also suppose that there is a risk-free asset with rate of return r, using which present values are determined. Without loss of generality, we can suppose that the monetary unit of value is scaled such that at time 0 the risk-free asset has value 1, so that at time 1 it has value $(1+r) \cdot 1 = 1+r$. A total of c units of the risk-free asset is then worth c at the start and $c(1+r)$ after one time period.

The presence of the risk-free asset and the picture on the right of Figure 1.10 suggest that a possible way of valuing the derivative at time 0 is just the present value of its expected claim value, that is:

$$V_0 = (1+r)^{-1} \left(p \cdot V_u + (1-p)V_d \right) = E_p \left[(1+r)^{-1} V_1 \right], \qquad (1.56)$$

where V_1 is the random variable that models the value of the derivative at time 1. But we do not have any justification for this yet. In fact, the fascinating conclusion of this section is that this formula is incorrect, but only by a little.

First we need a consequence of the assumption of no arbitrage.

Theorem 1. (Law of One Price) In a financial market with a risk-free asset with rate r, unlimited borrowing and short-selling, and no arbitrage opportunities, if two assets A and B are guaranteed to have the same final values $A_1 = B_1$ for all possible outcomes, then they must have the same initial values $A_0 = B_0$.

Proof. We argue by contradiction. Suppose that the two assets do not have the same initial value, for instance, $A_0 < B_0$. Then we can construct arbitrage by short-selling a unit of asset B at the start, using it to buy a unit of asset A, and investing the leftover amount $B_0 - A_0$ in the risk-free asset. After the time unit passes, sell off asset A at A_1, and since this has the same value B_1 as asset B, the short sale can be repaid exactly, and with certainty, leaving $(1+r)(B_0 - A_0)$ in our hands with no initial net investment and no risk. The same argument can be applied if asset B has the smaller value at time 0. This contradiction of the no-arbitrage assumption implies that we must have $A_0 = B_0$. ∎

The Law of One Price will enable us to uniquely value the derivative in Figure 1.10 at time 0 under the assumption of no arbitrage (and free borrowing and short-selling). In the rest of this discussion we assume these things without explicit reference.

To get started, we try to construct an asset which is a portfolio of the underlying asset and the risk-free asset, which replicates the values V_u and V_d of the derivative at time 1. Suppose that the portfolio is to consist of Δ units of the underlying asset and c units of the risk-free asset. Then its value at time 1 is:

$$\Delta S_1 + c(1+r),$$

and recall that the random asset price S_1 can either be S_u or S_d. Matching the up and down values of the portfolio with those of the derivative, we want to choose Δ and c to solve the system of linear equations:

$$\begin{cases} \Delta \cdot S_u + c(1+r) = V_u \\ \Delta \cdot S_d + c(1+r) = V_d. \end{cases} \qquad (1.57)$$

Subtracting the bottom equation from the top gives:

$$\Delta\,(S_u - S_d) = V_u - V_d$$

so that the number of units Δ of the risky asset to use in the portfolio is the ratio of the difference in final derivative values to the difference in final underlying asset values. This value of Δ can be substituted into the bottom equation in (1.57) to obtain:

$$\frac{V_u - V_d}{S_u - S_d} \cdot S_d + c(1+r) = V_d,$$

from which it is easy to solve for the number of units c of the risk-free asset. In summary,

$$\Delta = \frac{V_u - V_d}{S_u - S_d}, \quad c = (1+r)^{-1}\left(V_d - \frac{V_u - V_d}{S_u - S_d} \cdot S_d\right). \qquad (1.58)$$

Take a look back at what we have done. There is a portfolio of risky asset and risk-free asset which is guaranteed in both the up and down cases to have the same values V_u and V_d as the derivative has at time 1. Therefore, the Law of One Price shows that this replicating portfolio must have the same initial value V_0 as the derivative. So the derivative value at time 0 must be:

$$V_0 = \Delta S_0 + c = \frac{V_u - V_d}{S_u - S_d} \cdot S_0 + (1+r)^{-1}\left(V_d - \frac{V_u - V_d}{S_u - S_d} \cdot S_d\right). \qquad (1.59)$$

This formula uniquely values the derivative in terms of the known possible claim values, the possible values of the underlying asset, and the risk-free rate of return, assuming no arbitrage and unlimited borrowing and short-selling (because it can well be that Δ or c are negative).

The second term in formula (1.59) can be rewritten somewhat more appealingly as in the next theorem by getting a common denominator. We leave the details to the reader. This formula for the initial value of a derivative is the form that we will use from here on, and refer to as the *replicating portfolio formula*.

Theorem 2. Under the assumptions and model structure described above, the initial value of the derivative is:

$$V_0 = \frac{V_u - V_d}{S_u - S_d} \cdot S_0 + (1+r)^{-1} \cdot \frac{S_u \cdot V_d - S_d \cdot V_u}{S_u - S_d}. \ \blacksquare \qquad (1.60)$$

An unexpected consequence of what we have accomplished is that the up probability p does not figure into the expression at all.

Example 1. A numerical example will be helpful to understand the replication idea more concretely. In a single-period binomial model, an underlying asset has the behavior on the left of Figure 1.11, and a derivative on that asset has final claim values 5 and 0 as shown on the right of the figure. There is a risk-free asset with per-period rate $r = .01$. Find a portfolio of the risky asset and the risk-free asset that replicates the derivative. What is the initial value V_0 of the derivative?

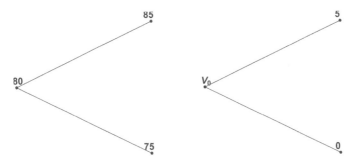

FIGURE 1.11 A particular binomial branch process.

Solution. Let Δ be the number of units of the underlying asset and let c be the number of units of the risk-free asset in the portfolio. Then in both the up and down cases we want to match the portfolio value with the derivative value, which produces the system of equations:

$$\begin{cases} \$85 \cdot \Delta + c(1.01) = \$5 \\ \$75 \cdot \Delta + c(1.01) = \$0 \end{cases}$$

Subtracting the two equations yields:

$$\$10 \cdot \Delta = \$5 \Longrightarrow \Delta = .5.$$

Substitution of this value of Δ into the second equation produces:

$$\$75 \cdot (.5) + c(1.01) = \$0 \Longrightarrow c = \frac{-\$37.5}{1.01} = -37.1287.$$

So we must borrow 37.1287 units of the risk-free asset (that is, take out a loan) and purchase .5 unit of the underlying risky asset to construct a replicating portfolio. Since this portfolio has the same final values as the derivative, to avoid arbitrage it must have the same initial value as the derivative. Therefore, we can price the derivative at time 0 as:

$$V_0 = \Delta \cdot S_0 + c = .5 \cdot 80 - 37.1287 = \$2.8713. \blacksquare$$

Example 2. A stock initially priced at $20 per share follows a one-step bi-nomial branch process with up rate $b = .04$ and down rate $a = -.04$. The risk-free rate per period is $r = .01$. Find the current value of a European call option on this stock with strike price $20.50 and expiration time $n = 1$. What portfolio of risk-free asset and stock replicates the option?

Solution. Since $S_0 = \$20$, recall from the reparameterization described above, the possible time 1 values of the stock are:

$$S_u = S_0(1 + b) = \$20(1.04) = \$20.80;$$

$$S_d = S_0(1 + a) = \$20(.96) = \$19.20.$$

The call option is worth executing only in the up case, in which case $V_u = \max\{\$20.80 - \$20.50, 0\} = \$.30$, and $V_d = 0$. From the replicating portfolio formula, we obtain:

$$
\begin{aligned}
V_0 &= \frac{V_u - V_d}{S_u - S_d} \cdot S_0 + (1 + r)^{-1} \cdot \frac{S_u \cdot V_d - S_d \cdot V_u}{S_u - S_d} \\
&= \frac{\$.30 - \$0}{\$20.80 - \$19.20} \cdot \$20 + 1.01^{-1} \cdot \frac{\$20.80 \cdot (\$0) - \$19.20(\$.30)}{\$20.80 - \$19.20} \\
&\approx \$.186.
\end{aligned}
$$

Therefore the value of the single option is a mere 18.6 cents. (However options are often sold in lots, for example 50 or 100, so investors pay more than pennies for them.) The formulas for Δ and c allow us to find the number of shares of stock and the amount to hold or to borrow in the replicating portfolio:

$$\Delta = \frac{V_u - V_d}{S_u - S_d} = \frac{\$.30 - \$0}{\$20.80 - \$19.20} = .1875,$$

$$
\begin{aligned}
c &= (1 + r)^{-1} \left(\frac{S_u \cdot V_d - S_d \cdot V_u}{S_u - S_d} \right) \\
&= 1.01^{-1} \cdot \frac{\$20.80(\$0) - \$19.20(\$.30)}{\$20.80 - \$19.20} \\
&= -\$3.56436.
\end{aligned}
$$

Thus, the single option can be replicated by holding .1875 shares of stock and borrowing $3.56. ∎

Roughly, replication implies that:

Δ units of underlying asset $+ c$ units of riskless asset $= 1$ unit of derivative,

which is equivalent to:

c units of riskless asset $= 1$ unit of derivative $- \Delta$ units of underlying asset.

$$(1.61)$$

Hence, if an investor is interested in "hedging away" all risk, he should short sell Δ units of the underlying asset for each unit of the derivative that he holds. This is called the **Δ-hedge**. Of course, the investor can choose to hedge only partway to reduce risk somewhat while still having the potential to benefit by the upside case of the derivative. Qualitatively, if an investor has some calls on an asset for example, he can purchase the asset at the fixed exercise price if it is beneficial, and short-selling the asset allows him to gain capital initially, and gives him a place to dump the assets he buys at the exercise price if the market price of the asset has risen, or leave the options unexecuted and purchase the asset in the market at the lower price to resolve the short sale, if that is better.

We illustrate hedging in the next example.

Example 3. Suppose that an investor Kate holds 1000 European puts on an asset whose current value is $100. The strike price on the puts is $95. There is a risk-free asset with one-period rate of return .03, and the asset either goes up by 10% or down by 10% in this period. Analyze Kate's possible outcomes in the up and down cases if: (a) she does not hedge; (b) she uses the full Δ-hedge, purchasing 1000Δ units of the asset; (c) she uses a 50% Δ-hedge, purchasing 500Δ units.

Solution. (a) The initial price of the asset is $S_0 = \$100$, and the possible next prices are $S_u = \$110$ and $S_d = \$90$. Each individual put is worth nothing at time 1 in the up case, and is worth $\$95 - \$90 = \$5$ in the down case, that is, $V_u = 0, V_d = \$5$. So there are possible claim values for all of the puts of 0 or $5000. Kate must have paid the following for each put:

$$
\begin{aligned}
V_0 &= \tfrac{V_u - V_d}{S_u - S_d} \cdot S_0 + (1+r)^{-1} \cdot \tfrac{S_u \cdot V_d - S_d \cdot V_u}{S_u - S_d} \\
&= \tfrac{\$0 - \$5}{\$110 - \$90} \cdot \$100 + 1.03^{-1} \cdot \tfrac{\$110 \cdot (\$5) - \$90(\$0)}{\$110 - \$90} \\
&\approx \$1.699.
\end{aligned}
$$

So, her overall initial investment in the puts was $1699.03. If the asset moves up, the puts are worthless and she loses this $1699.03. If the asset moves down, then the puts are worth $5000, which gives her a net profit of $5000 - \$1699.03 = \$3300.97$.

(b) From formula (1.61), the number of shares of the asset that she should use to hedge each put is:

$$
-\Delta = -\frac{V_u - V_d}{S_u - S_d} = -\frac{\$0 - \$5}{\$110 - \$90} = .25
$$

This implies that she needs 250 shares of the underlying asset, at $100 per share, for an additional investment of $25,000. In total, she has paid $25,000 + \$1699.03 = \$26,699.03$. Although this seems counterintuitive from the point of view of safety, let us consider carefully the two scenarios, where the asset goes up and where it goes down.

In the up case, the puts are worthless, but the 250 shares that she has bought are now worth:

$$250 \cdot \$110 = \$27,500.$$

Therefore her profit in the up case is $\$27,500 - \$26,699.03 = \$800.97$. You can check that if she had simply invested $26.699.03 in the risk-free asset at rate .03, she would have ended with the value $27,500 that the shares are worth in the up case, verifying the hedging relationships. Specifically, 1000 put options and 250 shares of stock give the same final value and profit as $26,699.03 invested in the risk-free asset.

In the case that the asset moves down, the 250 shares are each worth $90 for a total value of $22,500, and the 1000 put options are each worth $5, for a total of $5000. Therefore, again the final value of the portfolio is $22,500+\$5000 = \$27,500$, which illustrates the riskless nature of the portfolio of options and shares of the asset.

(c) To use a 50% hedging strategy would mean to purchase $250/2 = 125$ shares of the asset, costing $12,500 in addition to the $1699.03 that Kate paid for the put options. Her total investment is then $14,199.03. If the asset goes up, then the options are worthless but she owns 125 shares of the asset for a value of $125(\$110) = \$13,750$. In this case she has a modest loss of $\$14,199 - \$13,750 = \$449.03$, much less than if she had not hedged at all. If the asset goes down, then each option is worth $5 and each share is worth $90; hence her total return is:

$$1000(\$5) + 125(\$90) = \$16,250$$

In the down case she profits by an amount $\$16,250 - \$14,199.03 = \$2050.97$. This is not as much as in part (a) where she does not hedge, but she still has the ability to make a substantial profit in this case without the possibility of a very large loss. ∎

1.6.2 Risk-Neutral Valuation

There is a way to rewrite the portfolio replication formula (1.60) more simply and intuitively, which gives a surprising result about what the initial value of a derivative really is. It is an expected present value, but not the one we guessed in formula (1.56) at the start of this section. To begin, we want a present value expression, so factor the term $(1+r)^{-1}$ out of the expression for V_0 to get:

$$
\begin{aligned}
V_0 &= \tfrac{V_u - V_d}{S_u - S_d} \cdot S_0 + (1+r)^{-1} \cdot \tfrac{S_u \cdot V_d - S_d \cdot V_u}{S_u - S_d} \\
&= (1+r)^{-1} \left((1+r)\tfrac{V_u - V_d}{S_u - S_d} S_0 + \tfrac{S_u \cdot V_d - S_d \cdot V_u}{S_u - S_d} \right).
\end{aligned}
$$

Now bring all terms in parentheses that have to do with V_u together, and all terms that involve V_d together:

$$
\begin{aligned}
V_0 &= (1+r)^{-1}\left((1+r)\tfrac{V_u-V_d}{S_u-S_d}S_0 + \tfrac{S_u\cdot V_d - S_d\cdot V_u}{S_u-S_d}\right) \\
&= (1+r)^{-1}\left(V_u\left(\tfrac{S_0(1+r)}{S_u-S_d} - \tfrac{S_d}{S_u-S_d}\right) + V_d\left(\tfrac{S_u}{S_u-S_d} - \tfrac{S_0(1+r)}{S_u-S_d}\right)\right) \\
&= (1+r)^{-1}\left(\left(\tfrac{S_0(1+r)-S_d}{S_u-S_d}\right)V_u + \left(\tfrac{S_u-S_0(1+r)}{S_u-S_d}\right)V_d\right).
\end{aligned}
$$

(1.62)

It is easy to see that the two coefficients of V_u and V_d sum to 1; call them q and $1-q$, respectively. For binomial branch processes, it is convenient to write them in the reparameterized form with up and down rates of return b and a as:

$$
q = \frac{S_0(1+r)-S_d}{S_u-S_d} = \frac{S_0(1+r)-S_0(1+a)}{S_0(1+b)-S_0(1+a)} = \frac{r-a}{b-a};
$$

(1.63)

$$
1-q = \frac{S_u-S_0(1+r)}{S_u-S_d} = \frac{S_0(1+b)-S_0(1+r)}{S_0(1+b)-S_0(1+a)} = \frac{b-r}{b-a}.
$$

(1.64)

In Exercise 5 you will argue that the risk-free rate r cannot be less than or equal to the asset down rate a, or else arbitrage would result. Thus, $r > a$ and so $q > 0$. But also, if r is at least as large as the up rate b for the risky asset, then arbitrage can be constructed. To do this, sell short one share of the asset at S_0, investing the proceeds in the risk-free asset. There is no net initial outlay on the part of the arbitrageur. This investment will be worth $S_0(1+r)$ at time 1, which is at least as large as the highest possible asset price $S_0(1+b)$, and strictly larger than the lowest possible price $S_0(1+a)$. This arbitrage possibility contradicts the assumption that $r \geq b$, hence $r < b$, and thus $1-q > 0$. This establishes that q is a valid event probability, which we call the **risk-neutral probability**, so that the expression on the right side of formula (1.62) is an expectation. We have now derived the following result.

Theorem 3. Under the assumptions of no arbitrage, and no limits to borrowing or short selling, and the binomial branch model structure, the initial value of the derivative is:

$$
V_0 = (1+r)^{-1}\left(q\cdot V_u + (1-q)V_d\right) = E_q\left[(1+r)^{-1}V_1\right],
$$

(1.65)

that is, the expected present value of the final claim value of the derivative, using the risk-neutral measure q of formula (1.63). ∎

So we now know that the derivative is indeed an expected present value of the final claim value at time 1, not using the original up probability p for the underlying asset, but instead a constructed probability q. Formula (1.65) implements what we refer to as the **risk-neutral valuation** technique for derivatives.

Our final two examples show how simple it is to carry out risk-neutral valuation.

Example 4. Redo the valuation of the derivative of Example 1 using the risk-neutral approach.

Solution. In Example 1, the parameters were $S_0 = \$80$, $S_u = \$85$, $S_d = \$75$, $r = .01$, $V_u = \$5$, and $V_d = 0$. The risk-neutral probability is:

$$q = \frac{S_0(1+r) - S_d}{S_u - S_d} = \frac{\$80 \cdot 1.01 - \$75}{\$85 - \$75} = .58.$$

Therefore the time 0 value of the option is:

$$V_0 = (1+r)^{-1} \left(q \cdot V_u + (1-q)V_d \right) = 1.01^{-1}(.58 \cdot \$5 + .42 \cdot 0) = \$2.8713,$$

which is the same price as we found in the earlier example. ■

Example 5. Suppose that a one-period put option exists on an underlying asset following the binomial branch model with up and down rates $b = .07$, $a = -.02$, and initial price \$200. The option has exercise price \$198, and the risk-free rate is $r = .04$. Find the initial price of the put option.

Solution. We can use the form of q that depends on a, b, and r:

$$q = \frac{r-a}{b-a} = \frac{.04 - (-.02)}{.07 - (-.02)} = \frac{.06}{.09} = \frac{2}{3}.$$

From the problem statement, $V_u = \max\{E - S_u, 0\} = \max\{\$198 - \$200(1.07), 0\} = \$0$, and

$$V_d = \max\{E - S_d, 0\} = \max\{\$198 - \$200(.98)\} = \max\{\$198 - \$196, 0\} = \$2.$$

Thus, risk-neutral valuation gives:

$$V_0 = (1+r)^{-1} \left(q \cdot V_u + (1-q)V_d \right) = 1.04^{-1} \left(\frac{2}{3} \cdot \$0 + \frac{1}{3} \cdot \$2 \right) = \$.6410.\ ■$$

Before closing this section, we would like to derive an interesting property of the risk-neutral probability q relative to the underlying asset. Consider the expected present value of the time 1 asset value. We have:

$$
\begin{aligned}
E_q\left[(1+r)^{-1}S_1\right] &= (1+r)^{-1}\left(qS_u + (1-q)S_d\right) \\
&= (1+r)^{-1}\left(\frac{S_0(1+r)-S_d}{S_u-S_d} \cdot S_u + \frac{S_u-S_0(1+r)}{S_u-S_d} \cdot S_d\right).
\end{aligned}
$$

Note that there is a product $S_u S_d$ appearing in parentheses once with a positive sign and once with a negative sign, which then disappears. We can therefore simplify the expression as:

$$
\begin{aligned}
E_q\left[(1+r)^{-1} S_1\right] & = (1+r)^{-1}\left(\frac{S_0(1+r)}{S_u - S_d} \cdot S_u - \frac{S_0(1+r)}{S_u - S_d} \cdot S_d\right) \\
& = (1+r)^{-1}\left(\frac{S_0(1+r)(S_u - S_d)}{S_u - S_d}\right) \\
& = (1+r)^{-1}\left(S_0(1+r)\right) = S_0.
\end{aligned}
\tag{1.66}
$$

Thus, q is a probability that makes the expected present value of the asset at time 1 equal to its current value at time 0, which is the same thing that formula (1.65) says about the option values. Exercise 18 asks you to show that our q value is in fact the only one that has this property. Another way of looking at this is to write (1.66) equivalently as: $E_q\left[S_1\right] = (1+r) S_0$. Under q, the risky asset gives expected values that grow like the deterministic values of a risk-free asset with rate r. In later sections when we move to multiple time periods, we will show that this behavior continues, which allows valuation of derivatives to be done along the same lines as in the single period case.

Exercises 1.6

1. Find a portfolio of a risk-free asset with rate of return .03 and a risky asset with initial price \$50 and possible time 1 prices \$52 and \$47 that replicates a derivative on this risky asset that pays \$3 if the asset price is \$52 and −\$1 if the asset price is \$47. What is the arbitrage-free initial price of the derivative?

2. A risky asset follows a one-period binomial branch process with up rate $b = .05$ and down rate $a = -.04$. Its initial price is \$100. A risk-free asset is available with rate of return .02. Find a replicating portfolio for a put option with strike price \$98. Also, compute the arbitrage-free initial price of the put option.

3. An asset that is currently priced at \$120 will in one period either reach a price of \$122 or \$130. The risk-free rate is 4%. Find a replicating portfolio for a call option on the risky asset with exercise price \$125, and use it to find the initial value of this call.

4. Suppose that a stock initially priced at \$40 follows a binomial branch process with $b = .03$ and $a = -.01$. The risk-free rate per period is $r = .01$. A single-period European put option on this asset has strike price \$40. Find the current value of this option. (You may either use the replicating portfolio technique or the risk-neutral technique.)

5. Show, using an argument similar to the case $r > b$ in the section, that if the risk-free rate r was less than or equal to the down rate a in a binomial branch model, then arbitrage would result.

6. Find the initial value of a European call option with strike price $50 on an underlying asset whose time 0 price is $50 and which follows a binomial branch process with parameters $b = .04$, $a = -.02$. Assume that the risk-free rate is $r = .02$ per period. (You may either use the replicating portfolio technique or the risk-neutral technique.)

7. A risky asset follows the binomial branch model with initial value $200, up rate .08, and down rate $-.04$. A derivative on this asset pays $10 if the asset goes up, and $-$5 if the asset goes down. The risk-free rate is $r = .02$. Use the risk-neutral approach to give an initial value to this derivative.

8. In a market with a risk-free asset with rate of return .04, and a risky asset with initial value $40 and possible final values $40 and $44, find the initial value of a derivative that pays $10 if the asset goes up to $44 and $2 if the asset remains at $40. Use the risk-neutral approach.

9. Analyze Kate's possible outcomes in Example 3 if she uses: (a) a 75% hedge; (b) a 25% hedge. Compare each to the 50% hedge results from the example.

10. What would have to be the case in order for the replicating portfolio of a derivative to demand short-selling of the underlying asset? Give a numerical example.

11. The discussion before Example 3 suggested that if an investor holds calls, then hedging can be done by short-selling the asset on which the calls are based. The proceeds from the short sale will be invested in a risk-free asset. To try this out, suppose that Bob holds 500 call options on an underlying asset that is currently worth $50. The option exercise price is $52, and the asset satisfies a binomial branch model with $b = .10$ and $a = -.02$. The risk-free rate is $r = .02$. Analyze the two possible outcomes of Bob's investment in the cases that: (a) he does not hedge; (b) he uses a full 100% hedge; (c) he uses a 50% hedge.

12. Timid Tom owns 400 shares of a risky asset with current value $125 per share. The possible time 1 values of this asset are $130 and $120. Tom worries that the down case will prevail and therefore would like to purchase put options, available at an exercise price of $124, so that he can sell off his holdings if he needs to at a reasonable price. Assume a risk-free rate of $r = .01$. How should Tom act if he wants to use: (a) a full 100% hedge; (b) a 75% hedge. What are his two possible profits or losses in each case?

13. Suppose that a call option is initially priced at $1.50. The underlying asset has initial value $40, and possible time 1 values $S_u = $50 or $S_d = $30. The risk-free rate is $r = .04$. What must the exercise price E have been? (Assume that E lies between S_u and S_d.)

Exercises 14-16 lead you through an interesting alternative to the valuation of a call option. Exercises 14 and 15 can be done individually, but the full impact of the idea is only achieved through doing the whole group.

14. Use (a) replicating portfolios; (b) risk-neutral valuation to value a derivative which is a binary option that pays a constant amount $(1 + r)c$ dollars or nothing according to whether an underlying asset exceeds $100 or not. Assume that the asset has initial value $100, and up and down rates $b = .05$, $a = -.05$, and the risk-free rate is $r = .01$. (This kind of derivative is called a *cash or nothing derivative*.)

15. Assume again the financial situation in Exercise 14. Use risk-neutral valuation to value a derivative which pays a number Δ shares of stock if that stock exceeds $100 in value after one period, or zero otherwise. (This is called a *stock or nothing derivative*.)

16. Compare the final claim value of a portfolio consisting of the cash-or-nothing derivative of Exercise 14 plus the stock-or-nothing derivative of Exercise 15 to a call option on the same risky asset, with exercise price $100, where Δ and c are chosen as in the replicating portfolio formulas. What can you conclude?

17. We can think of a future as a derivative held by the buyer in the sense of this section, and use replicating portfolios or risk-neutral valuation to value it, but there is a twist. No money changes hands at the beginning, so the initial value for the buyer is $0. Check that if the futures price F is left general and the initial derivative price in terms of F is forced to be zero, then the formula of Section 1.5, $F = S_0 \cdot (1 + r)$, follows. Use risk-neutral valuation, and also, let the problem parameters r, S_0, b, and a be general.

18. Show that for the binomial branch asset model with general up and down rates b and a, the risk-neutral probability q is the unique probability such that $E_q \left[(1 + r)^{-1} S_1 \right] = S_0$.

2

Portfolio Selection and CAPM Theory

Chapter 1 introduced the problem of optimizing portfolios of assets. But there we were limited to simple problems with two assets. In this chapter we would like to expand on the ideas, considering in the first section the problem of choosing optimal portfolios of many assets. Multivariable calculus is essential here, and also the computations are made easier and more compact using ideas from matrix algebra. In Sections 2.2 and 2.3 we use the material to present the famous CAPM, or **Capital Asset Pricing Model**, which gives surprising relationships between holdings in individual assets and the market behavior of all assets. One of the implications of this model is that if a market contains a risk-free asset, then all investors will hold portfolios that balance the risky securities in the same proportion to one another. Mathematical economists William F. Sharpe, Harry Markowitz, and Merton Miller jointly received the 1990 Nobel Memorial Prize in Economics for the development of this theory. The fourth section defines a new way of considering the objective of portfolio construction, using so-called **utility functions**. Finally, Section 2.5 introduces the study of multiple time period portfolio optimization.

2.1   Portfolio Optimization with Multiple Assets

In Section 1.4, we observed that the problem of mean-variance portfolio maximimization was a multivariable calculus optimization problem of maximizing the difference between the mean portfolio rate of return and the risk aversion times the variance of the portfolio rate of return. In compact form, we are to maximize $\mu_p - a \cdot \sigma_p^2$, but in full form the function we must cope with is:

$$\max_{(w_1, w_2, \ldots, w_n)} \sum_{i=1}^{n} w_i \mu_i - a \left(\sum_{i=1}^{n} w_i^2 \sigma_i^2 + \sum_{j=1}^{n} \sum_{\substack{k=1 \\ j \neq k}}^{n} w_j w_k \text{Cov}\left(R_j, R_k\right) \right), \quad (2.1)$$

where we also have the constraint:

$$\sum_{i=1}^{n} w_i = 1. \tag{2.2}$$

We start with a few examples in which multivariable calculus optimization can be brought to bear on the problem.

Example 1. Let us first consider a three asset model in which asset 1 is non-risky and assets 2 and 3 have correlated rates of return with correlation $\rho = -.5$. Let the rate of return on the non-risky asset be denoted by $\mu_1 = r = .02$, and suppose that the means for the other two assets are $\mu_2 = .05$ and $\mu_3 = .08$, and the standard deviations are $\sigma_2 = .10$, $\sigma_3 = .18$. Find the optimal portfolio for an investor with risk aversion 10.

Solution. To formulate the problem, let w_1, w_2, and w_3 be the weights on the three assets. Since $\sigma_1^2 = 0$ and also asset 1 has no correlation with the two risky assets, the objective function in formula (2.1) becomes:

$$f(w_1, w_2, w_3) = \begin{aligned}[t] & .02w_1 + .05w_2 + .08w_3 \\ & - 10\left((.10)^2 w_2^2 + (.18)^2 w_3^2 + 2w_2 w_3 \text{Cov}(R_2, R_3)\right). \end{aligned}$$

We will eliminate w_1 from the expression, using the relationship $w_1 + w_2 + w_3 = 1$, and also recall that $\text{Cov}(R_2, R_3) = \rho\sigma_2\sigma_3$, to obtain a new version of the objective as:

$$f(w_2, w_3) = \begin{aligned}[t] & .02(1 - w_2 - w_3) + .05w_2 + .08w_3 \\ & - 10\left((.10)^2 w_2^2 + (.18)^2 w_3^2 + 2w_2 w_3(-.5)(.10)(.18)\right). \end{aligned}$$

This simplifies to:

$$f(w_2, w_3) = .02 + .03w_2 + .06w_3 - .1w_2^2 + .18w_2 w_3 - .324w_3^2.$$

To optimize the function, we must find the partial derivatives with respect to each variable and set both to zero. The partials are:

$$\begin{aligned} \frac{\partial f}{\partial w_2} &= .03 - .2w_2 + .18w_3 = 0 \\ \frac{\partial f}{\partial w_3} &= .06 + .18w_2 - .648w_3 = 0 \end{aligned}$$

Solving simultaneously, we find $w_2 = .311$, $w_3 = .179$, and hence $w_1 = 1 - w_2 - w_3 = .510$.

Although we will not typically do this, recall that there is a second derivative test that checks to be sure that the critical point is actually a maximum. Calculate the second-order partial derivatives in the two variables and the mixed partial from which the discriminant can be found:

$$\frac{\partial^2 f}{\partial w_2^2} = -.2, \frac{\partial^2 f}{\partial w_3^2} = -.648, \frac{\partial^2 f}{\partial w_2 \partial w_3} = .18$$

$$\implies D = \frac{\partial^2 f}{\partial w_2^2} \cdot \frac{\partial^2 f}{\partial w_3^2} - \left(\frac{\partial^2 f}{\partial w_2 \partial w_3}\right)^2 = (-.2)(-.648) - (.18)^2 = .0972$$

Since the second order partial in w_2 is negative and the discriminant is positive, the critical point is guaranteed to be a maximum. The graph of the objective function is shown in Figure 2.1, and this maximum point does seem consistent with that graph.

We can also check the mean and variance of the optimal portfolio, which are:

$$\mu_p = .02w_1 + .05w_2 + .08w_3 = .02(.510) + .05(.311) + .08(.179) = .040;$$

$$\sigma_p^2 = (.10)^2.311^2 + (.18)^2.179^2 + 2(.311)(.179)(-.5)(.10)(.18)$$
$$= .0010033.$$

This means that the mean rate of return on the portfolio of about .04 lies between the non-risky rate of .02 and the mean rate on asset 2, which is logical because the portfolio is a linear combination of the assets with the majority of its weight on the non-risky asset. The standard deviation of the portfolio comes out as $\sigma_p = \sqrt{.0010033} = .0317$, which is much lower than either of the standard deviations of the risky asset rates. ∎

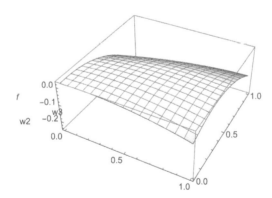

FIGURE 2.1 Portfolio objective function as a function of weights w_2 and w_3.

Example 2. The computations are rather easy when the assets are independent, because then there are no mixed product terms in the expression for the portfolio variance. Suppose now that there is a non-risky asset with constant rate of return r, there are three independent risky assets in the market with general means and variances of rates of return, and an investor has a general risk aversion a. Find expressions for the optimal portfolio weights.

Solution. Let the asset weights be denoted by w_i, $i = 1, 2, 3, 4$, where the non-risky asset is number 1. Observing that the covariance terms are all equal to zero, and writing $w_1 = 1 - w_2 - w_3 - w_4$, we obtain for the objective function the expression below.

$$f(w_2, w_3, w_4) = r(1 - w_2 - w_3 - w_4) + \mu_2 w_2 + \mu_3 w_3 + \mu_4 w_4 - a\left(\sigma_2^2 w_2^2 + \sigma_3^2 w_3^2 + \sigma_4^2 w_4^2\right).$$

We must again do the work of finding the critical point by setting the partial derivatives equal to zero, but we observe a very regular structure: each of the partials comes out as in formula (2.3) below:

$$\frac{\partial f}{\partial w_i} = (\mu_i - r) - 2a\sigma_i^2 w_i = 0 \implies w_i = \frac{\mu_i - r}{2a\sigma_i^2}, i = 2, 3, 4. \qquad (2.3)$$

There are some very interesting consequences of this representation that we will explore in the ensuing subsections. ∎

Example 3. Consider an extremely cautious investor for whom the minimization of variance means everything, regardless of mean return. In this market we will assume that there are three risky assets with standard deviations $\sigma_1 = .10$, $\sigma_2 = .15$, $\sigma_3 = .18$. Suppose that the correlations between rate of return pairs are $\rho_{1,2} = -.3$, $\rho_{1,3} = .2$, and $\rho_{2,3} = -.4$. What portfolio minimizes the portfolio variance?

Solution. The answer is not just obvious. Because of the asset correlations and the general benefit of diversification, it is possible that the best strategy is not simply to invest 100% of wealth in asset 1. For a portfolio with weights $w_1, w_2, w_3 = 1 - w_1 - w_2$, the variance is:

$$\begin{aligned} \sigma_p^2 &= \left(w_1{}^2\right)(.10)^2 + \left(w_2{}^2\right)(.15)^2 + (1 - w_1 - w_2)^2(.18)^2 \\ &\quad + 2(-.3)(.10)(.15)w_1 w_2 + 2(.2)(.10)(.18)w_1(1 - w_1 - w_2) \\ &\quad + 2(-.4)(.15)(.18)w_2(1 - w_1 - w_2) \\ &= .0324 - .0576 w_1 + .0352 w_1^2 - .0864 w_2 + .0765 w_2^2 + .0702 w_1 w_2, \end{aligned}$$

after tedious simplification. A plot of this variance as a function of the two portfolio weights is in Figure 2.2, and a minimum does seem to be indicated, occurring at a critical point.

With the variance σ_p^2 in this simplified form, it is easy to set the partial derivatives with respect to w_1 and w_2 equal to zero and solve simultaneously. Let $f(w_1, w_2)$ be the function that returns the portfolio variance; then:

$$\begin{aligned} \frac{\partial f}{\partial w_1} &= -.0576 + .0704 w_1 + .0702 w_2 = 0, \\ \frac{\partial f}{\partial w_2} &= -.0864 + .0702 w_1 + .153 w_2 = 0. \end{aligned}$$

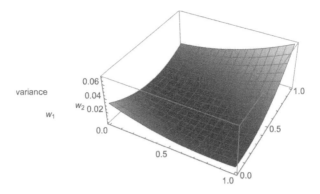

FIGURE 2.2 Portfolio variance as a function of weights w_1 and w_2.

The solution is $w_1 = .470; w_2 = .349$, and hence $w_3 = 1 - w_1 - w_2 = .181$. All of the assets should be held with positive weights in the minimum variance portfolio, but unsurprisingly, the highest proportion goes to asset 1, then asset 2, and then the most risky asset 3. If you substitute these values for w_1 and w_2 into the expression for σ_p^2, you get that the least possible variance itself is about .00378, which makes the minimum standard deviation σ_p equal to about $\sqrt{.00378} \approx .0615$. ∎

2.1.1 Lagrange Multipliers

Because the portfolio problem max $\mu_p - a\sigma_p^2$ is subject to the simple linear constraint $\sum_{i=1}^{n} w_i = 1$, it falls into the domain of the Lagrange multiplier technique. Let us do a quick review of how that is implemented, and see how a couple of our previous examples play out using this method.

To maximize a function of several variables $f(\boldsymbol{x}) = f(x_1, x_2, ..., x_n)$ subject to a single constraint on the variables, which can be written in the form $g(\boldsymbol{x}) = g(x_1, x_2, ..., x_n) = 0$, introduce a new real variable λ called the *Lagrange multiplier* and optimize $f(\boldsymbol{x}) - \lambda g(\boldsymbol{x})$ by setting the gradient below equal to zero:

$$\nabla(f(\boldsymbol{x}) - \lambda g(\boldsymbol{x})) = 0. \tag{2.4}$$

Recall that the gradient of a function j of several variables is the vector of partial derivatives:

$$\nabla j(\boldsymbol{x}) = \nabla j(x_1, x_2, ..., x_n) = \left(\frac{\partial j}{\partial x_1}, \frac{\partial j}{\partial x_2}, ..., \frac{\partial j}{\partial x_n} \right). \tag{2.5}$$

Formula (2.4) also translates to:

$$\nabla f(\boldsymbol{x}) = \lambda \nabla g(\boldsymbol{x}), \tag{2.6}$$

which is a system of equations in the variables that can be solved simultaneously with $g(x) = 0$. The additional condition is necessary to obtain a unique solution, since we have introduced the extra variable λ.

Example 4. In Example 1 we had a non-risky asset with rate of return $r = .02$, and two risky assets with means $\mu_2 = .05$ and $\mu_3 = .08$, and standard deviations $\sigma_2 = .10$, and $\sigma_3 = .18$. Their rates of return had correlation $-.5$. Use the Lagrange technique to recalculate the optimal portfolio for an investor with risk aversion $a = 10$.

Solution. Here we do not solve for one variable in terms of the others, so our problem is:

$$\max f\,(w_1, w_2, w_3) \;=\; \begin{aligned}[t] &.02w_1 + .05w_2 + .08w_3 \\ &-10\left((.10)^2 w_2^2 + (.18)^2 w_3^2 + 2w_2 w_3(-.5)(.10)(.18)\right)\end{aligned}$$

$$\text{subject to: } g\,(w_1, w_2, w_3) = w_1 + w_2 + w_3 - 1 = 0.$$

Notice how the constraint that the weight variables w_i sum to 1 is translated into the form $g(w) = 0$ by subtracting 1 from both sides. The key Lagrange relation $\nabla f(w) = \lambda \nabla g(w)$ becomes the system:

$$\begin{cases} .02 & = \lambda \cdot 1 \\ .05 - 20(.10)^2 w_2 + 10(.10)(.18)w_3 & = \lambda \cdot 1 \\ .08 - 20(.18)^2 w_3 + 10(.10)(.18)w_2 & = \lambda \cdot 1. \end{cases}$$

Thus the Lagrangian variable λ has the value .02, which is the same as r. The second two equations can now be rewritten as:

$$\begin{cases} .03 & = 20(.10)^2 w_2 - 10(.10)(.18)w_3 = .2w_2 - .18w_3 \\ .06 & = 20(.18)^2 w_3 - 10(.10)(.18)w_2 = .648w_3 - .18w_2. \end{cases}$$

These are the same linear equations in w_2 and w_3 that we derived in Example 1, so we confirm that $w_2 = .311$, $w_3 = .179$, and by the side constraint $g\,(w_1, w_2, w_3) = w_1 + w_2 + w_3 - 1 = 0$, it follows that again $w_1 = 1 - w_2 - w_3 = .510$. ∎

Example 5. Next let us use Lagrange multipliers to rederive formula (2.3) in the context of Example 2 in which there was a general non-risky asset and three general independent risky assets, and also a general risk aversion.

Solution. We must maximize the following:

$$f(w_1, w_2, w_3, w_4) = rw_1 + \mu_2 w_2 + \mu_3 w_3 + \mu_4 w_4 - a\left(\sigma_2^2 w_2^2 + \sigma_3^2 w_3^2 + \sigma_4^2 w_4^2\right),$$

subject to the condition:

$$g(w_1, w_2, w_3, w_4) = w_1 + w_2 + w_3 + w_4 - 1 = 0.$$

The gradient of f is:

$$\left(r, \mu_2 - 2a\sigma_2^2 w_2, \mu_3 - 2a\sigma_3^2 w_3, \mu_4 - 2a\sigma_4^2 w_4\right),$$

and the gradient of g is simply $(1, 1, 1, 1)$. This gives the system of equations $\nabla f(\boldsymbol{w}) = \lambda \nabla g(\boldsymbol{w})$ as below:

$$\begin{cases} r & = \lambda \cdot 1 \\ \mu_2 - 2a\sigma_2^2 w_2 & = \lambda \cdot 1 \\ \mu_3 - 2a\sigma_3^2 w_3 & = \lambda \cdot 1 \\ \mu_4 - 2a\sigma_4^2 w_4 & = \lambda \cdot 1. \end{cases}$$

Since $\lambda = r$ from the top equation, the rest are of the form $\mu_i - 2a\sigma_i^2 w_i = r$, whose solution, as in Example 2, is $w_i = \frac{\mu_i - r}{2a\sigma_i^2}, i = 2, 3, 4$. ■

2.1.2 Qualitative Behavior

It is easy to see from the work that we did in Examples 2 and 5 that in the general case of one risk-free asset with rate of return r and $n - 1$ independent risky assets with means μ_i and variances σ_i^2, $i = 2, 3, ..., n$, the optimal portfolio weights for investors with risk aversion a are:

$$w_i = \frac{\mu_i - r}{2a\sigma_i^2}, i = 2, 3, ..., n, \text{ and } w_1 = 1 - (w_2 + w_3 + \cdots + w_n). \qquad (2.7)$$

This formula allows us to see some simple, intuitively reasonable qualitative relationships between the weights and the individual problem parameters:

1. As the mean μ_i increases in relation to the risk-free rate r, the weight on asset i increases (proportionately);

2. As the risk aversion a increases, the weight on asset i decreases (in inverse proportion);

3. As the variance σ_i^2 increases, the weight on asset i decreases (in inverse proportion).

All of these are in the direction that you would expect for a rational investor.

Example 6. The following numerical example and related graphs illustrate the three relationships above. Suppose that in a market with a risk-free asset with rate of return $r = .04$, there are several independent risky assets including one specific asset whose historical mean rate of return is $\mu = .07$, and whose historical variance is $\sigma^2 = .015$. Consider an investor with risk aversion $a = 8$. If r, σ^2, and a remain the same, the relationship between the weight w on this asset and its mean rate of return is:

$$w = \frac{\mu - r}{2a\sigma^2} = \frac{\mu - .04}{16 \cdot .015} = \frac{\mu - .04}{.24}.$$

Figure 2.3 displays this relationship for values of μ from the non-risky rate of .04 through .10. If, all other parameters being constant, the mean rate of return on this asset grows, then it becomes more desirable to the investor, who will devote a higher proportion of wealth to it. In order that an investor of risk aversion 8 will give 20% of his wealth to it for example, the mean must satisfy:

$$\frac{\mu - .04}{.24} = .20 \Longrightarrow \mu = .04 + .24(.20) = .088.$$

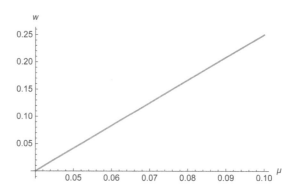

FIGURE 2.3 Portfolio weight as a function of mean rate of return.

Now suppose that the mean and variance are still at their historical levels of $\mu = .07$ and $\sigma^2 = .015$. The dependence of the portfolio weight on the investor's risk aversion a is below, and is graphed in Figure 2.4 for values of a from 2 through 20.

$$w = \frac{\mu - r}{2a\sigma^2} = \frac{.07 - .04}{2a \cdot .015} = \frac{.03}{.03a} = \frac{1}{a}.$$

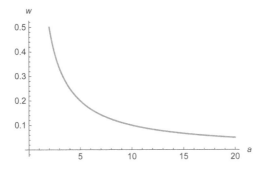

FIGURE 2.4 Portfolio weight as a function of risk aversion.

At lower risk aversions such as 2, $w = .5$ and the investor is more daring and more willing to devote wealth to the risky asset as opposed to the non-risky asset, while at higher risk aversions such as 20, $w = .05$ and the cautious investor much prefers the risk-free asset. For example, the risk aversion at which an investor would devote 10% of his wealth to this risky asset satisfies:

$$w = \frac{1}{a} = .10 \Longrightarrow a = 10.$$

Finally, suppose that we again look at an investor with risk aversion 8, suppose that the mean is still at its historical value of .07, but the variance may have changed. The relationship between the portfolio weight for the risky asset and its variance is:

$$w = \frac{\mu - r}{2a\sigma^2} - \frac{.07 - .04}{16 \cdot \sigma^2} = \frac{.03}{16 \cdot \sigma^2}.$$

The higher the variance, the lower the proportion devoted to the risky asset, as shown in Figure 2.5 for values of σ^2 ranging from .001 through .05. In this plot we see that for some very low variances, the portfolio weight for the risky asset is more than 1, requiring borrowing on the risk-free asset. The largest value of σ^2 for which this happens satisfies the equation:

$$1 = w = \frac{.03}{16 \cdot \sigma^2} \Longrightarrow \sigma^2 = \frac{.03}{16} = .001875. \blacksquare$$

2.1.3   Correlated Assets

Example 7. The convenient formulas for the optimal portfolio weights in the independent case are not as pleasant when the risky assets are correlated. Take the case of a non-risky asset and two correlated assets to begin to see if the qualitative behavior holds to the patterns of the independent case. Let

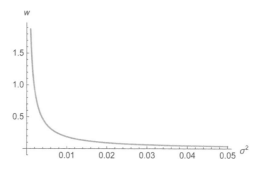

FIGURE 2.5 Portfolio weight as a function of variance.

the risk aversion be a general value a, let the non-risky rate be r, and suppose that risky assets 2 and 3 have general means μ_2 and μ_3, variances σ_2^2 and σ_3^2, and correlation ρ. Derive formulas for the portfolio weights w_2 and w_3.

Solution. We proceed using Lagrange multipliers. The objective function to maximize is:

$$f(w_1, w_2, w_3) = rw_1 + \mu_2 w_2 + \mu_3 w_3 - a\left(\sigma_2^2 w_2^2 + \sigma_3^2 w_3^2 + 2w_2 w_3 \rho \sigma_2 \sigma_3\right),$$

subject to:

$$g(w_1, w_2, w_3, w_4) = w_1 + w_2 + w_3 - 1 = 0.$$

The gradient of f is:

$$\left(r, \mu_2 - 2a\left(\sigma_2^2 w_2 + w_3 \rho \sigma_2 \sigma_3\right), \mu_3 - 2a\left(\sigma_3^2 w_3 + w_2 \rho \sigma_2 \sigma_3\right)\right),$$

and the gradient of g is $(1, 1, 1)$. The system $\nabla f(\boldsymbol{w}) = \lambda \nabla g(\boldsymbol{w})$ can be written out as:

$$\begin{cases} r & = \lambda \cdot 1 \\ \mu_2 - 2a\left(\sigma_2^2 w_2 + w_3 \rho \sigma_2 \sigma_3\right) & = \lambda \cdot 1 \\ \mu_3 - 2a\left(\sigma_3^2 w_3 + w_2 \rho \sigma_2 \sigma_3\right) & = \lambda \cdot 1. \end{cases}$$

Hence $\lambda = r$ as in the independent case, and the other two equations in w_2 and w_3 become:

$$\begin{cases} \frac{\mu_2 - r}{2a} & = \left(\sigma_2^2 w_2 + w_3 \rho \sigma_2 \sigma_3\right) \\ \frac{\mu_3 - r}{2a} & = \left(\sigma_3^2 w_3 + w_2 \rho \sigma_2 \sigma_3\right). \end{cases} \tag{2.8}$$

We could tediously solve for w_2 and w_3 by the usual elementary solve-and-substitute methods, but to preview what we will do later, view system (2.8) as a matrix linear system $\Sigma \boldsymbol{w} = \boldsymbol{c}$, where:

$$\Sigma = \begin{pmatrix} \sigma_2^2 & \rho\sigma_2\sigma_3 \\ \rho\sigma_2\sigma_3 & \sigma_3^2 \end{pmatrix}, \boldsymbol{w} = \begin{pmatrix} w_2 \\ w_3 \end{pmatrix}, \boldsymbol{c} = \begin{pmatrix} \frac{\mu_2 - r}{2a} \\ \frac{\mu_3 - r}{2a} \end{pmatrix} = \frac{1}{2a}\begin{pmatrix} \mu_2 - r \\ \mu_3 - r \end{pmatrix}.$$
$$(2.9)$$

Matrix Σ can be recognized as the **covariance matrix** of the random rates of return R_2 and R_3, with the individual variances on the diagonal and the covariance of the rates off the diagonal. The solution of the linear system is $\boldsymbol{w} = \Sigma^{-1}\boldsymbol{c}$. Fortunately, there is a convenient formula for the inverse of a 2×2 matrix:

$$A^{-1} = \begin{pmatrix} a_{11} & a_{12} \\ a_{21} & a_{22} \end{pmatrix}^{-1} = \frac{1}{a_{11}a_{22} - a_{21}a_{12}} \cdot \begin{pmatrix} a_{22} & -a_{12} \\ -a_{21} & a_{11} \end{pmatrix}. \quad (2.10)$$

Recall that the expression in the denominator of (2.10) is the **determinant** of the matrix A. Applying this to our coefficient matrix Σ,

$$\begin{aligned} \Sigma^{-1} = \begin{pmatrix} \sigma_2^2 & \rho\sigma_2\sigma_3 \\ \rho\sigma_2\sigma_3 & \sigma_3^2 \end{pmatrix}^{-1} &= \frac{1}{\sigma_2^2\sigma_3^2 - (\rho\sigma_2\sigma_3)^2} \cdot \begin{pmatrix} \sigma_3^2 & -\rho\sigma_2\sigma_3 \\ -\rho\sigma_2\sigma_3 & \sigma_2^2 \end{pmatrix} \\ &= \frac{1}{(1-\rho^2)\sigma_2^2\sigma_3^2} \cdot \begin{pmatrix} \sigma_3^2 & -\rho\sigma_2\sigma_3 \\ -\rho\sigma_2\sigma_3 & \sigma_2^2 \end{pmatrix}. \end{aligned}$$
$$(2.11)$$

Thus, the solution vector $\boldsymbol{w}$ is:

$$\begin{aligned} \boldsymbol{w} = \begin{pmatrix} w_2 \\ w_3 \end{pmatrix} &= \Sigma^{-1}\boldsymbol{c} \\ &= \frac{1}{(1-\rho^2)\sigma_2^2\sigma_3^2} \cdot \begin{pmatrix} \sigma_3^2 & -\rho\sigma_2\sigma_3 \\ -\rho\sigma_2\sigma_3 & \sigma_2^2 \end{pmatrix} \frac{1}{2a}\begin{pmatrix} \mu_2 - r \\ \mu_3 - r \end{pmatrix} \\ &= \frac{1}{2a(1-\rho^2)\sigma_2^2\sigma_3^2} \begin{pmatrix} \sigma_3^2(\mu_2 - r) - \rho\sigma_2\sigma_3(\mu_3 - r) \\ -\rho\sigma_2\sigma_3(\mu_2 - r) + \sigma_2^2(\mu_3 - r) \end{pmatrix}. \end{aligned}$$
$$(2.12)$$

Written in full and slightly simplified, we have that:

$$\begin{aligned} w_2 &= \frac{1}{2a(1-\rho^2)} \cdot \frac{\sigma_3^2(\mu_2-r)-\rho\sigma_2\sigma_3(\mu_3-r)}{\sigma_2^2\sigma_3^2} \\ &= \frac{1}{2a(1-\rho^2)} \cdot \left(\frac{(\mu_2-r)}{\sigma_2^2} - \frac{\rho(\mu_3-r)}{\sigma_2\sigma_3} \right); \\ w_3 &= \frac{1}{2a(1-\rho^2)} \cdot \frac{-\rho\sigma_2\sigma_3(\mu_2-r)+\sigma_2^2(\mu_3-r)}{\sigma_2^2\sigma_3^2} \\ &= \frac{1}{2a(1-\rho^2)} \cdot \left(\frac{(\mu_3-r)}{\sigma_3^2} - \frac{\rho(\mu_2-r)}{\sigma_2\sigma_3} \right). \end{aligned}$$
$$(2.13)$$

It is interesting to see that these formulas generalize formulas (2.3) in the independent case; specifically, they reduce to those formulas when $\rho = 0$. The role of the risk aversion a remains the same. All other parameters being unchanged, increasing a reduces the proportion of wealth in each risky asset. Also, increasing a mean rate of return μ_i, keeping all else unchanged, will increase the proportion of wealth in risky asset i, which is the same effect that we observed in the independent case. The effect of increasing the variance σ_i^2 of asset i is harder to see. The left term in parentheses in (2.13) will decrease, but so will the right term that is subtracted from it. But the net effect should still be to decrease w_i in most cases. If we take, for instance, the parameters in Example 1, namely $r = .02$, $\mu_2 = .05$, $\mu_3 = .08$, $\sigma_2 = .10$, $\sigma_3 = .18$, $\rho = -.5$, and $a = 10$, and consider what happens to w_2 as σ_2 increases from .05 to .3 we get the curve sketched in Figure 2.6. The expected decreasing relationship does hold here. In fact, you are asked to show in Exercise 19 that w_2 must decrease with an increase in σ_2 when the correlation ρ is negative, and also in the case that ρ is positive but ρ satisfies:

$$\rho < \frac{2\left(\mu_2 - r\right)/\sigma_2}{\left(\mu_3 - r\right)/\sigma_3}. \quad \blacksquare \tag{2.14}$$

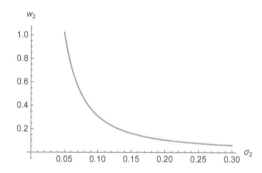

FIGURE 2.6 Portfolio weight as a function of standard deviation, dependent case.

2.1.4   Portfolio Separation and the Market Portfolio

In the case that a market contains a risk-free asset, there is a very interesting consequence of formula (2.3) in the independent case and formulas (2.13) in the case of two correlated risky assets that serves as a precursor for the CAPM theory to come. Again number the risk-free asset as asset 1, and suppose that there are $n - 1$ risky assets in the market as well, numbered as $2, 3, ..., n$. We have seen that if the risky assets are mutually independent, then the optimal portfolio weights for an investor with risk aversion a are:

$$w_i = \frac{\mu_i - r}{2a\sigma_i^2}, i = 2, 3, ..., n.$$

But then look at the wealth devoted to asset i as a proportion of the total part of the investor's wealth that goes to all risky assets:

$$w_i^* = \frac{w_i}{\sum_{j=2}^n w_j} = \frac{\frac{\mu_i - r}{2a\sigma_i^2}}{\sum_{j=2}^n \frac{\mu_j - r}{2a\sigma_j^2}} = \frac{\frac{\mu_i - r}{\sigma_i^2}}{\sum_{j=2}^n \frac{\mu_j - r}{\sigma_j^2}}, i = 2, 3, ..., n. \qquad (2.15)$$

The common factors 2 and a have divided away and so the quantity w_i^* defined above does not depend on the risk aversion. Put another way, regardless of risk aversion, all investors will apportion the wealth that they give to the risky assets in the same way; that is, asset i gets a share w_i^* of the wealth at risk. For example, if a risk-free asset exists and there are three risky assets that a particular investor holds in a 3:2:3 ratio, then all investors do the same, that is, their risky wealth is divided such that 3/8 goes to the first risky asset, 2/8 to the second, and 3/8 to the third. The exact fractions are determined by the problem parameters and formula (2.15). We give this collection of weights a name.

Definition 1. The ***market portfolio*** is the portfolio in a market with a risk-free asset and $n - 1$ independent risky assets whose relative portfolio weights (i.e. proportions of risky wealth devoted to each asset) are given by (2.15). ∎

Notice also that in our two risky asset correlated case, the formulas:

$$w_2 = \frac{1}{2a(1-\rho^2)} \cdot \left(\frac{(\mu_2 - r)}{\sigma_2^2} - \frac{\rho(\mu_3 - r)}{\sigma_2 \sigma_3} \right);$$

$$w_3 = \frac{1}{2a(1-\rho^2)} \cdot \left(\frac{(\mu_3 - r)}{\sigma_3^2} - \frac{\rho(\mu_2 - r)}{\sigma_2 \sigma_3} \right)$$

imply that the relative risky proportions $w_2^* = w_2 / (w_2 + w_3)$ and $w_3^* = w_3 / (w_2 + w_3)$ do not depend on a, since the same cancellation of the common factors 2 and a occurs as in the independent case. This does turn out to be true in general, so the existence of a market portfolio does not hinge on the independence assumption. We will see this in the next section.

We close with a computational example.

Example 8. Find the market portfolio in a market that has a risk-free asset with rate .03, and three independent risky assets with mean rates of return $\mu_2 = .06$, $\mu_3 = .08$, and $\mu_4 = .10$, and variances $\sigma_2^2 = .05$, $\sigma_3^2 = .15$, and $\sigma_4^2 = .25$. Find also the overall proportion of wealth devoted to all risky assets by investors with risk aversion 8 and 15.

Solution. Because of the second question, it will be useful to compute the denominator in formula (2.15). It is:

$$\sum_{j=2}^4 \frac{\mu_j - r}{\sigma_j^2} = \frac{.06 - .03}{.05} + \frac{.08 - .03}{.15} + \frac{.10 - .03}{.25} = 1.21333.$$

The market portfolio weights are just the individual terms that make up this sum, divided by the sum, which are:

$$w_2^* = \frac{\frac{\mu_2 - r}{\sigma_2^2}}{\sum_{j=2}^{4} \frac{\mu_j - r}{\sigma_j^2}} = \frac{\frac{.06 - .03}{.05}}{1.21333} = .4945$$

$$w_3^* = \frac{\frac{\mu_3 - r}{\sigma_3^2}}{\sum_{j=2}^{4} \frac{\mu_j - r}{\sigma_j^2}} = \frac{\frac{.08 - .03}{.15}}{1.21333} = .2747$$

$$w_4^* = \frac{\frac{\mu_4 - r}{\sigma_4^2}}{\sum_{j=2}^{4} \frac{\mu_j - r}{\sigma_j^2}} = \frac{\frac{.10 - .03}{.25}}{1.21333} = .2308.$$

For the second question, since the portfolio weights themselves are given by:

$$w_i = \frac{\mu_i - r}{2a\sigma_i^2},$$

the total proportion of wealth devoted to risky assets is:

$$\sum_{j=2}^{4} w_j = \sum_{j=2}^{4} \frac{\mu_j - r}{2a\sigma_j^2} = \frac{1}{2a} \sum_{j=2}^{4} \frac{\mu_j - r}{\sigma_j^2}.$$

But the sum is just the value 1.21333 computed above. Hence the results are:

risk aversion $8 : \dfrac{1}{2 \cdot 8} \cdot 1.21333 = .0758$; risk aversion $15 : \dfrac{1}{2 \cdot 15} \cdot 1.21333 = .0404$.

For these risk aversions, the investor spends a large majority of his wealth on the non-risky asset, apparently because the variances of the risky assets are too large for the investor to tolerate. Taking it a bit farther, we can plot the function $P(a) = \frac{1}{2a} \cdot 1.21333$, which gives the proportion of wealth given to all risky assets as a function of risk aversion, to produce the graph in Figure 2.7. The proportion at risk decreases in an inverse proportion to the risk aversion.
∎

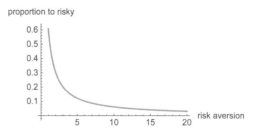

FIGURE 2.7 Proportion of wealth given to risky assets as a function of risk aversion.

Exercises 2.1

1. Suppose that there is a risk-free asset in the market with rate of return $r = .01$, and also there are two risky assets with mean rates of return $\mu_2 = .04$, $\mu_3 = .08$, variances $\sigma_1^2 = .1$, $\sigma_2^2 = .3$, and correlation $\rho = .2$. Use calculus to find the optimal portfolio for an investor with risk aversion $a = 4$. What are the mean and variance of the rate of return on this portfolio?

2. Use calculus optimization to solve the portfolio problem with the following information. There are four independent risky assets in the market with mean rates of return .03, .05, .07, and .09, and variances .08, .10, .12, .14, and the investor has risk aversion $a = 6$.

3. A market has four assets, one being non-risky with rate .02, and the others being independent risky assets with means .04, .06, and .08 and standard deviations .05, .07, and .09. Without using formula (2.3), for an investor with risk aversion 8: (a) find the optimal portfolio by calculus; (b) find the optimal portfolio using the Lagrange multiplier technique.

4. Derive a general system of equations for the optimal portfolio consisting of n independent risky assets (there is no non-risky asset) with general mean and variance parameters and general risk aversion. Use the Lagrange multiplier method to set up the equations, then use the constraint $w_1 + w_2 + \cdots + w_n = 1$ to solve for the Lagrange multiplier variable λ, and thereby to find explicit formulas for the optimal portfolio weights.

5. In the situation of Exercise 3, how large must the risk aversion be so that no more than .2 of the investor's wealth is devoted to the riskiest asset?

6. In the situation of Exercise 2, and again assuming a risk aversion of 6, how large must the mean on the second least risky asset be in order that it receives weight of at least .3? (You may need a computer algebra system to solve the necessary equations.)

7. In the situation of Exercise 1, suppose that the risk-free rate r is general and the other parameters remain the same. How large must the risk-free rate be so that the proportion of wealth at risk is no more than 4%?

8. Consider a market with three risky assets, with mean rates of return $\mu_1 = .06$, $\mu_2 = .08$, and $\mu_3 = .10$, and with standard deviations $\sigma_1 = .10$, $\sigma_2 = .15$, $\sigma_3 = .18$. Suppose that the correlations between rate of return pairs are $\rho_{1,2} = -.3$, $\rho_{1,3} = .2$, and $\rho_{2,3} = -.4$. Use the Lagrange multiplier

technique to derive the optimal portfolio for an investor with risk aversion $a = 3$. (You may need a computer algebra system to solve the necessary equations.)

9. Apply the Lagrange multiplier method to derive the optimal portfolio in a general case of two correlated risky assets, with general mean and variance parameters and general risk aversion.

10. Use the Lagrange multiplier method to find the optimal portfolio of an investor with risk aversion 12, if there is a non-risky asset with rate of return .01, and two independent risky assets with means $\mu_2 = .05, \mu_3 = .09$ and standard deviations $\sigma_2 = .08, \sigma_3 = .12$. Check to see that your result is consistent with what you get when you plug the parameters into formula (2.3).

11. Use the Lagrange multiplier method to find the optimal portfolio of an investor with risk aversion 6 in the case that there are three independent risky assets (no non-risky asset) with means .04, .06, and .08 and variances .03, .09, .12, respectively.

12. Set up, but don't solve, the Lagrange system of equations in the general case of one non-risky asset and $n - 1$ correlated assets. What does the system look like when expressed in matrix form?

13. In the context of Exercise 3: (a) find the market portfolio; (b) other parameters being equal, by how much must the mean rate of return μ_3 on the second risky asset increase such that its market portfolio weight w_3 is at least .36?

14. Suppose that, in addition to the three independent risky assets in Exercise 11, there is a non-risky asset with rate of return .01. (a) Find the market portfolio; (b) How small must the variance of the first risky asset be in order that its market portfolio weight is at least .5?

15. Referring to Exercise 14, compute the overall optimal portfolio weights for all four assets for an investor with risk aversion 2 and another with risk aversion 18. Check that the ratios of the weights of asset 3 to asset 2 and asset 4 to asset 2 are the same for each investor.

16. Find the market portfolio, and the optimal portfolio for an investor with risk aversion 10, in a market with a risk-free asset with rate of return .03, and two risky assets with mean rates of return .07 and .09, standard deviations .05 and .10, and correlation $-.2$.

17. Suppose that there are eight possible outcomes $\omega_1, ..., \omega_8$ and three risky assets in a market, whose values at each outcome are given in the table below. A non-risky asset with rate of return .005 is available. Find the market portfolio.

outcome	$P[\omega]$	value of R_2	value of R_3	value of R_4
ω_1	1/8	−.01	0	−.02
ω_2	1/8	−.01	0	.08
ω_3	1/8	−.01	.05	−.02
ω_4	1/8	−.01	.05	.08
ω_5	1/8	.03	0	−.02
ω_6	1/8	.03	0	.08
ω_7	1/8	.03	.05	−.02
ω_8	1/8	.03	.05	.08

18. In the introduction of the section we anticipated the existence of the market portfolio by saying: "all investors will hold portfolios that balance the risky securities in the same proportion to one another." Explain carefully how the work of this section justifies this claim, and make note of underlying assumptions.

19. As in Example 7, with problem parameters $r = .02$, $\mu_2 = .05$, $\mu_3 = .08$, σ_2 general, $\sigma_3 = .18$, $\rho = -.5$, and $a = 10$, analyze the first derivative of w_2 with respect to σ_2 to show that w_2 is a decreasing function of σ_2 when ρ satisfies:

$$\rho < \frac{2(\mu_2 - r)/\sigma_2}{(\mu_3 - r)/\sigma_3}.$$

Using this formula, construct a set of problem parameters and a range of values of σ_2 such that w_2 is not a decreasing function of σ_2 over that range.

20. (Technology required) This exercise previews the geometric approach to portfolio optimization and the market portfolio that we will take in the next section. Consider a market with three independent assets, with mean rates of return $\mu_1 = .06$, $\mu_2 = .08$, and $\mu_3 = .10$, and standard deviations $\sigma_1 = .10$, $\sigma_2 = .15$, $\sigma_3 = .18$. Write expressions for the possible portfolio mean and standard deviations in terms of the first two weights w_1 and w_2. Plot a large number of the potential pairs (σ_p, μ_p) for values of w_1, w_2 ranging through non-negative values such that $w_1 + w_2 \leq 1$. Describe the shape of the region of the (σ_p, μ_p) plane that is produced. On what part of that region are the optimal portfolios located? Approximate the smallest possible standard deviation.

2.2   Capital Market Theory, Part I

2.2.1   Linear Algebraic Approach

The purpose of the last section was to show standard multivariate calculus techniques, and also the Lagrange multiplier method, in the solution of instances of the general portfolio problem:

$$
\max_{(w_1,w_2,\dots,w_n)} \sum_{i=1}^{n} w_i \mu_i - a \left(\sum_{i=1}^{n} w_i^2 \sigma_i^2 + \sum_{\substack{j=1 \\ j\neq k}}^{n} \sum_{k=1}^{n} w_j w_k \mathrm{Cov}\left(R_j, R_k\right) \right), \quad (2.16)
$$

subject to:

$$
\sum_{i=1}^{n} w_i = 1. \tag{2.17}
$$

This sometimes entailed a great deal of computational struggle. We begin the section by bringing matrix methods into play to simplify these computations and to find a general solution to the portfolio problem that can also shed light on other issues. One of those issues is the extension of the market portfolio idea to the case of correlated assets. Another is the minimization of portfolio variance subject to a fixed portfolio mean. We continue the section on this note, and then go on to look at a geometric interpretation of the market portfolio. This geometry leads to some relationships between individual risky assets and the market to be pursued in the next section, which are collectively called the ***Capital Market Theory*** (or ***Capital Asset Pricing Model***, CAPM for short).

We will be assuming that asset 1 is a risk-free asset with deterministic rate of return $\mu_1 = r$, and that the market also has $n - 1$ risky assets with random rates of return $R_2, R_3, \dots, R_n$. The means of these random variables are denoted by $\mu_2, \mu_3, \dots, \mu_n$, the variances are denoted by $\sigma_2^2, \sigma_3^2, \dots, \sigma_n^2$, and the covariances between asset rates of return will be denoted by $\sigma_{jk} = \mathrm{Cov}\left(R_j, R_k\right)$. Note that $\sigma_{jj} = \sigma_j^2$ and that $\sigma_{ij} = \sigma_{ji}$ by the definition of covariance. The variances and covariances of all assets can be assembled into the ***covariance matrix***:

$$
\Sigma = \begin{pmatrix}
\sigma_{11} & \sigma_{12} & \sigma_{13} & \cdots & \sigma_{1n} \\
\sigma_{21} & \sigma_{22} & \sigma_{23} & \cdots & \sigma_{2n} \\
\sigma_{31} & \sigma_{32} & \sigma_{33} & \cdots & \sigma_{3n} \\
\vdots & \vdots & \vdots & \ddots & \vdots \\
\sigma_{n1} & \sigma_{n2} & \sigma_{n3} & \cdots & \sigma_{nn}
\end{pmatrix} \tag{2.18}
$$

But since asset 1 is non-random, all elements σ_{i1} of the first column and all elements σ_{1j} of the first row are equal to 0. We will use Σ_r to indicate the variance-covariance matrix for the risky assets only, with the first row and column of Σ removed. To form a matrix expression for the objective function, we can introduce column vectors for the mean rates of return and the portfolio weights, and also use $\mathbf{1}$ to denote a column vector of n ones:

$$\boldsymbol{\mu} = \begin{pmatrix} \mu_1 = r \\ \mu_2 \\ \mu_3 \\ \vdots \\ \mu_n \end{pmatrix}, \ \boldsymbol{w} = \begin{pmatrix} w_1 \\ w_2 \\ w_3 \\ \vdots \\ w_n \end{pmatrix}, \ \mathbf{1} = \begin{pmatrix} 1 \\ 1 \\ 1 \\ \vdots \\ 1 \end{pmatrix}. \tag{2.19}$$

In this book we will use a superscript ^{t} to denote transpose of a matrix, and by default, we suppose that vectors are column vectors, so that if $\boldsymbol{v}$ is a column vector then $\boldsymbol{v}^t$ is a row vector. You can check (see Exercise 4) that the variance part of the portfolio objective function is actually the quadratic form:

$$\begin{aligned} \sigma_p^2 &= \left(\sum_{i=1}^n w_i^2 \sigma_i^2 + \sum_{\substack{j=1 \\ j \neq k}}^n \sum_{k=1}^n w_j w_k \mathrm{Cov}\left(R_j, R_k\right) \right) \\ &= \sum_{j=1}^n \sum_{k=1}^n w_j w_k \mathrm{Cov}\left(R_j, R_k\right) \\ &= \boldsymbol{w}^t \Sigma \boldsymbol{w}. \end{aligned} \tag{2.20}$$

Because of the fact that asset 1 is deterministic, the indices in the double sum actually begin at 2 rather than 1. The mean part of the objective function is clearly the vector product $\boldsymbol{\mu}^t \cdot \boldsymbol{w}$, and the sum of the weights can be written as $\mathbf{1}^t \cdot \boldsymbol{w}$, so that in vector form our problem reduces to:

$$\text{maximize}_{\boldsymbol{w}} f(\boldsymbol{w}) = \boldsymbol{\mu}^t \cdot \boldsymbol{w} - a \cdot \boldsymbol{w}^t \Sigma \boldsymbol{w} \text{ subject to} : g(\boldsymbol{w}) = \mathbf{1}^t \cdot \boldsymbol{w} - 1 = 0. \quad (2.21)$$

Example 7 of Section 2.1 pointed the way to what we will do next. The gradient of a constant vector dotted with a vector of variables is easily seen to be the constant vector, and in Exercise 9 you will check that the gradient of the quadratic form $\boldsymbol{w}^t \Sigma \boldsymbol{w}$ (in column form) is $2\Sigma \boldsymbol{w}$. So, the Lagrange multiplier relationship leads to the equation:

$$\nabla f(\boldsymbol{w}) = \lambda \nabla g(\boldsymbol{w}) \Longleftrightarrow \boldsymbol{\mu} - 2a\Sigma \boldsymbol{w} = \lambda \cdot \mathbf{1}, \tag{2.22}$$

together, of course, with the constraint $\mathbf{1}^t \cdot \boldsymbol{w} = 1$, which says that the non-risky weight w_1 is one minus the total of the risky weights.

Because of the fact that the first row of Σ consists entirely of 0's, the first equation in system (2.22) takes the form $\mu_1 = r = \lambda$. In terms of the subvector $\boldsymbol{w}_r = (w_2, w_3, ..., w_n)^t$ of weights for the $n-1$ risky assets and the corresponding subvector $\boldsymbol{\mu}_r$ of means, system (2.22) now becomes:

$$\Sigma_r \boldsymbol{w}_r = \frac{1}{2a} \left(\boldsymbol{\mu}_r - r \cdot \mathbf{1} \right), \tag{2.23}$$

which, written in full, is:

$$
\begin{pmatrix}
\sigma_{22} & \sigma_{23} & \cdots & \sigma_{2n} \\
\sigma_{32} & \sigma_{33} & \cdots & \sigma_{3n} \\
\vdots & \vdots & \ddots & \vdots \\
\sigma_{n2} & \sigma_{n3} & \cdots & \sigma_{nn}
\end{pmatrix}
\begin{pmatrix}
w_2 \\
w_3 \\
\vdots \\
w_n
\end{pmatrix}
= \frac{1}{2a}
\begin{pmatrix}
\mu_2 - r \\
\mu_3 - r \\
\vdots \\
\mu_n - r
\end{pmatrix}. \tag{2.24}
$$

Assuming that the matrix Σ_r is invertible, the formal solution of vector equation (2.23) is:

$$\boldsymbol{w}_r = \frac{1}{2a} \Sigma_r^{-1} \left(\boldsymbol{\mu}_r - r \cdot \mathbf{1} \right) \Longleftrightarrow$$

$$
\begin{pmatrix}
w_2 \\
w_3 \\
\vdots \\
w_n
\end{pmatrix}
= \frac{1}{2a}
\begin{pmatrix}
\sigma_{22} & \sigma_{23} & \cdots & \sigma_{2n} \\
\sigma_{32} & \sigma_{33} & \cdots & \sigma_{3n} \\
\vdots & \vdots & \ddots & \vdots \\
\sigma_{n2} & \sigma_{n3} & \cdots & \sigma_{nn}
\end{pmatrix}^{-1}
\begin{pmatrix}
\mu_2 - r \\
\mu_3 - r \\
\vdots \\
\mu_n - r
\end{pmatrix}. \tag{2.25}
$$

Something very important follows from this. Let x_i be the entry in row i of the product $\Sigma_r^{-1} \left(\boldsymbol{\mu}_r - r \cdot \mathbf{1} \right)$; then x_i has nothing to do with the risk aversion constant a, and moreover, $1/(2a)$ is a common factor of each row in the solution vector $\boldsymbol{w}_r$. The ratios w_i/w_j therefore do not depend on the investor's risk aversion, which is the general market portfolio result. The market portfolio weights would again be:

$$w_i^* = \frac{w_i}{\sum_{j=2}^{n} w_j}, i = 2, 3, ..., n. \tag{2.26}$$

By formula (2.25), since the total of the optimal portfolio weights for the risky assets can be written in vector form as $\mathbf{1}^t \cdot \boldsymbol{w}_r$, upon cancelling the common factor $1/(2a)$, the vector of market portfolio weights would be:

$$\boldsymbol{w}^* = \frac{\Sigma_r^{-1} \left(\boldsymbol{\mu}_r - r \cdot \mathbf{1} \right)}{\left(\mathbf{1}^t \cdot \Sigma_r^{-1} \left(\boldsymbol{\mu}_r - r \cdot \mathbf{1} \right) \right)}. \tag{2.27}$$

Our hard work has given us a very powerful result.

Theorem 1. Suppose that a market has a risk-free asset with rate r and $n - 1$ risky assets whose covariance matrix Σ_r is invertible. Let $\boldsymbol{\mu}_r$ be the vector of mean risky asset returns. Then the optimal portfolio for an investor with risk aversion a has risky weight vector:

$$\boldsymbol{w}_r = \frac{1}{2a} \Sigma_r^{-1} \left(\boldsymbol{\mu}_r - r \cdot \mathbf{1} \right),$$

and the optimal non-risky weight w_1 satisfies $w_1 = 1 - (w_2 + w_3 + \cdots + w_n)$. Moreover, there exists a market portfolio such that every investor holds risky assets in proportions w_i^*, where the vector of these proportions $\boldsymbol{w}^*$ is given by

$$\boldsymbol{w}^* = \frac{\Sigma_r^{-1}(\boldsymbol{\mu}_r - r \cdot \mathbf{1})}{\mathbf{1}^t \cdot \Sigma_r^{-1}(\boldsymbol{\mu}_r - r \cdot \mathbf{1})}. \quad \blacksquare$$

We now know that even in correlated cases, and with arbitrarily many risky assets, there is such a thing as a market portfolio as in formula (2.27), and the optimal portfolio problem has an easy solution given by formula (2.25), although technology is needed to invert the covariance matrix.

Example 1. As a first example, let us find the optimal portfolio for an investor with risk aversion $a = 10$, and the market portfolio, for the situation in Example 1 of Section 2.1: $r = .02$, two risky assets with means $\mu_2 = .05$ and $\mu_3 = .08$ and standard deviations $\sigma_2 = .10$, and $\sigma_3 = .18$, and correlation $-.5$ between asset rates of return.

Solution. The covariance matrix for the risky assets is:

$$\Sigma_r = \begin{pmatrix} .10^2 & (-.5)(.10)(.18) \\ (-.5)(.10)(.18) & .18^2 \end{pmatrix} = \begin{pmatrix} .01 & -.009 \\ -.009 & .0324 \end{pmatrix}.$$

Its inverse is:

$$\Sigma_r^{-1} = \begin{pmatrix} .01 & -.009 \\ -.009 & .0324 \end{pmatrix}^{-1} = \begin{pmatrix} 133.333 & 37.037 \\ 37.037 & 41.1523 \end{pmatrix}.$$

By Theorem 1, the optimal risky weights for the investor whose risk aversion is $a = 10$ are:

$$\begin{aligned} \boldsymbol{w}_r &= \tfrac{1}{2a}\Sigma_r^{-1}(\boldsymbol{\mu}_r - r \cdot \mathbf{1}) \\ &= \tfrac{1}{20}\begin{pmatrix} 133.333 & 37.037 \\ 37.037 & 41.1523 \end{pmatrix}\begin{pmatrix} .05 - .02 \\ .08 - .02 \end{pmatrix} \\ &= \begin{pmatrix} .311 \\ .179 \end{pmatrix}, \end{aligned}$$

that is, $w_2 = .311, w_3 = .179$, and hence $w_1 = 1 - w_2 - w_3 = .510$. These answers match those from the earlier example.

For the market portfolio, since the risk aversion doesn't matter, we may either compute directly:

$$w_2^* = \frac{w_2}{w_2 + w_3} = \frac{.311}{.311 + .179} = .635; w_3^* = \frac{w_3}{w_2 + w_3} = \frac{.179}{.311 + .179} = .365,$$

or we may employ formula (2.27) to obtain:

$$w^* = \frac{\Sigma_r^{-1}(\mu_r - r \cdot 1)}{1^t \cdot \Sigma_r^{-1}(\mu_r - r \cdot 1)}$$

$$= \frac{\begin{pmatrix} 133.333 & 37.037 \\ 37.037 & 41.1523 \end{pmatrix} \begin{pmatrix} .05 - .02 \\ .08 - .02 \end{pmatrix}}{(1 \quad 1) \begin{pmatrix} 133.333 & 37.037 \\ 37.037 & 41.1523 \end{pmatrix} \begin{pmatrix} .05 - .02 \\ .08 - .02 \end{pmatrix}}$$

$$= \begin{pmatrix} 6.2222 \\ 3.5803 \end{pmatrix} / 9.8025$$

$$= \begin{pmatrix} .635 \\ .365 \end{pmatrix}. \ \blacksquare$$

Example 2. It is no more difficult to deal with a market in which there are four correlated risky assets than it is to deal with a market that has two. Suppose again that the risk-free rate is .02, and that the risky assets have means $\mu_2 = .04$, $\mu_3 = .045$, $\mu_4 = .05$, and $\mu_5 = .055$. Assume that asset 2 has correlation $\rho_{24} = .2$ with asset 4, asset 3 has correlation $\rho_{35} = -.4$ with asset 5, and otherwise all correlations are zero. The standard deviations are $\sigma_2 = .1$, $\sigma_3 = .3$, $\sigma_4 = .5$, and $\sigma_5 = .7$. Find the market portfolio, and also its mean and variance.

Solution. We need the covariance matrix, which is:

$$\Sigma_r = \begin{pmatrix} \sigma_2^2 & \rho_{23}\sigma_2\sigma_3 & \rho_{24}\sigma_2\sigma_4 & \rho_{25}\sigma_2\sigma_5 \\ \rho_{23}\sigma_2\sigma_3 & \sigma_3^2 & \rho_{34}\sigma_3\sigma_4 & \rho_{35}\sigma_3\sigma_5 \\ \rho_{24}\sigma_2\sigma_4 & \rho_{34}\sigma_3\sigma_4 & \sigma_4^2 & \rho_{45}\sigma_4\sigma_5 \\ \rho_{25}\sigma_2\sigma_5 & \rho_{35}\sigma_3\sigma_5 & \rho_{45}\sigma_4\sigma_5 & \sigma_5^2 \end{pmatrix}$$

$$= \begin{pmatrix} .1^2 & 0 & .2(.1)(.5) & 0 \\ 0 & .3^2 & 0 & (-.4)(.3)(.7) \\ .2(.1)(.5) & 0 & .5^2 & 0 \\ 0 & (-.4)(.3)(.7) & 0 & .7^2 \end{pmatrix}$$

$$= \begin{pmatrix} .01 & 0 & .01 & 0 \\ 0 & .09 & 0 & -.084 \\ .01 & 0 & .25 & 0 \\ 0 & -.084 & 0 & .49 \end{pmatrix}.$$

Using technology, the inverse matrix is:

$$\Sigma_r^{-1} = \begin{pmatrix} 104.167 & 0 & -4.16667 & 0 \\ 0 & 13.2275 & 0 & 2.26757 \\ -4.16667 & 0 & 4.16667 & 0 \\ 0 & 2.26757 & 0 & 2.42954 \end{pmatrix}.$$

Then the market portfolio vector is computed from formula (2.27) as:

$$\boldsymbol{w}^* = \frac{\Sigma_r^{-1}(\boldsymbol{\mu}_r - r \cdot \mathbf{1})}{\mathbf{1}^t \cdot \Sigma_r^{-1}(\boldsymbol{\mu}_r - r \cdot \mathbf{1})}$$

$$= \frac{\begin{pmatrix} 104.167 & 0 & -4.167 & 0 \\ 0 & 13.228 & 0 & 2.268 \\ -4.167 & 0 & 4.167 & 0 \\ 0 & 2.268 & 0 & 2.43 \end{pmatrix} \begin{pmatrix} .02 \\ .025 \\ .03 \\ .035 \end{pmatrix}}{\begin{pmatrix} 1 & 1 & 1 & 1 \end{pmatrix} \begin{pmatrix} 104.167 & 0 & -4.167 & 0 \\ 0 & 13.228 & 0 & 2.268 \\ -4.167 & 0 & 4.167 & 0 \\ 0 & 2.268 & 0 & 2.43 \end{pmatrix} \begin{pmatrix} .02 \\ .025 \\ .03 \\ .035 \end{pmatrix}}$$

$$= \begin{pmatrix} .767 \\ .161 \\ .016 \\ .056 \end{pmatrix}.$$

The mean of the market portfolio is found as usual, by computing:

$$\mu^* = \boldsymbol{\mu}_r^t \cdot \boldsymbol{w}^* = \begin{pmatrix} .04 & .045 & .05 & .055 \end{pmatrix} \cdot \begin{pmatrix} .767 \\ .161 \\ .016 \\ .056 \end{pmatrix} = .0418.$$

The variance of the market portfolio is:

$$(\boldsymbol{w}^*)^t \cdot \Sigma_r \cdot \boldsymbol{w}^*$$

$$= \begin{pmatrix} .767 & .161 & .016 & .056 \end{pmatrix} \cdot \begin{pmatrix} .01 & 0 & .01 & 0 \\ 0 & .09 & 0 & -.084 \\ .01 & 0 & .25 & 0 \\ 0 & -.084 & 0 & .49 \end{pmatrix} \cdot \begin{pmatrix} .767 \\ .161 \\ .016 \\ .056 \end{pmatrix}$$

$$= .008547. \ \blacksquare$$

2.2.2 Efficient Mean-Standard Deviation Frontier

If you did Exercise 20 of Section 2.1, you saw a preview of what we are about to look at, which is the geometry of the region of feasible portfolios. This study will shed light on the meaning of a market portfolio and will enable us to relate the parameters of individual assets or portfolios of assets to the parameters of the market as a whole.

Example 3. As in the exercise referred to above, we suppose that there are three risky assets (and for the time being no non-risky asset), with mean rates of return $\mu_1 = .06$, $\mu_2 = .08$, and $\mu_3 = .10$, and standard deviations $\sigma_1 = .10$, $\sigma_2 = .15$, $\sigma_3 = .18$. But this time we suppose that the assets are correlated, with $\rho_{12} = .3$, $\rho_{13} = -.4$, and $\rho_{23} = -.2$. Let us graph the set of possible pairs (σ_p, μ_p) of portfolio standard deviations and means ranging over a broad selection of weight values such that $w_1 + w_2 + w_3 = 1$.

Solution. The portfolio mean and standard deviation can be viewed as parametric functions of w_1, w_2, and w_3, which would be:

$$\mu_p = w_1(.06) + w_2(.08) + w_3(.10);$$

$$\sigma_p = \sqrt{\begin{array}{l} w_1^2(.10)^2 + w_2^2(.15)^2 + w_3^2(.18)^2 + 2\,(w_1 w_2(.3)(.1)(.15) \\ + w_1 w_3(-.4)(.1)(.18) + w_2 w_3(-.2)(.15)(.18)) \end{array}}.$$

Weight w_3 can be replaced by $1 - w_1 - w_2$ in these functions. Technology is definitely necessary here to compute a collection of these (σ_p, μ_p) pairs for a large range of values of w_1 and w_2.

The plot of these pairs is in Figure 2.8. The boundary of the region has a bullet-nosed shape. Where can the optimal portfolios lie? Certainly not in the interior of this region, because, as the black line segment shows, a point like the right endpoint of the segment is dominated by the left endpoint on the boundary, since the left endpoint has the same mean as the right and smaller standard deviation. Moreover, only points on the northern boundary of this region could be optimal portfolios, since, south of the leftmost tip of the region at about $(.07, .075)$, portfolios are dominated by those above them that have the same standard deviation and higher mean. Since optimal portfolios for more cautious investors will have lower standard deviations and lower means, we surmise that points that are down and to the left on this northern boundary correspond to optimal portfolios of investors with higher risk aversions, and points that are up and to the right on the boundary are optimal portfolios of investors with lower risk aversions. ■

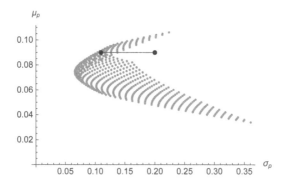

FIGURE 2.8 Possible pairs (σ_p, μ_p) of portfolio standard deviation and mean.

The northern boundary of the region of feasible standard deviation-mean pairs is important enough to deserve its own name.

Definition 1. The (***mean-standard deviation***) ***efficient frontier*** for a portfolio problem with risky assets only is the northern boundary of the feasible region of points (σ_p, μ_p) that are attainable using these assets. Algebraically, if (σ^{mv}, μ^{mv}) is the feasible point with minimum standard deviation, the efficient frontier is the set of all points (σ_p, μ_p) such that $\mu_p \geq \mu^{mv}$, and σ_p is the minimum portfolio standard deviation subject to fixed mean μ_p. ■

The characterization of the efficient frontier in the second sentence of this definition follows by consideration of the line segment in Figure 2.8, or more precisely, the collection of all such horizontal line segments from the interior of the feasible region to the northern boundary. If the mean μ_p is fixed, then the boundary point for that mean has coordinates (σ_p, μ_p) for which σ_p is as far to the left as possible; that is, σ_p is minimized. If we wished, by considering vertical line segments whose upper endpoints lie on the efficient frontier, we could also characterize the curve as the set of points such that for each fixed value of σ_p, μ_p is maximized.

It can be proved that the efficient frontier, viewed as the graph of a function $\mu = h(\sigma)$, is a concave function as Figure 2.8 suggests. That is, line segments connecting two points on the frontier lie entirely below the frontier, except for the endpoints.

For the rest of this subsection we consider markets of risky assets only, and we attempt a more quantitative identification of points on the efficient frontier. To do this, we first have to find the lowest point, which minimizes portfolio standard deviation regardless of mean. Then, to find the other points on the frontier, we must solve the problem of minimizing standard deviation subject not only to the constraint that the sum of the weights equals 1, but also to a second constraint that the portfolio mean is a given value. For this, we need the following extension of the Lagrange multiplier technique to the case of two constraints.

For a single constraint, to optimize a function of several variables $f(\boldsymbol{x}) = f(x_1, x_2, ..., x_n)$ subject to $g(\boldsymbol{x}) = g(x_1, x_2, ..., x_n) = 0$, we define one new real variable λ as the Lagrange multiplier and equate the gradient $\nabla(f(\boldsymbol{x}) - \lambda g(\boldsymbol{x}))$ to zero. Similarly, if there is a second constraint, $h(\boldsymbol{x}) = h(x_1, x_2, ..., x_n) = 0$, introduce a second Lagrange multiplier γ and set

$$\nabla(f(\boldsymbol{x}) - \lambda g(\boldsymbol{x}) - \gamma h(\boldsymbol{x})) = 0, \tag{2.28}$$

together with the two constraints. In total there are n equations generated by system (2.28) plus the two constraints in the total of $n + 2$ variables $x_1, x_2, ..., x_n$ and λ and γ.

First let us address the question of the smallest standard deviation. In Example 3 of Section 2.1 we showed the computational details in one case of variance minimization. Since standard deviation is an increasing function, namely the square root, of variance, we may find the minimum variance first, and then its square root will be the minimum standard deviation. The portfolio weights that achieve the minimum can then be substituted into the mean function $\mu^t \cdot \boldsymbol{w}$ to get the y-coordinate of the lowest point (σ^*, μ^*) on the efficient frontier. Let us try to use matrix algebra to simplify the derivation and to produce a general result.

Playing off of formula (2.21), our problem is:

$$\text{minimize}_{\boldsymbol{w}}\, f(\boldsymbol{w}) = \boldsymbol{w}^t \Sigma_r \boldsymbol{w} \text{ subject to}: g(\boldsymbol{w}) = \mathbf{1}^t \cdot \boldsymbol{w} - 1 = 0. \qquad (2.29)$$

Introducing the Lagrange multiplier λ, we find:

$$\nabla f(\boldsymbol{w}) = \lambda \nabla g(\boldsymbol{w}) \iff 2\Sigma_r \boldsymbol{w} = \lambda \cdot \mathbf{1}, \qquad (2.30)$$

and hence the optimal portfolio weight vector, which we now denote by $\boldsymbol{w}^{mv}$ (for minimum variance), satisfies:

$$\boldsymbol{w}^{mv} = \frac{1}{2}\lambda \cdot \Sigma_r^{-1} \cdot \mathbf{1}. \qquad (2.31)$$

The constraint $\mathbf{1}^t \cdot \boldsymbol{w} = 1$ allows us to eliminate λ:

$$\mathbf{1}^t \cdot \boldsymbol{w}^{mv} = 1 \implies \mathbf{1}^t \cdot \frac{1}{2}\lambda \cdot \Sigma_r^{-1} \cdot \mathbf{1} = 1 \implies \frac{1}{2}\lambda = \frac{1}{\mathbf{1}^t \cdot \Sigma_r^{-1} \cdot \mathbf{1}}. \qquad (2.32)$$

Substitution of this result into formula (2.31) gives:

$$\boldsymbol{w}^{mv} = \frac{1}{2}\lambda \cdot \Sigma_r^{-1} \cdot \mathbf{1} = \frac{\Sigma_r^{-1} \cdot \mathbf{1}}{\mathbf{1}^t \cdot \Sigma_r^{-1} \cdot \mathbf{1}}. \qquad (2.33)$$

Hence we have derived the following result about the minimum variance portfolio.

Theorem 2. Suppose that a market has n risky assets whose covariance matrix Σ_r is invertible. Then the portfolio that achieves minimum variance has weight vector:

$$\boldsymbol{w}^{mv} = \frac{\Sigma_r^{-1} \cdot \mathbf{1}}{\mathbf{1}^t \cdot \Sigma_r^{-1} \cdot \mathbf{1}}.$$

Moreover, if μ_r is the vector of mean rates of return of market assets, then the mean rate of return on the minimum variance portfolio is $\mu^{mv} = \mu_r^t \cdot \boldsymbol{w}^{mv}$ and the minimum variance is:

$$(\sigma^{mv})^2 = (\boldsymbol{w}^{mv})^t \Sigma_r \boldsymbol{w}^{mv}. \quad \blacksquare$$

Exercise 13 asks you to derive the following general matrix formulas for the mean and variance of the minimum variance portfolio:

$$\mu^{mv} = \frac{\mu^t \cdot \Sigma_r^{-1} \cdot \mathbf{1}}{\mathbf{1}^t \cdot \Sigma_r^{-1} \cdot \mathbf{1}}, \quad (\sigma^{mv})^2 = \frac{1}{\mathbf{1}^t \cdot \Sigma_r^{-1} \cdot \mathbf{1}}. \qquad (2.34)$$

Example 4. Compute the coordinates of the westernmost point on the efficient frontier in Figure 2.8 and the portfolio weights that achieve minimum variance.

Solution. Recall that the mean rates of return for the three risky assets were $\mu_1 = .06$, $\mu_2 = .08$, and $\mu_3 = .10$, the standard deviations were $\sigma_1 = .10$, $\sigma_2 = .15$, and $\sigma_3 = .18$, and the correlations were $\rho_{12} = .3$, $\rho_{13} = -.4$, and $\rho_{23} = -.2$. Then the covariance matrix is:

$$
\begin{aligned}
\Sigma_r &= \begin{pmatrix}
\sigma_1^2 & \rho_{12}\sigma_1\sigma_2 & \rho_{13}\sigma_1\sigma_3 \\
\rho_{12}\sigma_1\sigma_2 & \sigma_2^2 & \rho_{23}\sigma_2\sigma_3 \\
\rho_{13}\sigma_1\sigma_3 & \rho_{23}\sigma_2\sigma_3 & \sigma_3^2
\end{pmatrix} \\[2mm]
&= \begin{pmatrix}
.1^2 & (.3)(.1)(.15) & (-.4)(.1)(.18) \\
(.3)(.1)(.15) & .15^2 & (-.2)(.15)(.18) \\
(-.4)(.1)(.18) & (-.2)(.15)(.18) & .18^2
\end{pmatrix} \\[2mm]
&= \begin{pmatrix}
.01 & .0045 & -.0072 \\
.0045 & .0225 & -.0054 \\
-.0072 & -.0054 & .0324
\end{pmatrix}.
\end{aligned}
$$

The inverse of this matrix turns out to be:

$$
\Sigma_r^{-1} = \begin{pmatrix}
126.649 & -19.3492 & 24.9194 \\
-19.3492 & 49.2524 & 3.90892 \\
24.9194 & 3.90892 & 37.0533
\end{pmatrix}.
$$

The portfolio weight vector that we seek is, by Theorem 2,

$$
\begin{aligned}
\boldsymbol{w}^{mv} &= \frac{\Sigma_r^{-1} \cdot \mathbf{1}}{\mathbf{1}' \cdot \Sigma_r^{-1} \cdot \mathbf{1}} \\[2mm]
&= \frac{\begin{pmatrix} 126.649 & -19.3492 & 24.9194 \\ -19.3492 & 49.2524 & 3.90892 \\ 24.9194 & 3.90892 & 37.0533 \end{pmatrix}\begin{pmatrix} 1 \\ 1 \\ 1 \end{pmatrix}}{\begin{pmatrix} 1 & 1 & 1 \end{pmatrix}\begin{pmatrix} 126.649 & -19.3492 & 24.9194 \\ -19.3492 & 49.2524 & 3.90892 \\ 24.9194 & 3.90892 & 37.0533 \end{pmatrix}\begin{pmatrix} 1 \\ 1 \\ 1 \end{pmatrix}} \\[2mm]
&= \begin{pmatrix} 132.219 \\ 33.8122 \\ 65.8816 \end{pmatrix} \bigg/ 231.913 = \begin{pmatrix} .570 \\ .146 \\ .284 \end{pmatrix}.
\end{aligned}
$$

The mean rate of return on the minimum variance portfolio becomes:

$$
\mu^{mv} = \boldsymbol{\mu}^t \cdot \boldsymbol{w}^{mv} = (.06, .08, .10)\begin{pmatrix} .570 \\ .146 \\ .284 \end{pmatrix} = .0743,
$$

and the variance and standard deviation are:

$$
\begin{aligned}
(\sigma^{mv})^2 &= (\boldsymbol{w}^{mv})^t \, \Sigma_r \boldsymbol{w}^{mv} \\
&= \begin{pmatrix} .570 & .146 & .284 \end{pmatrix}
\begin{pmatrix}
.01 & .0045 & -.0072 \\
.0045 & .0225 & -.0054 \\
-.0072 & -.0054 & .0324
\end{pmatrix}
\begin{pmatrix} .570 \\ .146 \\ .284 \end{pmatrix} \\
&= .00431,
\end{aligned}
$$

$$
\sigma^{mv} = \sqrt{(\sigma^{mv})^2} = \sqrt{.00431} = .0657.
$$

So the westernmost point has coordinates $(.0657, .0743)$. These values seem to be consistent with Figure 2.8. ∎

Now we move to the second, more complicated question of finding the rest of the efficient frontier. Given a target portfolio mean μ_p, our new objective is to minimize:

$$
\begin{aligned}
&\text{minimize}_{\boldsymbol{w}} \, f(\boldsymbol{w}) = \boldsymbol{w}^t \Sigma_r \boldsymbol{w} \\
&\text{subject to: } g(\boldsymbol{w}) = \mathbf{1}^t \cdot \boldsymbol{w} - 1 = 0 \text{ and } h(\boldsymbol{w}) = \boldsymbol{\mu}^t \cdot \boldsymbol{w} - \mu_p = 0.
\end{aligned}
\tag{2.35}
$$

To solve the problem, we introduce a second Lagrange multiplier γ, and try to solve:

$$
\nabla(f(\boldsymbol{x}) - \lambda g(\boldsymbol{x}) - \gamma h(\boldsymbol{x})) = \mathbf{0},
\tag{2.36}
$$

together with the two constraints $\mathbf{1}^t \cdot \boldsymbol{w} = 1$ and $\boldsymbol{\mu}^t \cdot \boldsymbol{w} = \mu_p$. The gradient condition translates to:

$$
2\Sigma_r \boldsymbol{w} - \lambda \cdot \mathbf{1} - \gamma \cdot \boldsymbol{\mu} = \mathbf{0}.
\tag{2.37}
$$

Then,

$$
\boldsymbol{w} = \frac{1}{2} \Sigma_r^{-1} (\lambda \cdot \mathbf{1} + \gamma \cdot \boldsymbol{\mu}).
\tag{2.38}
$$

We can find a system of two linear equations for the Lagrange multipliers λ and γ by substituting into the constraints:

$$
\begin{cases}
\mathbf{1}^t \cdot \boldsymbol{w} = 1 & \implies 1 = \frac{1}{2} \cdot \mathbf{1}^t \cdot \Sigma_r^{-1}(\lambda \cdot \mathbf{1} + \gamma \cdot \boldsymbol{\mu}) = \frac{\lambda}{2} \mathbf{1}^t \cdot \Sigma_r^{-1} \cdot \mathbf{1} + \frac{\gamma}{2} \mathbf{1}^t \cdot \Sigma_r^{-1} \cdot \boldsymbol{\mu}; \\
\boldsymbol{\mu}^t \cdot \boldsymbol{w} = \mu_p & \implies \mu_p = \frac{1}{2} \cdot \boldsymbol{\mu}^t \Sigma_r^{-1}(\lambda \cdot \mathbf{1} + \gamma \cdot \boldsymbol{\mu}) = \frac{\lambda}{2} \boldsymbol{\mu}^t \cdot \Sigma_r^{-1} \cdot \mathbf{1} + \frac{\gamma}{2} \boldsymbol{\mu}^t \cdot \Sigma_r^{-1} \cdot \boldsymbol{\mu}.
\end{cases}
\tag{2.39}
$$

Once λ and γ are known, then $\boldsymbol{w}$ is known, and the minimum variance σ_p^2 for a given mean μ_p is equal to $\boldsymbol{w}^t \Sigma_r \boldsymbol{w}$ as before. (Since the portfolio vector depends on $\lambda/2$ and $\gamma/2$, when doing problems you may find it useful to solve system (2.39) for these two instead of λ and γ themselves.)

We now have the following result.

Theorem 3. Suppose that a market has n risky assets whose covariance matrix Σ_r is invertible. Then the portfolio that achieves minimum variance for a given mean return μ_p has weight vector:

$$w = \frac{1}{2}\Sigma_r^{-1}(\lambda \cdot \mathbf{1} + \gamma \cdot \mu)$$

where the real constants λ and γ satisfy the linear system:

$$\begin{cases} \mathbf{1}^t \cdot w = 1 & \Longrightarrow 1 = \frac{\lambda}{2}\mathbf{1}^t \cdot \Sigma_r^{-1} \cdot \mathbf{1} + \frac{\gamma}{2}\mathbf{1}^t \cdot \Sigma_r^{-1} \cdot \mu; \\ \mu^t w = \mu_p & \Longrightarrow \mu_p = \frac{\lambda}{2}\mu^t \cdot \Sigma_r^{-1} \cdot \mathbf{1} + \frac{\gamma}{2}\mu^t \cdot \Sigma_r^{-1} \cdot \mu. \end{cases}$$

The minimum variance corresponding to mean return μ_p is:

$$\sigma_p^2 = w^t \Sigma_r w. \quad \blacksquare$$

These computations are of course best not done by hand. A rough algorithm for mapping out the efficient frontier is then: (1) Use Theorem 1 to compute the coordinates of the extreme leftmost point (σ^{mv}, μ^{mv}) on the boundary of the feasible region; and (2) for a selection of portfolio mean values μ_p greater than μ^{mv}, use Theorem 2 to find the associated σ_p for each mean, and plot the points (σ_p, μ_p). This strategy can be implemented in any computer system that supports matrix multiplication, inversion, and solution of linear systems.

Example 5. Let's graph an efficient frontier. For simplicity, we begin with just two risky assets, with parameters $\mu_1 = .05$, $\mu_2 = .08$, $\sigma_1 = .4$, $\sigma_2 = .6$, and $\rho = -.5$. The covariance matrix is:

$$\Sigma_r = \begin{pmatrix} .4^2 & (.5)(.4)(.6) \\ (-.5)(.4)(.6) & .6^2 \end{pmatrix} = \begin{pmatrix} .16 & -.12 \\ -.12 & .36 \end{pmatrix},$$

whose inverse is:

$$\Sigma_r^{-1} = \begin{pmatrix} 8.3333 & 2.7778 \\ 2.7778 & 3.7037 \end{pmatrix}.$$

Then the portfolio that gives lowest variance is:

$$w^{mv} = \frac{\Sigma_r^{-1} \cdot \mathbf{1}}{\mathbf{1}^t \cdot \Sigma_r^{-1} \cdot \mathbf{1}} = \frac{\begin{pmatrix} 8.3333 & 2.7778 \\ 2.7778 & 3.7037 \end{pmatrix}\begin{pmatrix} 1 \\ 1 \end{pmatrix}}{(\begin{matrix} 1 & 1 \end{matrix})\begin{pmatrix} 8.3333 & 2.7778 \\ 2.7778 & 3.7037 \end{pmatrix}\begin{pmatrix} 1 \\ 1 \end{pmatrix}} = \begin{pmatrix} .6316 \\ .3684 \end{pmatrix}.$$

Using these portfolio weights, the mean return on the minimum variance portfolio is:

$$\mu^{mv} = \mu^t \cdot w^{mv} = (\begin{matrix} .05 & .08 \end{matrix})\begin{pmatrix} .6316 \\ .3684 \end{pmatrix} = .0611,$$

and the variance and standard deviation are:

$$(\sigma^{mv})^2 \;=\; (\boldsymbol{w}^{mv})^t \, \Sigma_r \boldsymbol{w}^{mv}$$

$$= \; (\begin{array}{cc} .6316 & .3684 \end{array}) \left(\begin{array}{cc} .16 & -.12 \\ -.12 & .36 \end{array} \right) \left(\begin{array}{c} .6316 \\ .3684 \end{array} \right)$$

$$= \; .0568;$$

$$\sigma^{mv} = \sqrt{(\sigma^{mv})^2} = \sqrt{.0568} = .2384.$$

Hence, the coordinates of the westernmost point on the efficient frontier are $(.2384, .0611)$.

We cannot compute all of the other points on the frontier by hand, but to illustrate the process, we continue by finding the point (σ_p, μ_p) on the frontier for which the target mean is $\mu_p = .07$. To do this, we start by solving the system below for the Lagrange multipliers λ and γ:

$$\begin{cases} \frac{1^t \cdot \Sigma_r^{-1} \cdot 1}{2}\lambda + \frac{1^t \cdot \Sigma_r^{-1} \cdot \mu}{2}\gamma &= 1 \\ \frac{\mu^t \Sigma_r^{-1} \cdot 1}{2}\lambda + \frac{\mu^t \Sigma_r^{-1} \cdot \mu}{2}\gamma &= \mu_p \end{cases} \implies \begin{cases} 8.7963\lambda + .537038\gamma &= 1 \\ .537038\lambda + .0333797\gamma &= .07 \end{cases}$$

The coefficients are found by carrying out the matrix multiplications $1^t \cdot \Sigma_r^{-1} \cdot 1$, $1^t \cdot \Sigma_r^{-1} \cdot \mu$ (which is the same as $\mu^t \cdot \Sigma_r^{-1} \cdot 1$) and $\mu^t \cdot \Sigma_r^{-1} \cdot \mu$, where $\mu = \left(\begin{array}{c} .05 \\ .08 \end{array} \right)$. The solution of this system is:

$$\lambda = -.808934, \gamma = 15.1118.$$

The next step is to compute the portfolio weights for the portfolio with smallest variance and mean of .07, by the calculation:

$$\boldsymbol{w} \;=\; \tfrac{1}{2}\Sigma_r^{-1}(\lambda \cdot 1 + \gamma \cdot \mu)$$

$$= \; \tfrac{1}{2} \left(\begin{array}{cc} 8.3333 & 2.7778 \\ 2.7778 & 3.7037 \end{array} \right) \left((-.808934) \left(\begin{array}{c} 1 \\ 1 \end{array} \right) + 15.1118 \left(\begin{array}{c} .05 \\ .08 \end{array} \right) \right)$$

$$= \; \left(\begin{array}{c} .3333 \\ .6667 \end{array} \right).$$

The variance and standard deviation corresponding to the portfolio mean of .07 are:

$$\sigma_p^2 = \boldsymbol{w}^t \Sigma_r \boldsymbol{w} = (\begin{array}{cc} .3333 & .6667 \end{array}) \left(\begin{array}{cc} .16 & -.12 \\ -.12 & .36 \end{array} \right) \left(\begin{array}{c} .3333 \\ .6667 \end{array} \right) = .12446;$$

$$\sigma_p = \sqrt{\sigma_p^2} = \sqrt{.12446} = .3528.$$

A full graph of the frontier produces the curve in Figure 2.9, which is consistent with the computations above of the minimum standard deviation and mean, and the minimum standard deviation for a fixed mean of .07. ■

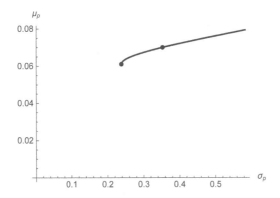

FIGURE 2.9 Efficient frontier of pairs (σ_p, μ_p).

It is interesting to look at the graphical relationship between the region of possible portfolio combinations (σ_p, μ_p) and the efficient frontier. Remember that Figure 2.8 contained a set of possible standard deviation-mean pairs using three risky assets and the parameters from Example 3: $\mu_1 = .06$, $\mu_2 = .08$, $\mu_3 = .10$, $\sigma_1 = .10$, $\sigma_2 = .15$, $\sigma_3 = .18$, $\rho_{12} = .3$, $\rho_{13} = -.4$, and $\rho_{23} = -.2$. The coordinates of the minimum variance portfolio were $\mu^* = .0743$, and $\sigma^* = .0657$. We have used *Mathematica* to combine the graph of that figure with a portion of the efficient frontier computed using the algorithm; the result is in Figure 2.10. The westernmost point as well as the rest of the curve fits the region of feasible portfolios very well.

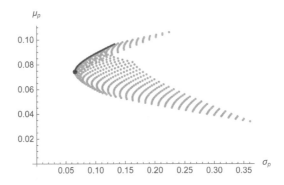

FIGURE 2.10 Efficient frontier with collection of possible pairs (σ_p, μ_p).

Exercises 2.2

1. Suppose that a market consists of a non-risky asset with rate of return $r = \mu_1 = .01$, and three risky assets with mean rates of return $\mu_2 = .04$, $\mu_3 = .07$, and $\mu_4 = .08$, and standard deviations $\sigma_2 = .06$, $\sigma_3 = .09$, and $\sigma_4 = .11$, respectively. Assets 3 and 4 are independent, as are assets 2 and 4, but assets 2 and 3 have correlation $\rho_{2,3} = -.3$. Use matrix techniques to find the optimal portfolio for an investor with risk aversion $a = 8$, and also find the market portfolio of risky assets.

2. Assume a non-risky rate of $r = .02$ in a market with five independent risky assets with means .05, .06, .07, .08, and .09 and variances .01, .04, .08, .10, and .15. Find the market portfolio and its mean and variance by matrix techniques. Also find the optimal portfolio for an investor with risk aversion $a = 15$. Compare the results to the explicit formulas of the last section.

3. Use matrix techniques to find the market portfolio if the risk-free rate is $r = .03$, and there are two risky assets with mean rates of return .06 and .08, standard deviations .05 and .07, and correlation .2. What are the mean and standard deviation of the market?

4. Check formula (2.20), that is, verify that if $R_p = \sum_{i=1}^{n} w_i R_i$ is the rate of return on a portfolio with n assets, weights w_i, and individual asset rates of return R_i, then

$$\text{Var}\,(R_p) = \boldsymbol{w}^t \Sigma \boldsymbol{w},$$

where $\boldsymbol{w}$ is the vector of portfolio weights.

5. Show in the case of two risky assets with standard deviations $\sigma_1 < \sigma_2$ and correlation ρ that the minimum possible portfolio variance is:

$$(\sigma^{mv})^2 = \frac{\sigma_1^2 \sigma_2^2 \left(1 - \rho^2\right)}{\left(\sigma_1^2 - 2\rho\sigma_1\sigma_2 + \sigma_2^2\right)}.$$

6. Suppose that there are five possible outcomes $\omega_1, ..., \omega_5$ with given probabilities, and three risky assets in a market, whose values at each outcome are given in the table below. A non-risky asset with rate of return .002 is available. Find the market portfolio. Comment on what is unusual about the solution, which may imply that the data in the table is not realistic.

outcome	$P[\omega]$	value of R_1	value of R_2	value of R_3
ω_1	1/5	$-.01$	.07	.03
ω_2	1/5	$-.01$	.05	.01
ω_3	1/5	0	.01	0
ω_4	1/5	.02	$-.01$	$-.01$
ω_5	1/5	.03	$-.04$	$-.015$

7. The following market data is from Exercise 2.1-16: Risk-free rate of return: .03, two risky assets with mean rates of return .07 and .09, standard deviations .05 and .10, and correlation $-.2$.

(a) Using Theorem 1, verify the solution for the market portfolio weights $w_2^* = .708$, $w_3^* = .292$ that were obtained in that exercise by longhand methods;

(b) Find the minimum variance portfolio of risky assets and the corresponding standard deviation-mean pair (σ^{mv}, μ^{mv});

(c) Find the minimum variance portfolio of risky assets subject to mean return $\mu_p = .075$. (Is there really any optimization that needs to be done?) What are the variance and standard deviation of that portfolio?

8. (Technology required) Consider a market with two risky assets. Assume that the mean returns are .05 and .08, and the standard deviations are .03 and .06. Write expressions for the portfolio mean and standard deviation, and show that if the correlation between the rates of return is $+1$ and there is no borrowing or short-selling, then the efficient frontier (which is, in fact, the collection of all possible portfolios in this case) graphs as a line segment. More generally, suppose that the correlation is ρ; sketch graphs of the set of possible portfolios when $\rho = .3, .7$, and $-.7$.

9. Check that the gradient of the quadratic form $\mathbf{w}^t \Sigma \mathbf{w}$ (in column form) is $2\Sigma \mathbf{w}$.

10. In Example 1, how low must the risk-free rate of return be so that no more than 40% of the investor's wealth is held in that asset?

11. In Example 2, how large must the mean rate of return on asset 4 be so that its market portfolio weight is at least 3%?

12. Argue that the market portfolio lies on the efficient frontier.

13. For the minimum variance portfolio of Theorem 2, derive the expressions below for the mean rate of return and variance of the rate of return:

$$\mu^{mv} = \frac{\boldsymbol{\mu}^t \cdot \Sigma_r^{-1} \cdot \mathbf{1}}{\mathbf{1}^t \cdot \Sigma_r^{-1} \cdot \mathbf{1}}, \quad (\sigma^{mv})^2 = \frac{1}{\mathbf{1}^t \cdot \Sigma_r^{-1} \cdot \mathbf{1}}.$$

14. When the market consists of two risky assets whose parameters are: $\mu_1 = .05$, $\mu_2 = .08$, $\sigma_1 = .4$, $\sigma_2 = .6$, and $\rho = -.5$, is there more than one portfolio that achieves a target mean of $\mu_p = .07$? How do you find the variance of such a portfolio? What does this mean relative to the geometry of the feasible region and the efficient frontier?

15. In Exercise 2 of Section 2.1 we had a market with four independent risky assets with mean rates of return .03, .05, .07, and .09, and variances .08, .10,

.12, .14. Find the portfolio of minimum variance and its mean and standard deviation, and find the portfolio on the efficient frontier whose mean return is .06 and its standard deviation.

16. Assume that a market has three risky assets with standard deviations $\sigma_1 = .06$, $\sigma_2 = .09$, $\sigma_3 = .12$ and correlations $\rho_{1,2} = 0$, $\rho_{1,3} = 0$, and $\rho_{2,3} = .5$. The mean rates of return are $\mu_1 = .03$, $\mu_2 = .04$, $\mu_3 = .05$ and the risk-free rate is $r = .01$.

   (a) Find the coordinates in the (σ, μ) plane of the minimum variance portfolio of risky assets;

   (b) Find the coordinates of the point on the efficient frontier with $\mu_p = .035$;

   (c) Find the market portfolio;

   (d) What combination of an efficient portfolio and the risk-free asset has mean rate of return .025?

17. In Exercise 3 of Section 2.1 there was a market with a non-risky asset with rate .02, and three independent risky assets with means .04, .06, and .08 and standard deviations .05, .07, and .09.

   (a) Find the coordinates in the (σ, μ) plane of the minimum variance portfolio;

   (b) Find the coordinates of the point on the efficient frontier with $\mu_p = .05$;

   (c) Find the market portfolio;

   (d) What combination of an efficient portfolio and the risk-free asset has mean rate of return .03?

18. Express Theorem 3 in matrix form, and give the solution vector (λ, γ) in matrix form.

19. Check in Theorem 3 that if μ_p is chosen to be μ^{mv}, then σ_p reduces to σ^{mv}. (Hint, in the system of equations for λ and γ, express the coefficients $\boldsymbol{\mu}^t \cdot \Sigma_r^{-1} \cdot \mathbf{1}$ and $\mathbf{1}^t \cdot \Sigma_r^{-1} \cdot \boldsymbol{\mu}$ in terms of μ^{mv}, and show that $\gamma = 0$.)

20. Characterize the efficient frontier in a general, two-asset independent case by finding an algebraic relationship between the portfolio mean μ_p and standard deviation σ_p. Show that the relationship has the form:

$$a\mu_p^2 + b\mu_p + c = d\sigma_p^2,$$

and then apply the quadratic formula to solve for μ_p. (Note that there will be two branches to the solution curve.)

2.3 Capital Market Theory, Part II

We continue our study of Capital Market Theory in this section, beginning with a characterization of the optimal portfolios in a market which has several risky assets and a non-risky asset. It turns out that all such portfolios have standard deviation-mean pairs (σ_p, μ_p) that lie on a line called the *Capital Market Line*.

2.3.1 Capital Market Line

Example 6. Reconsider the two risky assets of Example 1 of Section 2.2 for a moment. The parameters were: $\mu_2 = .05$, $\mu_3 = .08, \sigma_2 = .10$, $\sigma_3 = .18$, and $\rho_{23} = -.5$. We are able to use the methods of the last subsection to find the efficient frontier for those assets in isolation. Now suppose that in addition to these two assets, there is a risk-free asset with rate of return $r = \mu_1 = .02$. There exists a market portfolio, and we found in Example 1 that it has weights $w_2^* = .635$ and $w_3^* = .365$. The mean, variance, and standard deviation corresponding to the market portfolio are:

$$\mu^* = (.635)(.05) + (.365)(.08) = .06095;$$

$$(\sigma^*)^2 = (.635)^2(.10)^2 + (.365)^2(.18)^2 + 2(.635)(.365)(-.5)(.10)(.18) = .00418;$$

$$\sigma^* = \sqrt{(\sigma^*)^2} = .06463.$$

Now let's look at a sketch in which the point $(0, r) = (0, .02)$, corresponding to the portfolio consisting of only the non-risky asset, and the point $(\sigma^*, \mu^*) = (.06463, .06095)$ for the market portfolio are superimposed on the efficient frontier for the two risky assets, and are joined by a line. This is shown in Figure 2.11. The market portfolio does lie on the efficient frontier. (You were asked for a general proof of this fact in Exercise 2.2-12) Also, we will show shortly that due to the concave shape of the frontier, the line is tangent to the frontier at the point (σ^*, μ^*).

We know that investors who choose portfolios based on mean-variance optimization will select some portfolio, according to their risk aversion value, that devotes a proportion p of wealth to the market portfolio and $1 - p$ to the risk-free asset. The mean and standard deviation of such a portfolio, in general, are (assuming $p > 0$ for the moment):

$$\mu = (1 - p)r + p \cdot \mu^*, \sigma = \sqrt{\sigma^2} = \sqrt{0 + p^2 (\sigma^*)^2} = p \cdot \sigma^*. \qquad (2.40)$$

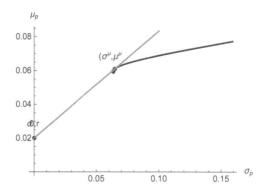

FIGURE 2.11 Line connecting risk-free asset with market portfolio, two-asset case.

But notice that points lying on the segment connecting $(0, r)$ with (σ^*, μ^*) are of the form:

$$(1 - t)(0, r) + t\, (\sigma^*, \mu^*) = (t\sigma^*, (1 - t)r + t\mu^*)\,.$$

Hence the optimal portfolios lie on the segment. Points on the line extending northeast of the market portfolio would be those for which $p > 1$, so that short-selling of the non-risky asset (i.e. borrowing) is necessary to achieve higher portfolio mean than the market at the expense of higher standard deviation. (See, for instance, Exercise 4). In Exercise 23 you are asked to consider the case where the market portfolio is short-sold, i.e. $p < 0$. It turns out that no investor would choose such a portfolio, so we may ignore this case. ∎

The general situation is shown in Figure 2.12.

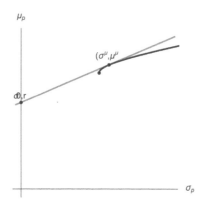

FIGURE 2.12 Capital Market Line.

Definition 1. The *capital market line* is the line passing through the points $(0, r)$ and (σ^*, μ^*), which is also the tangent line to the efficient frontier at the point (σ^*, μ^*). The *efficient portfolios* are those whose standard deviation-mean pairs fall on this line, i.e. they are simple linear combinations of the market portfolio (σ^*, μ^*) with the risk-free asset. The *market price of risk* or *risk premium* is the slope of the capital market line. ∎

We now check that the capital market line is indeed tangent to the efficient frontier. If not, then there would be a second point (σ_p, μ_p) on the frontier that the line cuts through, as shown in Figure 2.13, for some efficient portfolio of risky assets. (The case shown is the case that the market mean μ^* exceeds the portfolio mean μ_p; the argument in the other case is the same.) But if this were so, then portfolios whose (σ, μ) pairs lie on the line segment connecting (σ_p, μ_p) and (σ^*, μ^*) would also be efficient. But these points correspond to portfolios of risky assets that are in the interior of the feasible region, not the frontier; hence these portfolios are not in the ratios dictated by the market portfolio, contradicting the market portfolio theorem. Thus, there cannot be a second intersection point between the capital market line and the efficient frontier. It is also easy to see from Figure 2.12 that the line connecting $(0, r)$ and (σ^*, μ^*) has the largest slope among all lines joining $(0, r)$ with feasible (σ, μ) combinations, a fact that we will use in the proof of Theorem 4 later.

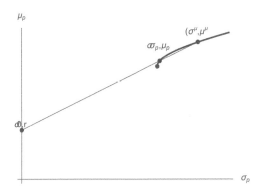

FIGURE 2.13 Spurious Capital Market Line, case $\mu^* > \mu_p$.

It is easy to find the equation of the capital market line, and therefore an expression for the market price of risk. Since two points on the line are $(0, r)$ and (σ^*, μ^*), the slope is:

$$\text{market price of risk} = \frac{\mu^* - r}{\sigma^*}, \qquad (2.41)$$

and since the vertical intercept is $(0, r)$, the equation of the line is:

$$\text{capital market line equation}: \ \mu = r + \frac{\mu^* - r}{\sigma^*}\sigma. \tag{2.42}$$

The reason for the term "market price of risk" should now be clear. For each unit increase in the standard deviation σ of an asset, in order to be efficient, the mean of that asset must increase by $(\mu^* - r)/\sigma^*$. You can think of it as the price that an asset must pay in mean return to compensate for increased risk.

Example 7. In the context of Example 2 of Section 2.2, find an efficient portfolio that has mean return .03. What is the variance of that portfolio?

Solution. In that problem we were given that the risk-free rate was $r = .02$, and that the mean vector and covariance matrix of the risky assets were:

$$\mu_r = \begin{pmatrix} .04 \\ .045 \\ .05 \\ .055 \end{pmatrix}, \quad \Sigma_r = \begin{pmatrix} .01 & 0 & .01 & 0 \\ 0 & .09 & 0 & -.084 \\ .01 & 0 & .25 & 0 \\ 0 & -.084 & 0 & .49 \end{pmatrix}.$$

The market portfolio came out to be:

$$w^* = \begin{pmatrix} .767 \\ .161 \\ .016 \\ .056 \end{pmatrix}.$$

The mean and variance of the market portfolio were:

$$\mu^* = \mu_r^t \cdot w^* = .0418; \ (\sigma^*)^2 = (w^*)^t \Sigma_r w^* = .008547.$$

Let p be the proportion of wealth in the efficient portfolio that we seek that is given to the market, so that $1 - p$ is given to the risk-free asset. In order for the mean return of this efficient portfolio to equal .03, we must have

$$.03 = (1 - p)(.02) + p(.0418) \implies .01 = .0218p \implies p = \frac{.01}{.0218} = .4587.$$

The variance of this portfolio is

$$(1 - p)^2 \cdot 0 + p^2 \cdot (\sigma^*)^2 = .4587^2(.008547) = .00180. \ \blacksquare$$

Example 8. Suppose that the market mean and standard deviation are .08 and .10 respectively, and the risk-free rate is .04. Is an asset with mean return .10 and standard deviation .20 by itself an efficient investment?

Solution. If the asset were itself efficient, then its standard deviation-mean pair would lie on the capital market line. Using the given parameters, the right side of the equation is:

$$r + \frac{\mu^* - r}{\sigma^*}\sigma = .04 + \frac{.08 - .04}{.10} \cdot .20 = .12.$$

Since the mean asset return is just .10, it does not give enough expected return to be efficient.

Actually, the question answers itself. The only efficient portfolio that uses none of the risk-free asset is the market portfolio itself, and the given asset does not have mean and standard deviation parameters that match those of the market. ∎

Example 9. Consider a market in which the market mean return is .05, the market standard deviation is .04, and the risk-free rate is .01. Verify algebraically that there is no portfolio of the risk-free asset and two independent assets with mean returns .04 and .06 and standard deviations .08 and .10 that is efficient. Assume no short selling or borrowing.

Solution. Within the risky assets, suppose that w_1 is the portion of wealth given to the first asset and $w_2 = 1 - w_1$ the portion to the second. Overall, suppose that a proportion p is invested in risky assets, leaving $1 - p$ in the risk-free asset. Then we have:

$$
\begin{aligned}
\mu_p &= (1-p)(.01) + p\left(w_1(.04) + w_2(.06)\right) \\
&= (1-p)(.01) + p\left(w_1(.04) + (1-w_1)(.06)\right) \\
&= (1-p)(.01) + p\left(.06 - .02w_1\right) \\
&= .01 + .05p - .02pw_1;
\end{aligned}
$$

$$\sigma_p^2 = (pw_1)^2(.08)^2 + (p(1-w_1))^2(.10)^2 = p^2\left(.01 - .02w_1 + .0164w_1^2\right).$$

Since we are assuming that p is between 0 and 1 to ensure no borrowing or short-selling, the standard deviation is:

$$\sigma_p = \sqrt{\sigma_p^2} = \sqrt{p^2(.01 - .02w_1 + .0164w_1^2)} = p\sqrt{.01 - .02w_1 + .0164w_1^2}.$$

We must decide if there are values p and w_1 that satisfy the capital market line equation:

$$
\begin{aligned}
\mu_p &= r + \frac{\mu^*-r}{\sigma^*}\sigma_p \\
&\Longrightarrow .01 + .05p - .02pw_1 = .01 + \frac{.05-.01}{.04} \cdot p\sqrt{.01 - .02w_1 + .0164w_1^2}.
\end{aligned}
$$

Subtracting the common terms of .01 from both sides, dividing both sides by p, and simplifying the quotient gives the equation:

$$.05 - .02w_1 = \sqrt{.01 - .02w_1 + .0164w_1^2} \iff .016w_1^2 - .018w_1 + .0075 = 0.$$

The last equation follows from squaring both sides and collecting terms. You can check that the discriminant of this quadratic equation in w_1 is negative; consequently there are no real solutions for w_1 that lie on the capital market line. ∎

2.3.2 CAPM Formula; Asset β

It should not come as too much of a surprise that rates of return on individual assets tend to move with the rate of return on the market. Stock indices such as Standard & Poors 500 and the Dow Jones Industrial Average can serve as rough approximators of the market, and it is surprising how simple and tight the relationship can be between companies such as General Electric and Caterpillar and these market representatives. Data on weekly rates of return for GE, CAT, and the S&P 500 were gathered during the 2016 calendar year, and scatterplots of the asset rates of return against the $S\&P$ are shown in Figure 2.14. It appears as if there is a rough linear relationship between the rates of return on the individual assets and the market, with slope and variability depending on the asset. The CAPM theory actually predicts this, and identifies the slope, or rate of change, of an asset's return relative to that of the market, which is a key economic quantity called the asset's *β-coefficient*, or *β-value*. This subsection elaborates on this idea.

We continue to assume that investors are interested only in the means, variances, and covariances of assets when they make their optimal decisions. Again, σ^* and μ^* denote the standard deviation and mean return of the market portfolio. There are a finite number $n-1$ of risky assets in the market, with parameters (σ_i, μ_i) and a known correlation structure. And there is a risk-free asset with rate of return r. We can prove the following relationship between the mean return on a single asset and the mean return on the market portfolio.

Theorem 4. Let asset i have covariance with the market denoted by σ_{mi}, that is, $\sigma_{mi} = \mathrm{Cov}(R_i, R_m)$ where R_i is the random rate of return on the asset and R_m is the random rate of return on the market portfolio. Then,

$$\mu_i = r + \beta_i (\mu^* - r), \text{ where } \beta_i = \frac{\sigma_{mi}}{(\sigma^*)^2} \tag{2.43}$$

Remark. Viewing this equation instead as a relationship between the mean return μ_i and a security's β_i, we see that the relationship is linear. The graph of this line is called the *securities market line*. The securities market line goes though $(0, r)$ and $(1, \mu^*)$, the former corresponding to the risk-free asset that is not correlated with the market, and the latter corresponding to the market portfolio which is perfectly correlated with itself. This is displayed in

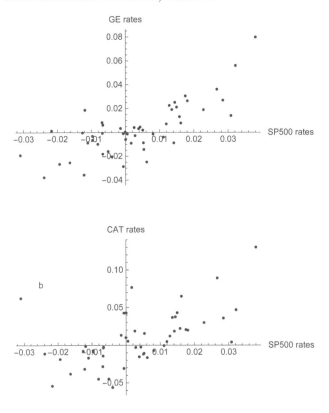

FIGURE 2.14 Rates of return of (a) General Electric; (b) Caterpillar vs. Standard & Poors 500 index.

Figure 2.15. When an asset has a β-value less than 1, the mean excess return over the risk-free asset $\mu_i - r$ will be less than the mean excess return of the market portfolio over the risk-free asset $\mu_m - r$. The reverse is true if $\beta > 1$.

Proof of Theorem 4. Because of the fact that the capital market line is tangent to the efficient frontier, it has the largest possible slope of any line segment connecting the risk-free asset $(0, r)$ with a feasible portfolio (σ_p, μ_p). Thus,

$$\frac{\mu^* - r}{\sigma^*} = \max_p \frac{\mu_p - r}{\sigma_p}. \tag{2.44}$$

Let σ_{ij} be the asset covariance between assets i and j as before, let w_i^* be the weights in the market portfolio of risky assets for assets $i = 2, 3, ..., n$, and let w_i be the weights in an arbitrary portfolio. By our earlier expressions for portfolio mean and variance, for the market portfolio,

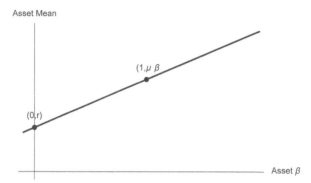

FIGURE 2.15 Securities market line.

$$\mu^* = \sum_{i=2}^{n} w_i^* \mu_i, \, (\sigma^*)^2 = \sum_{i=2}^{n} \sum_{j=2}^{n} w_i^* w_j^* \sigma_{ij},$$

and similarly for arbitrary portfolios; hence:

$$
\begin{aligned}
\frac{\mu^*-r}{\sigma^*} &= \max_p \frac{\sum_{i=2}^{n} w_i \mu_i + r\left(1-\sum_{i=2}^{n} w_i\right) - r}{\left(\sum_{i=2}^{n} \sum_{j=2}^{n} w_i w_j \sigma_{ij}\right)^{1/2}} \\
&= \max_p \frac{\sum_{i=2}^{n} w_i (\mu_i - r)}{\left(\sum_{i=2}^{n} \sum_{j=2}^{n} w_i w_j \sigma_{ij}\right)^{1/2}}.
\end{aligned}
\tag{2.45}
$$

Since the maximum occurs at the market portfolio weights, the partial derivative of the function being maximized with respect to each w_i must be zero when evaluated at the weights of the market portfolio vector $\boldsymbol{w}^*$. The function is a quotient, and to set its derivative to zero we only need be concerned with the numerator of the partial derivative. By the quotient rule, the numerator of the partial derivative with respect to w_i is:

$$(\mu_i - r) \left(\sum_{i=2}^{n} \sum_{j=2}^{n} w_i w_j \sigma_{ij}\right)^{1/2}$$
$$- \left(\sum_{i=2}^{n} w_i (\mu_i - r)\right) \cdot \frac{1}{2} \cdot \left(\sum_{i=2}^{n} \sum_{j=2}^{n} w_i w_j \sigma_{ij}\right)^{-1/2} \cdot \left(2 \sum_{j=2}^{n} w_j \sigma_{ij}\right).$$

We now use the result (see Exercise 10) that $\sigma_{mi} = \sum_{j=2}^{n} w_j^* \sigma_{ij}$. Evaluating at the market portfolio and noting that $\sum_{i=2}^{n} w_i^* = 1$, setting the numerator to zero gives:

$$
0 = (\mu_i - r) \left((\sigma^*)^2\right)^{1/2} - \frac{(\mu^*-r)}{\sigma^*} \sigma_{mi} \implies (\mu_i - r) \sigma^* = \frac{(\mu^*-r)}{\sigma^*} \sigma_{mi}
$$
$$
\implies \mu_i - r = \frac{\sigma_{mi}}{(\sigma^*)^2} (\mu^* - r),
$$

which establishes the formula. ∎

Example 10. For the four risky assets in Example 2 of Section 2.2, find the β-coefficient of asset number 2, and verify that formula (2.43) holds true for that asset.

Solution. Recall that the risk-free rate is .02, the means of the risky assets are $\mu_2 = .04$, $\mu_3 = .045$, $\mu_4 = .05$, and $\mu_5 = .055$. The covariance matrix is:

$$\Sigma_r = \begin{pmatrix} .01 & 0 & .01 & 0 \\ 0 & .09 & 0 & -.084 \\ .01 & 0 & .25 & 0 \\ 0 & -.084 & 0 & .49 \end{pmatrix}.$$

We noted in Example 7 that the market portfolio vector is:

$$w^* = \begin{pmatrix} .767 \\ .161 \\ .016 \\ .056 \end{pmatrix}.$$

We computed in that example that the market mean is $\mu^* = .0418$, and the market variance is $(\sigma^*)^2 = .008547$. We must next compute $\sigma_{m2} = \text{Cov}(R_m, R_2)$, where R_m is the random rate of return on the market portfolio. By linearity in the first argument of the covariance, we have:

$$\begin{aligned} \sigma_{m2} = \text{Cov}(R_m, R_2) &= \text{Cov}(w_2^* R_2 + w_3^* R_3 + w_4^* R_4 + w_5^* R_5, R_2) \\ &= w_2^* \text{Var}(R_2) + w_3^* \text{Cov}(R_3, R_2) + w_4^* \text{Cov}(R_4, R_2) \\ &\quad + w_5^* \text{Cov}(R_5, R_2) \\ &= (.767)(.01) + (.161)(0) + (.016)(.01) + (.056)(0) \\ &= .00783. \end{aligned}$$

Therefore the β-coefficient of asset 2 is:

$$\beta_2 = \frac{\sigma_{m2}}{(\sigma^*)^2} = \frac{.00783}{.008547} = .9161.$$

We are asked to check the CAPM formula:

$$\mu_2 = r + \beta_2 (\mu^* - r) \iff .04 = .02 + .9161(.0418 - .02).$$

Up to some small rounding error, the right side of the equation does agree with the left; that is, the right side is about .04. ∎

2.3.3 Systematic and Non-Systematic Risk; Pricing Using CAPM

We will close our treatment of the Capital Asset Pricing Model with two ideas, the first of which gives a breakdown of the risk of an asset into two components and the second of which suggests how CAPM can be used in pricing assets.

The CAPM formula is:

$$\mu_i - r = \beta_i \left(\mu^* - r \right) \text{ where } \beta_i = \frac{\sigma_{mi}}{\left(\sigma^* \right)^2}. \tag{2.46}$$

Defining $R_i^e = R_i - r; R_m^e = R_m - r$, the (random) excess rates of return beyond the risk-free return, on an asset and on the market, respectively, CAPM therefore poses a linear relationship between the means of the excess rates of return $\mu_i - r$ and $\mu^* - r$. This also suggests a special linear regression relationship between the excess returns themselves, with slope coefficient β_i and constant term $a = 0$:

$$R_i^e = \beta_i \cdot R_m^e + \epsilon_i \tag{2.47}$$

So, actually define $\epsilon_i = R_i^e - \beta_i \cdot R_m^e$ to make the relationship in formula (2.47) true. By the CAPM formula, ϵ_i must have mean 0. In Exercise 18, you are to show that

$$\text{Cov} \left(R_i^e, R_m^e \right) = \sigma_{im} = \beta_i \left(\sigma^* \right)^2 + \text{Cov} \left(\epsilon_i, R_m^e \right). \tag{2.48}$$

Then the error terms are uncorrelated with the market excess return, because, using the definition of β_i:

$$\text{Cov} \left(\epsilon_i, R_m^e \right) = \sigma_{im} - \beta_i \left(\sigma^* \right)^2 = \sigma_{im} - \frac{\sigma_{im}}{\left(\sigma^* \right)^2} \cdot \left(\sigma^* \right)^2 = 0.$$

Thus, a linear regression model with intercept forced to 0 does apply here, although the normality of errors that a regression model also assumes is not necessarily in evidence. Also, because of the lack of correlation between ϵ_i and R_m^e, we observe that:

$$\text{Var} \left(R_i^e \right) = \beta_i^2 \cdot \text{Var} \left(R_m^e \right) + \text{Var} \left(\epsilon_i \right).$$

But since the only difference between R_i^e and the original rate R_i, and between R_m^e and R_m, is the subtraction of the constant r, it also follows that:

$$\sigma_i^2 = \text{Var} \left(R_i \right) = \beta_i^2 \cdot \text{Var} \left(R_m \right) + \text{Var} \left(\epsilon_i \right) = \beta_i^2 \cdot \left(\sigma^* \right)^2 + \text{Var} \left(\epsilon_i \right). \tag{2.49}$$

This is the risk decomposition that we sought. The first term $\beta_i^2 \cdot \left(\sigma^* \right)^2$ on the right side of (2.49) is called the **market risk**, or **systematic risk**, while the second term $\text{Var} \left(\epsilon_i \right)$ is called the **specific risk** or **unsystematic risk** of the asset. Assets with high market risks (high betas) correspond to high mean rates of return, roughly to compensate for the fact that market risk cannot be reduced by diversification as specific risk can. Systematic risk is tied to the volatility of the market as a whole.

It can also be useful to know not just the amount of market risk that an asset has, which is influenced by its own variance, but the share that market risk has of the total variance of the asset. We refer to this as **relative market risk**, and calculate it as:

$$\text{relative market risk} = \frac{\beta_i^2 \cdot (\sigma^*)^2}{\sigma_i^2}. \tag{2.50}$$

These ideas are illustrated in the next example.

Example 11. In Example 10 for instance, the market risk of asset number 2 would be:

$$\beta_2^2 \cdot (\sigma^*)^2 = (.9161)^2 (.008547) = .00717,$$

and the specific risk is:

$$\text{Var}(\epsilon_2) = \text{Var}(R_2) - \beta_2^2 \cdot (\sigma^*)^2 = .01 - .00717 = .00283.$$

Remember that the market portfolio was heavily weighted on asset 2, so that the market rate of return is highly correlated with the rate of return on asset 2, which in turn makes the market risk of asset 2 much larger than its specific risk. The relative market risk, that is the ratio of market risk to overall variance, is:

$$\frac{\beta_2^2 \cdot (\sigma^*)^2}{\sigma_2^2} = \frac{.00717}{.01} = .717 = 71.7\%.$$

So a very high percentage of the variability of asset 2 comes as a result of its close tie to the market portfolio. ∎

Moving to our last topic, the acronym CAPM stands for "Capital Asset Pricing Model", so in what sense can the CAPM theory be used for pricing, that is for comparing a current market situation to one that is theoretically expected in equilibrium? Can we tell if an asset is over- or underpriced?

Let $S(0)$ and $S(1)$ denote the prices of an asset at the initial and terminal times, respectively. We assume that $S(0)$ is known, while $S(1)$ is a random variable. Since the rate of return on an asset is $\frac{S(1)-S(0)}{S(0)}$, its expected value is:

$$\mu = E\left[\frac{S(1) - S(0)}{S(0)}\right] = \frac{1}{S(0)} E[S(1)] - 1.$$

On the other hand, the CAPM formula says that:

$$\mu = r + \beta \left(\mu^* - r \right)$$

so that combining the two gives:

$$r + \beta \left(\mu^* - r \right) = \frac{1}{S(0)} E[S(1)] - 1 \Longrightarrow S(0) = \frac{E[S(1)]}{1 + r + \beta \left(\mu^* - r \right)}. \qquad (2.51)$$

For example, if an asset is currently priced at \$30 in the market, if an investor thinks that its value at the next period is expected to be 1.06 times its current value, and if the market portfolio return is .05, the riskless rate is .03, and the asset's β is .4, then the CAPM initial price is:

$$S(0) = \frac{E[S(1)]}{1 + r + \beta \left(\mu^* - r \right)} = \frac{1.06(30)}{1 + .03 + .4(.05 - .03)} = 30.636.$$

The current market price is slightly under the CAPM value, so this might be a good investment; at least, we have no evidence that the asset is over-valued.

Example 12. Another way of using CAPM theory to tell if assets are overperforming or underperforming is to apply a linear regression model:

$$R_i^e = a + \beta_i \cdot R_m^e + \epsilon_i \qquad (2.52)$$

to the excess rate of return on the asset and on the market. CAPM theory asserts that the constant a is zero. Standard regression testing techniques can tell us if it is reasonable to assume that this is the case. If a is significantly positive, then the asset is outperforming the market, and if a is significantly less than zero it is underachieving, according to what the theory predicts. Let us see how this works on one of the assets that was considered earlier, General Electric.

We will suppose that the risk-free rate was about .5% = .005 in 2016, and since our asset rates of return were computed weekly, we will translate this risk-free rate to a weekly rate by dividing by 52: $r = .005/52 \approx .0001$.

Using the Standard & Poors 500 as an approximator of the market, the 52 week sample of data that yielded the rates of return graphed in Figure 2.14 gave sample mean and variance:

$$\bar{X}_m \approx .00286, \, S_m^2 \approx .0002679,$$

which we use to estimate μ^* and $(\sigma^*)^2$, respectively. Similarly, for the GE weekly rates, we use the sample mean and variance to estimate:

$$\mu_i \approx .002542, \, \sigma_i^2 \simeq .0005668.$$

The covariance can be estimated by the sample covariance:

$$S_{mi} \approx \sigma_{mi} \approx .0003042,$$

from which we get the approximate β-value of GE as $\beta_i = \frac{\sigma_{mi}}{(\sigma^*)^2} = \frac{.0003042}{.0002679} = 1.1357$. Unsurprisingly, the output parameter table that we get when we perform linear regression on the excess rates $R_i^e = R_i - .0001$ and $R_m^e = R_m - .0001$ shows that the estimated slope coefficient is the same $\beta_i = 1.1357$, and this has a very small p-value and so is highly significant. The intercept estimate $a \approx -.0007$ is slightly negative, so GE may be underperforming; however the very high p-value of .745 for the constant term indicates that it is not significantly different from zero, so that we do not have evidence against the CAPM model in this particular data set. ∎

	Estimate	Standard Error	t-Statistic	P-Value
1	-0.000690948	0.00211364	-0.3269	0.745108
x	1.13557	0.128551	8.83354	8.78888×10^{-12}

FIGURE 2.16 Regression output for excess returns of GE against market.

Exercises 2.3

1. Reconsider the market of one non-risky asset and three risky assets in Exercise 2.2-1, in which the parameters were: $r = \mu_1 = .01$, asset means $\mu_2 = .04$, $\mu_3 = .07$, $\mu_4 = .08$, asset standard deviations $\sigma_2 = .06$, $\sigma_3 = .09$, and $\sigma_4 = .11$. The only non-zero correlation was $\rho_{2,3} = -.3$. Build an efficient portfolio with mean rate of return $\mu_p = .05$, and find the standard deviation of that portfolio.

2. In Exercise 2.2-15 we had a market of four independent risky assets with mean rates of return .03, .05, .07, and .09, and variances .08, .10, .12, .14. Suppose that there is a risk-free asset with rate of return .02. Build an efficient portfolio whose mean rate of return is $\mu_p = .04$, and find the variance of that portfolio.

3. Referring to the situation in Exercise 2, for what risk aversion is the portfolio that you computed the optimal portfolio?

4. In Example 6, what efficient portfolio has a mean of $\mu_p = .07$, and what is the standard deviation of that portfolio? If an investor short sells the risk-free asset so as to invest an extra 30% of wealth in the market portfolio, what are the mean and standard deviation of the rate of return on that portfolio?

5. Suppose that there is a non-risky asset with rate of return $r = .04$, an asset with mean return .07 and standard deviation .12, and a second risky asset with mean .10, standard deviation .15, and correlation .6 with the first risky

asset. Is a portfolio with half of its wealth reserved for the non-risky asset and 1/4 for each of the risky assets an efficient portfolio?

6. Suppose that a market has three risky assets with mean rates of return $\mu_2 = .05$, $\mu_3 = .07$ and $\mu_4 = .10$ and variances $\sigma_2^2 = .01$, $\sigma_3^2 = .04$, $\sigma_4^2 = .09$. Assets 2 and 4 are correlated with correlation $\rho_{24} = -.4$, but asset 3 is independent of assets 2 and 4. The risk-free rate is $r = .03$. Find an efficient portfolio with mean rate of return .09, find its variance, and compare to the variance of the market. Is borrowing on the risk-free asset necessary to form this portfolio?

7. In the case of three independent risky assets, find a specific relation between the point on the capital market line between $(0, r)$ and (σ^*, μ^*) and the risk aversion of the investor.

8. Explain algebraically and intuitively why the securities market line goes through the points $(0, r)$ and $(1, \mu^*)$.

9. Consider a situation in which the risk-free asset has rate of return $r = .03$ and the market portfolio has mean return $\mu^* = .05$.

(a) At least how large must any asset mean be whose β-value is at least 1.2?

(b) For an asset whose mean is less than .04, how small must its β-value be?

10. Show the fact used in the proof of Theorem 4: $\sigma_{mi} = \sum_{j=2}^{n} w_j^* \sigma_{ij}$.

11. Consider the market described in Exercise 6. (Three risky assets, means $\mu_2 = .05$, $\mu_3 = .07$ and $\mu_4 = .10$; variances $\sigma_2^2 = .01$, $\sigma_3^2 = .04$, $\sigma_4^2 = .09$; $\rho_{24} = -.4$, and other ρ values 0; $r = .03$.) Compute the β-coefficients of each risky asset.

12. For the market in Exercise 5 ($r = .04$; two risky assets with mean returns .07, .10; standard deviations .12, .15; correlation $\rho_{23} = .6$) compute the β-value of each risky asset and check each asset to see that the security market line equation is satisfied.

13. Let the market be as in Exercise 2 (four independent risky assets numbered $2, 3, 4, 5$ with means .03, .05, .07, and .09; variances .08, .10, .12, .14; $r = .02$). Find the β-value of asset 4 and check that the securities market line equation is satisfied.

14. Show that the β-value of the market portfolio equals 1 using the definition of β.

15. Two risky assets have rates of return R_1 and R_2 and β-coefficients β_1 and β_2. If a portfolio is formed from these with weights w_1 and w_2, show that the β coefficient for the portfolio is $w_1\beta_1 + w_2\beta_2$. Is a similar result true for portfolios of more than two assets?

16. Calculate the systematic and specific risks of assets 3, 4, and 5 in Example 10. For which asset is the amount of market risk lowest, and for which is it highest? For which asset is the relative market risk lowest, and for which is it highest?

17. For the market of Exercise 13 find the market and specific risks of all assets. Which asset has the highest and which the lowest relative market risk?

18. In the subsection on Systematic and Non-systematic Risk, derive formula (2.48), that is, show that

$$\text{Cov}\left(R_i^e, R_m^e\right) = \sigma_{im} = \beta_i \left(\sigma^*\right)^2 + \text{Cov}\left(\epsilon_i, R_m^e\right),$$

where $R_i^e = \beta_i \cdot R_m^e + \epsilon_i$.

19. Suppose that in a market with risk-free rate of .01, the market portfolio has mean rate of return of .07 and a standard deviation of .2. Emma suspects that stock in company A will grow by 5% from its current value of $100 per share. Emma has used recent price behavior to estimate that the covariance between company A and the market is about .03. Does CAPM pricing suggest that Emma should invest in A's stock?

20. An asset in a market follows a single period binomial branch process with initial price $75, up rate $b = .10$, down rate $a = -.02$, up probability $p = .6$ and down probability $1 - p = .4$. In the market, there is a risk-free asset with rate of return $r = .015$, and a market portfolio with mean $\mu^* = .03$, and variance $(\sigma^*)^2 = .25$ whose covariance with the risky asset is .08. Is the binomial branch model under- or overpriced relative to the CAPM price?

21. Playing off the idea of the market price of risk, the **Sharpe index** of an asset or portfolio with mean return μ and standard deviation σ is defined by $S = \frac{\mu-r}{\sigma}$, where r is the risk-free rate, as usual. This is one way of measuring and comparing investment performance; the higher the Sharpe index the better.
  (a) Find the Sharpe indices of all assets in Example 7;
  (b) Find the Sharpe index of a portfolio that splits wealth evenly between the first and second risky asset of Example 7;
  (c) If the CAPM theory is correct, argue that the largest possible Sharpe index is that of the market portfolio.

22. (Technology required) Using weekly rates of return on Caterpillar stock and the Standard & Poors 500 for the year 2016, perform a regression of excess

rates for Caterpillar on excess rates for the S&P. Is the intercept coefficient significantly different from zero? What is the sign, and what does that mean for the question of whether Caterpillar is over- or underperfoming?

23. In this section, the capital market line was graphed in Figure 2.12 for proportions $p \geq 0$. What does the line look like in the case $p < 0$? (Hint: Does the expression for σ in the parameterized form $\sigma = \sqrt{\sigma^2} = \sqrt{0 + p^2(\sigma^*)^2} = p \cdot \sigma^*$ change?) What does this say about whether there could be optimal portfolios that short-sell the market portfolio?

2.4   Utility Theory

The idea of characterizing investors' aversion to risk using a single number, as we have done, raises legitimate objections. Perhaps investor attitudes are more complicated than that. In this section we will briefly introduce an alternative approach to portfolio optimization in which the investor will maximize a function called a **utility function** of the final wealth achieved at the end of the investment time period. The modeling power given to us by using functions rather than single numbers may help to generalize investor behavior, although as we will see, closed form calculation becomes harder.

To be more specific, the goals of this section are to: (a) introduce the concept of a utility function of the wealth received; (b) characterize **risk aversion** in the context of utility functions and define the **certainty equivalent** of a risky gamble; (c) exhibit example utilities and illustrate some of their general properties regarding risk; (d) show the connection to the mean-variance portfolio optimization problem covered earlier; and (e) solve some simple single-period utility optimization problems.

2.4.1   Securities and Axioms for Investor Behavior

We consider discrete securities that, in one time period, return an amount of wealth X bounded between two numbers m and M, where $m < M$. Specifically, the wealth received on investing in a share of the security is a discrete random variable X taking on possible values $x_1, x_2, ..., x_n \in [m, M]$ with probabilities $p_1, p_2, ..., p_n$, respectively. Figure 2.17(a) depicts the probabilistic behavior of such a security. In the finance literature these random variables are sometimes referred to as **gambles**. Two special cases are important: the deterministic security $X(\omega) = x$ for all outcomes ω, and the simple security Y_h shown in Figure 2.17(b) taking on the maximum possible value M with probability h and the minimum possible value m with probability $1 - h$.

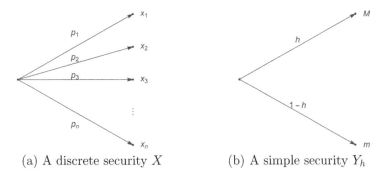

(a) A discrete security X (b) A simple security Y_h

FIGURE 2.17 Examples of securities.

Remark. This version of the idea of a risky asset is not really new. Previously, we have considered such assets in terms of their random rates of return R, but if R has probability mass function

$$p(r) = P[R = r], r = r_1, r_2, ..., r_n,\qquad(2.53)$$

then if the investor devotes an amount of wealth W_0 to this asset, the possible final wealth values are:

$$x_1 = (1 + r_1)W_0, x_2 = (1 + r_2)W_0, ..., x_n = (1 + r_n)W_0,\qquad(2.54)$$

occurring, respectively, with probabilities:

$$p_1 = p(r_1), p_2 = p(r_2), ..., p_n = p(r_n).\qquad(2.55)$$

Portfolios of assets can be considered as discrete securities in this sense as well, since the amount of wealth returned by a portfolio is just a discrete random variable.

Example 1. Consider a portfolio that devotes $1/4$ of an initial investment of $\$1000$ to risky asset 1 and $3/4$ to asset 2. The two assets have independent rates of return, with distributions as in the tables below. Model the portfolio as a discrete security in the sense described above, and draw a tree diagram indicating its possible outcomes.

asset 1 rates	.02	.05	asset 2 rates	0	.06
asset 1 probs	.5	.5	asset 2 probs	.4	.6

Solution. The investor will invest $\$250$ in asset 1 and $\$750$ in asset 2. There are four possible pairs of rates of return: $(.02, 0)$, $(.05, 0)$, $(.02, .06)$, and $(.05, .06)$.

The possible final wealth values and their probabilities are displayed in the table below, using the independence assumption to find the probabilities of each:

outcome	final wealth	probability
$(.02, 0)$	$1.02(\$250) + 1(\$750) = \$1005$	$(.5)(.4) = .2$
$(.05, 0)$	$1.05(\$250) + 1(\$750) = \$1012.50$	$(.5)(.4) = .2$
$(.02, .06)$	$1.02(\$250) + 1.06(\$750) = \$1050$	$(.5)(.6) = .3$
$(.05, .06)$	$1.05(\$250) + 1.06(\$750) = \$1057.50$	$(.5)(.6) = .3$

The tree diagram is in Figure 2.18. ■

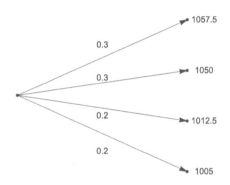

FIGURE 2.18 Portfolio viewed as a security.

Although we will not go into depth here on the topic of Utility Theory, in order to motivate the use of utility functions in finding optimal portfolios, we will state some reasonable assumptions about investor behavior. It is shown in the literature (see for example references[21] or [7]) that these axioms imply the so-called "Expected Utility Maxim". This maxim says that each investor has a **utility function** and will choose a security (or a portfolio of securities) so as to maximize the expected value of the utility of the final wealth that it returns.

Formally, the axioms are as follows:

Definition 1. (Axioms of investor preference)

Axiom 1 (Total ordering) Each investor has a total order relation $\gtrsim$ on the set of discrete securities, so that if X_1 and X_2 are securities, either $X_1 \gtrsim X_2$, or $X_2 \gtrsim X_1$, or if neither of these, we say that $X_1 \approx X_2$. (X_1 is **equivalent** to X_2)

Axiom 2 (Completeness) Given $x \in [m, M]$ there exists a simple security Y_h such that $x \approx Y_h$ (viewing x as a deterministic security).

Axiom 3 (Transitivity) If $X_1 \gtrsim X_2$ and $X_2 \gtrsim X_3$, then $X_1 \gtrsim X_3$.

Axiom 4 (Dominance) If h_1, h_2 are probabilities with $h_1 \geq h_2$, then the corresponding relation holds for the associated simple securities: $Y_{h_1} \gtrsim Y_{h_2}$.

Axiom 5 (Substitution) Let X be the security taking on possible values $x_1, x_2, ..., x_n$ with probabilities $p_1, p_2, ..., p_n$, respectively, assume that $x_i \approx Y_{h_i}$, and let Z be the security with the same probability distribution as X, except that in place of outcome x_i, Z takes on one of the values M or m with probabilities $p_i h_i$ or $p_i (1 - h_i)$, respectively. Then $X \approx Z$. ■

To interpret these, refer first to Axiom 1. We want the investor to be able to compare securities and be able to say that one or the other is better, or he is indifferent between them, in which case we call the securities equivalent. In particular, the investor should always be able to tell which is better: a deterministic security returning x or a simple security Y_h, and for which pairs (x, h) these two securities are equivalent. The completeness property (sometimes called the **continuity** property) in Axiom 2 should hold in the latter context: given a value of wealth $x \in [m, M]$ there should exist a simple security Y_h that is equivalent to it. The boundary cases $x = m$ and $x = M$, respectively correspond to $h = 0$ and $h = 1$. The investor's ordering of securities ought to be transitive as described by Axiom 3, so that if security A is at least as good as security B, and B is at least as good as C in the eyes of an investor, then A will be at least as good as C. In view of the fact that earning the highest possible amount of wealth M with higher probability is better than earning it with lower probability, it is reasonable to assume the dominance property in Axiom 4.

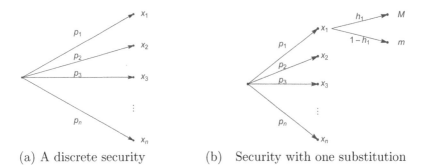

(a) A discrete security   (b)   Security with one substitution

FIGURE 2.19 Substitution of a simple security.

So far the axioms seem very intuitive and natural. The substitution property in Axiom 5 is mostly a technical necessity for the proof of the Expected Utility Maxim, which says that if a discrete security is transformed into a new security in which one or more of the outcomes x_i are replaced by simple securities Y_{h_i} equivalent to them, then the new security is equivalent to the old. The graph in Figure 2.19 illustrates the outcomes of the new security formed after substitution of a simple security for the first outcome x_1 of the original security. It can be proved by mathematical induction that substitution can be repeated for more than just one of the possible x_i. (See Exercise 5.) In fact if all of the possible security values x_i are so replaced, then the new security that is formed is a simple security Y_h in which the value M occurs with total probability $\sum_{i=1}^{n} p_i h_i$, and m occurs with the complementary probability. Therefore, any security can be replaced by an equivalent simple one.

2.4.2   Indifference Curves, Certainty Equivalent, Risk Aversion

For each pair (x, h), $x \in [m, M]$, $h \in [0, 1]$, by Axiom 1 an investor can compare the deterministic security with value x to the simple security Y_h that yields M with probability h and m with probability $1 - h$. In the graph of Figure 2.20, shade those points for which the deterministic security is preferred; otherwise leave the point unshaded. Clearly, if an investor prefers one sure amount x_1 to a gamble Y_h, that investor will prefer x_2 to Y_h for any $x_2 > x_1$; hence once a point is shaded, all points on a horizontal line to its right must be shaded. Also, if an investor prefers sure amount x to a gamble Y_{h_1}, then the investor will prefer x to another gamble Y_{h_2} where $h_2 < h_1$, since the chance of winning the maximum amount M is less for the second gamble. Therefore once a point is shaded, all points on a vertical line below it must be shaded.

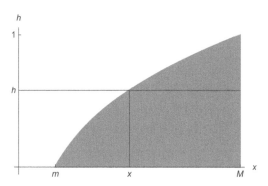

FIGURE 2.20 An indifference curve $h = f(x)$.

The analysis of the last paragraph indicates that the shaded region of pairs (x, h) for which x is preferred to Y_h is a continuous region bounded on the northwest by a curve. By Axiom 2, given x there is h so that the deterministic and simple securities are equivalent in the investor's eyes. The collection of all such (x, h) pairs forming the boundary of the shaded region is the so-called *indifference curve* for the investor. The indifference curve is the graph of an increasing function $h = f(x)$, since the higher the guaranteed reward x, the higher must be the probability h of receiving the amount M.

Notice that for each fixed probability h, the value of x corresponding to it on the indifference curve is the smallest amount of sure wealth that the investor will prefer to the gamble Y_h. Similarly, given a sure amount x, the associated $h = f(x)$ is the largest probability of receiving M (as opposed to m) such that the investor will still choose the sure amount x over the gamble Y_h.

We now generalize the idea of the equivalence of a sure amount of wealth to the wealth of a gamble, from simple securities to arbitrary ones.

Definition 2. The **certainty equivalent** of a security X is that amount of sure wealth, denoted by $\mathrm{CE}(X)$, that the investor considers equivalent to the expected return on the security. ∎

Roughly speaking, if an investor dislikes risk, that investor would trade a risky portfolio for a sure amount of money smaller than the expected wealth on that portfolio. Thus, risk attitudes can be characterized in terms of the certainty equivalent as follows:

$$\text{An investor is } \textbf{\textit{risk-averse}} \text{ if } \mathrm{CE}(X) \leq E[X] \text{ for all } X; \qquad (2.56)$$

$$\text{An investor is } \textbf{\textit{risk-neutral}} \text{ if } \mathrm{CE}(X) = E[X] \text{ for all } X; \qquad (2.57)$$

$$\text{An investor is } \textbf{\textit{risk-preferring}} \text{ if } \mathrm{CE}(X) \geq E[X] \text{ for all } X. \qquad (2.58)$$

Risk aversion implies something about the shape of the indifference curve $h = f(x)$. As we mentioned before, the curve is clearly increasing, since a bigger amount of sure wealth x demands a higher probability h of achieving wealth M in the simple security. We would like to show that the indifference curve is also concave.

Refer to Figure 2.21 and consider two arbitrary points x_1 and x_2 between m and M, which are certainty equivalents of simple gambles Y_{h_1} and Y_{h_2}, respectively. The graph shows the line segment connecting the points $(x_1, h_1 = f(x_1))$ and $(x_2, h_2 = f(x_2))$ on the indifference curve. An arbitrary point on that segment can be written as:

$$
\begin{aligned}
(x, h) &= (1 - t)(x_1, h_1) + t(x_2, h_2) \\
&= ((1 - t)x_1 + tx_2, (1 - t)h_1 + th_2), t \in [0, 1].
\end{aligned}
\qquad (2.59)
$$

Construct a gamble Y that returns x_1 with probability $1 - t$ and x_2 with probability t. In Figure 2.22 are three equivalent gambles: part (a) displays the possible outcomes of the gamble Y that we just constructed; part (b) uses substitution to replace x_1 and x_2 by their equivalent simple gambles with probabilities h_1 and h_2, and the simple security in part (c) just simplifies part (b) by combining the probabilities for M and m as $h = (1 - t)h_1 + th_2$ and $1 - h$, respectively.

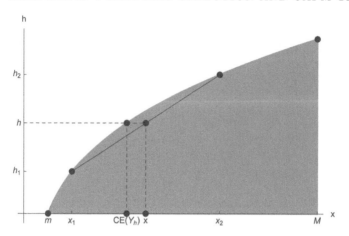

FIGURE 2.21 Concavity of the indifference curve $h = f(x)$.

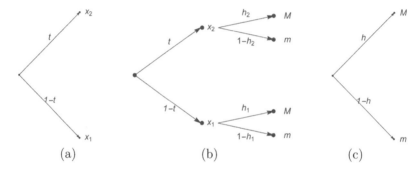

FIGURE 2.22 Three equivalent securities.

But that means that, by risk aversion, the certainty equivalent of the security in part (c) is less than or equal to the expected value of the gamble in part (a), i.e. $\mathrm{CE}\,(Y_h) \leq (1 - t)x_1 + tx_2 = x$. So Figure 2.21 has been drawn properly, i.e. the point at x on the line segment does lie to the right of the point at $\mathrm{CE}\,(Y_h)$, and it is therefore inside the shaded region. Since the original points x_1 and x_2, as well as the chosen point (x, h), were arbitrary, the shaded region is concave.

Remark. For a risk-neutral investor, since the expected value $E\,[Y_h]$ must coincide with the certainty equivalent $\mathrm{CE}\,(Y_h)$ for all probabilities h, it is apparent from the figure that the indifference curve must lie directly on the line segment connecting $(m, 0)$ and $(M, 1)$, i.e. the curve is linear in this case. In Exercise 6 you will argue that a risk-preferring individual has a concave up, or convex, indifference curve. Probably very few investors will be of either of these types, so we will focus on risk-averse investors.

Example 2. Suppose that an investor has an indifference curve given by $h = b\sqrt{x - 2}$, where b is a constant, and the maximum wealth received on a simple gamble is \$6. Find the certainty equivalents of the simple securities $Y_{.5}$ and $Y_{.3}$.

Solution. Since $h(2) = 0$, and $(m, 0)$ is always on the indifference curve, we must have $m = 2$. Since $(M, 1)$ is also always on the curve, we have the equation:

$$1 = b\sqrt{M - 2} = b\sqrt{6 - 2} = 2b.$$

Hence $b = \frac{1}{2}$. For $Y_{.5}$, the certain amount $x = \text{CE}\,(Y_{.5})$ satisfies:

$$.5 = \frac{1}{2}\sqrt{x - 2} \Longrightarrow 1 = \sqrt{x - 2} \Longrightarrow x = 3.$$

That is, our investor would be indifferent between receiving a certain amount of \$3 and playing the game in which either \$6 or \$2 is received with equal probability, which has expected value \$4. For $Y_{.3}$, the amount $x = \text{CE}\,(Y_{.3})$ satisfies:

$$.3 = \frac{1}{2}\sqrt{x - 2} \Longrightarrow .6 = \sqrt{x - 2} \Longrightarrow .36 = x - 2 \Longrightarrow x = 2.36.$$

So the investor is indifferent between receiving \$2.36 and playing the game with expected value $.3(6) + .7(2) = 3.2$. ∎

Example 3. If an investor has an indifference curve of the form $f(x) = b\log(x - 2)$, where b is a constant, and the certainty equivalent $\text{CE}\,(Y_{.2}) = \$4$ is known, then what are $E\,[Y_{.7}]$ and $\text{CE}\,(Y_{.7})$?

Solution. As before, let m and M be the low and high returns on simple securities. Since $f(m) = b\log(m - 2) = 0$, it must be that $m = \$3$. We would like to find M, and to do so, we will need b, since:

$$1 = f(M) = b\log(M - 2) \Longrightarrow \log(M - 2) = \frac{1}{b}.$$

We are given that the certainty equivalent of $Y_{.2}$ is 4; hence:

$$f(4) = b\log(4 - 2) = .2 \Longrightarrow b = \frac{.2}{\log(2)}.$$

Substitution of this expression into the equation for M yields:

$$\log(M - 2) = \frac{1}{b} = \frac{\log(2)}{.2} = 5\log(2) = \log(32)$$
$$\Longrightarrow M - 2 = 32 \Longrightarrow M = \$34.$$

Now that we have m and M, we can compute:

$$E\,[Y_{.7}] = .3(\$3) + .7(\$34) = \$24.7.$$

For the certainty equivalent of $Y_{.7}$, we set up the equation:

$$.7 = b \log\left(\text{CE}\left(Y_{.7}\right) - 2\right) = \frac{.2}{\log(2)} \cdot \log\left(\text{CE}\left(Y_{.7}\right) - 2\right)$$

$$\implies \tfrac{7}{2}\log(2) = \log\left(2^{7/2}\right) = \log\left(\text{CE}\left(Y_{.7}\right) - 2\right).$$

Thus, $\text{CE}\left(Y_{.7}\right) - 2 = 2^{7/2}$, and so $\text{CE}\left(Y_{.7}\right) = 2 + 2^{7/2} = \$13.31.$ ∎

Example 4. Suppose that an investor has an indifference curve of the form $h = f(x) = c\sqrt{x - 50}$ on domain $[m, M] = [\$50, \$150]$, where c is a constant. Find the certainty equivalent of the security X which returns either \$66, \$86, or \$114, with equal probabilities.

Solution. First we should solve for c. Since $f(M) = 1$, we have:

$$1 = f(150) = c\sqrt{150 - 50} = c\sqrt{100} = c \cdot 10 \implies c = .1.$$

Thus, the complete indifference curve is the graph of the function $h = f(x) = .1\sqrt{x - 50}$ on $[\$50, \$150]$. By the substitution axiom, a new security Z can be created from X by replacing each of the three possible values $x_1 = \$66$, $x_2 = \$86$, $x_3 = \$114$ by simple gambles Y_{h_i}, where $x_i = \text{CE}\left(h_i\right)$ for each $i = 1, 2, 3$. (See Figure 2.23) This new security will have three cases resulting in a payoff of $M = \$150$, and three resulting in $m = \$50$ which can be combined, so that Z will be a simple security. Now we have $h_i = f\left(x_i\right), i = 1, 2, 3$. For the given x values we find:

$$h_1 = .1\sqrt{x_1 - 50} = .1\sqrt{66 - 50} = .1\sqrt{16} = .4;$$
$$h_2 = .1\sqrt{x_2 - 50} = .1\sqrt{86 - 50} = .1\sqrt{36} = .6;$$
$$h_3 = .1\sqrt{x_3 - 50} = .1\sqrt{114 - 50} = .1\sqrt{64} = .8.$$

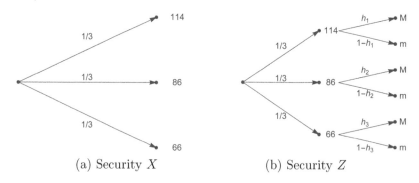

(a) Security X (b) Security Z

FIGURE 2.23 Finding an equivalent simple security.

Our new security Z can take on the value $M = \$150$ in three disjoint cases: $X = x_1$, $Y_{h_1} = M$, or $X = x_2$, $Y_{h_2} = M$, or $X = x_3$, $Y_{h_3} = M$. Since the three probabilities for X are all $1/3$, the total probability that $Z = M = \$150$ is:

$$h = p_1 h_1 + p_2 h_2 + p_3 h_3 = \frac{1}{3}(.4) + \frac{1}{3}(.6) + \frac{1}{3}(.8) = .6.$$

The complementary probability that $Z = m = \$50$ is therefore $1 - .6 = .4$. It could also be computed as $1 - h = p_1(1 - h_1) + p_2(1 - h_2) + p_3(1 - h_3)$. Since Z is now a simple gamble, the certainty equivalent of Z, which equals that of X, is:

$$\mathrm{CE}(Z) = \mathrm{CE}(X) = f^{-1}(h)$$

$$\begin{aligned}
\Longrightarrow \quad & h = f(\mathrm{CE}(X)) \\
\Longrightarrow \quad & .6 = .1\sqrt{\mathrm{CE}(X) - 50} \\
\Longrightarrow \quad & 6 = \sqrt{\mathrm{CE}(X) - 50} \\
\Longrightarrow \quad & 36 = \mathrm{CE}(X) - 50 \\
\Longrightarrow \quad & \mathrm{CE}(X) = \$86.
\end{aligned}$$

Notice that $\mathrm{CE}(X) \leq E[X] = \frac{1}{3}(66) + \frac{1}{3}(86) + \frac{1}{3}(114) = \$88.67.$ ∎

2.4.3 Examples of Utility Functions

The Expected Utility Maxim is the following:

Theorem 1. Assume Axioms 1-5, and let $h = f(x)$ be the indifference curve function of an investor. Then the investor will seek the security Z that maximizes $E[U(Z)]$, where $U(x) = a + bh = a + bf(x)$ is any linear transformation of the indifference curve function with constants $a \geq 0$ and $b > 0$. ∎

Because of the expected utility maxim, not only indifference curves with range $h \in [0, 1]$ are appropriate as utilities for risk-averse investors, but practically any increasing, concave down function due to the arbitrary (but positive) constants a and b.

Remark. A consequence of the theorem and our definition of certainty equivalent is that the certainty equivalent $\mathrm{CE}(X)$ of a security X satisfies the equation:

$$E[U[X]] = U(\mathrm{CE}(X)). \qquad (2.60)$$

You are guided through a proof of this fact in Exercise 17. This also provides an alternative method for solving for $\mathrm{CE}(X)$ for an arbitrary security X which avoids the need to convert the security to an equivalent simple one, as in Example 4.

Recall that an investor is risk averse if and only if $CE(X) \leq E[X]$ for all securities, and since the utility function is increasing, this happens if and only if $U(CE(X)) \leq U(E[X])$. Hence, by formula (2.60), the investor is risk averse if and only if

$$E[U[X]] \leq U(E[X]). \tag{2.61}$$

Here are a few simple and useful examples of utility functions.

Exponential Utility

A utility function is said to be an **exponential utility** if it takes the form:

$$U(x) = c - \frac{1}{a}e^{-ax}, x \geq 0. \tag{2.62}$$

Here we assume that the constant parameters c and a are positive. We have that:

$$U'(x) = -\frac{1}{a}(-a)e^{-ax} = e^{-ax} \tag{2.63}$$

which is always positive, so that U is increasing. Also,

$$U''(x) = -a \cdot e^{-ax} < 0; \tag{2.64}$$

hence U is concave down, as desired. In Figure 2.24 is a typical graph of an exponential utility function.

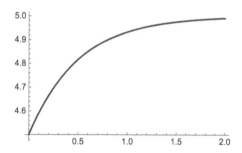

FIGURE 2.24 A utility function of exponential class.

Example 5. For an investor with utility function $U(x) = 40 - 2e^{-.5x}, x \geq 0$, find the certainty equivalent of a security with the probability distribution in the table below.

x	0	1	2	3
$p(x)$	1/4	1/4	1/4	1/4

Solution. To do this, we use formula (2.60):

$$
\begin{aligned}
U(\text{CE}(X)) &= E[U[X]] \\
40 - 2e^{-.5\text{CE}(X)} &= \tfrac{1}{4}(U(0) + U(1) + U(2) + U(3)) \\
&= \tfrac{1}{4}(38 + 38.7869 + 39.2642 + 39.5537) \\
&= 38.9012.
\end{aligned}
$$

Now we just solve algebraically for $\text{CE}(X)$:

$$
38.9012 = 40 - 2e^{-.5\text{CE}(X)} \implies 2e^{-.5\text{CE}(X)} = 40 - 38.9012
$$
$$
\implies e^{-.5\text{CE}(X)} = \tfrac{40 - 38.9012}{2}
$$
$$
\implies \text{CE}(X) = \log\left(\tfrac{40 - 38.9012}{2}\right) / (-.5) = 1.198. \ \blacksquare
$$

Power Utility

A utility function is a ***power utility*** if:

$$
U(x) = cx^{\gamma}; \ x \geq 0, \tag{2.65}
$$

where the parameter γ satisfies $\gamma \in (0,1)$ and $c > 0$. The first derivative of this power utility is:

$$
U'(x) = c\gamma x^{\gamma - 1}, \tag{2.66}
$$

which is positive for $x > 0$, and the second derivative is:

$$
U''(x) = c\gamma(\gamma - 1)x^{\gamma - 2}. \tag{2.67}
$$

Since the exponent γ was chosen to be between 0 and 1, this is less than zero. So again, U is increasing and concave down. A typical member of this class of utility functions is displayed in Figure 2.25.

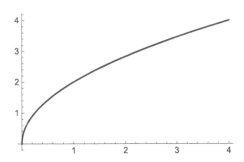

FIGURE 2.25 A power utility function.

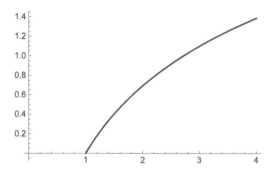

FIGURE 2.26 A logarithmic utility function.

Logarithmic Utility

We say that a utility function belongs to the **logarithmic utility** class if it is defined by the formula:

$$U(x) = b \cdot \log(x) + c, x > 0, \tag{2.68}$$

where b and c are assumed to be positive constants. In this case,

$$U'(x) = \frac{b}{x} \text{ and } U''(x) = \frac{-b}{x^2}. \tag{2.69}$$

Therefore members of this class, like the one plotted in Figure 2.26, are also increasing and concave down.

Quadratic Utility

As the name suggests, a **quadratic utility function** has the form:

$$U(x) = bx - cx^2. \tag{2.70}$$

Here, b and c are positive constants, and of course the domain of application must be limited to x such that the function U is increasing as is the case in the example of Figure 2.27. Since the function will be increasing up to the x-coordinate of its vertex, this means that $x \leq b/2c$. On this domain, the negativity of the square coefficient implies that the function is again concave down.

The problem of maximizing a quadratic utility function of wealth more or less reduces to the mean-variance optimization problem from before. To see this, consider a quadratic utility $U(x) = bx - cx^2$ and let $X = (1+R)W_0$ be the final wealth generated by a portfolio of n assets with weights $w_i, i = 1, 2, ..., n$, constant initial wealth W_0, and random portfolio rate of return R. Recall also the relationship $\text{Var}(Y) = E\left[Y^2\right] - (E[Y])^2$, which is true for any random

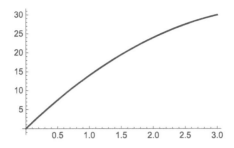

FIGURE 2.27 A quadratic utility function.

variable Y. The investor wants to maximize:

$$
\begin{aligned}
E[U(X)] &= bE[X] - cE\left[X^2\right] \\
&= bE[X] - c\left(\text{Var}(X) + (E[X])^2\right) \\
&= \left(bE[X] - cE[X]^2\right) - c\text{Var}(X) \\
&= \left(bE\left[W_0(1+R)\right] - cE\left[W_0(1+R)\right]^2\right) - c\text{Var}\left(W_0(1+R)\right) \\
&= \left(bE\left[W_0(1+R)\right] - cE\left[W_0(1+R)\right]^2\right) - cW_0^2\text{Var}(R)
\end{aligned}
$$

$$(2.71)$$

Remember that in reference to the third line of (2.71), over the range of possible wealth values x, $bx - cx^2$ increases from zero and is non-negative. So the form in line 3 increases with the mean of X and decreases with the variance of X, and in turn X is clearly an increasing linear function of the rate of return so that the form in line 5 increases in $E[R]$ and decreases in $\text{Var}(R)$. Hence the quadratic utility chooses the portfolio that makes the mean return as large as possible while making the variance of return as small as possible. The exact function being maximized in line 5 of the derivation is a little different than the earlier formulation, $(\max E[R] - a\text{Var}(R))$, but there are strong similarities. Also, if the maximum level of wealth w^* is considered a known constant, then the quadratic function $bw - cw^2$ reaches its maximum at $w^* = b/2c$, so that choosing c and w^* fixes b as well, and the utility optimization problem depends on only one constant c to capture the risk aversion properties of the investor.

2.4.4 Absolute and Relative Risk Aversion

Two measures of the degree of risk aversion for utility functions are as follows:

Definition 3. The *absolute risk aversion* of a utility function is:

$$A(x) = -\frac{U''(x)}{U'(x)} = -\frac{d\log\left(U'(x)\right)}{dx}.$$

$$(2.72)$$

This measures the amount of concavity relative to the rate of increase in utility. The *relative risk aversion* of a utility is defined by:

$$R(x) = -\frac{U''(x)}{U'(x)/x} = -\frac{U''(x)}{U'(x)}x = A(x)x.$$

$$(2.73)$$

Relative risk aversion compares the amount of concavity to the rate of increase in utility per dollar. ∎

Example 6. Show that exponential utilities have constant absolute risk aversion and linear relative risk aversion.

Solution. First, we can calculate the absolute risk aversion by:

$$U(x) = c - \frac{1}{a}e^{-ax} \implies A(x) = -\frac{U''(x)}{U'(x)} = -\frac{(-a)e^{-ax}}{e^{-ax}} = a.$$

Notice that since $U'(x) = e^{-ax}$, $d \log (U'(x))/ dx = d/dx(-ax) = -a$; hence the alternative formula in (2.72) is also very easy to use. The relative risk aversion is:

$$R(x) = -\frac{U''(x)}{U'(x)}x = A(x)x = ax. \ \blacksquare$$

Example 7. Consider a normalized power utility $U(x) = x^\gamma/\gamma$. This utility function has absolute risk aversion $A(x) = \frac{1-\gamma}{x}$ and constant relative risk aversion $R(x) = 1 - \gamma$, as shown below.

$$U(x) = \frac{x^\gamma}{\gamma} \implies A(x) = -\frac{U''(x)}{U'(x)} = -\frac{(\gamma-1)x^{\gamma-2}}{x^{\gamma-1}} = \frac{1-\gamma}{x}$$

$$R(x) = -\frac{U''(x)}{U'(x)}x = A(x)x = 1 - \gamma. \ \blacksquare$$

In Exercise 19 you are asked to show that logarithmic utilities also have constant (equal to 1) relative risk aversion.

Example 8. A broader class called the HARA (*hyperbolic absolute risk aversion*) utilities have the form:

$$U(x) = \frac{1-\gamma}{\gamma}\left(\frac{\beta x}{1-\gamma} + \eta\right)^\gamma, \tag{2.74}$$

where γ, β, and η are all positive constants with $\gamma \in (0,1)$. Notice that if $\eta = 0$, then a power utility results. An example, with parameters $\gamma = .5, \beta = 2$, and $\eta = 1$ is shown in Figure 2.28.

For members of this class, the following computation shows that the reciprocal of the absolute risk aversion, sometimes called the **risk tolerance**, is linear. First we compute the first derivative of U :

$$U'(x) = \frac{1-\gamma}{\gamma} \cdot \gamma \cdot \left(\frac{\beta x}{1-\gamma}+\eta\right)^{\gamma-1} \cdot \frac{\beta}{1-\gamma} = \beta\left(\frac{\beta x}{1-\gamma}+\eta\right)^{\gamma-1}.$$

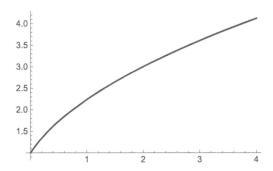

FIGURE 2.28 HARA utility function, $\gamma = .5, \beta = 2, \eta = 1$.

Next is the second derivative:

$$U''(x) = \beta(\gamma - 1)\left(\frac{\beta x}{1-\gamma} + \eta\right)^{\gamma-2} \cdot \frac{\beta}{1-\gamma} = -\beta^2 \left(\frac{\beta x}{1-\gamma} + \eta\right)^{\gamma-2}.$$

From these formulas it follows that the absolute risk aversion is:

$$A(x) = -\frac{U''(x)}{U'(x)} = \frac{\beta^2 \left(\frac{\beta x}{1-\gamma} + \eta\right)^{\gamma-2}}{\beta \left(\frac{\beta x}{1-\gamma} + \eta\right)^{\gamma-1}} = \frac{\beta}{\left(\frac{\beta x}{1-\gamma} + \eta\right)}.$$

So, as a function of x, the absolute risk aversion is of hyperbolic (that is, $1/x$) shape. The reciprocal of A, which is the risk tolerance, is a linear function of x:

$$\frac{1}{A(x)} = \frac{\left(\frac{\beta x}{1-\gamma} + \eta\right)}{\beta} = \frac{x}{1-\gamma} + \frac{\eta}{\beta}. \ \blacksquare$$

2.4.5 Utility Maximization

It is time to actually solve a few optimization problems with utility functions. The trouble is, when you move from objectives of the form $\max E[R_p] - \text{Var}(R_p)$, in which only means, variances, and covariances are involved, to objectives of the form $E[U(X)]$ for non-linear, non-quadratic utility functions U, closed form expressions for optimal portfolio weights become much harder to derive. Still, the approach is straightforward: express the final wealth X in terms of portfolio weights of assets, use the distribution of the rates of return to find the distribution of X, and set the derivatives of $E[U(X)]$ with respect to the portfolio weights to zero. We will be content with small examples. To begin to bridge to the next section on time dependent problems, we will now denote the final wealth X after one time period as W_1. Wealth at later times will be W_2, W_3, etc.

Example 9. Suppose that an investor's goal is formulated as maximizing the expected value of a power utility $U(x) = 2x^{1/2}$ evaluated at the final wealth at time 1. The investor begins with deterministic wealth W_0 at time 0, there is a risk-free asset with rate of return $r = .04$, and there is a risky asset A with possible rates of return $-.013$ and $.095$, occurring with equal probability. What portfolio maximizes the expected utility of final wealth? What changes if the utility is $U(x) = 3x^{1/3}$ instead?

Solution. Let W_1 denote the final wealth at the end of a period, let random variable R_1 be the rate of return on the risky asset, and let t be the proportion of initial wealth invested in the risky asset, so that a proportion $s = 1 - t$ is left for the non-risky asset. Then the final wealth is the random variable:

$$W_1 = W_0 \left((1+r)(1-t) + (1+R_1)t\right) = W_0 \left((1+r) + (R_1 - r)t\right).$$

This means that W_1 is a random variable with two equally likely possible values: $w_1 = W_0(1.04 - .053t)$ and $w_2 = W_0(1.04 + .055t)$. Thus,

$$
\begin{aligned}
E\left[U\left(W_1\right)\right] &= \tfrac{1}{2}\left(U\left(w_1\right) + U\left(w_2\right)\right) \\
&= \tfrac{1}{2} \cdot 2 \cdot \left(w_1{}^{1/2} + w_2{}^{1/2}\right) \\
&= \sqrt{W_0}\left((1.04 - .053t)^{1/2} + (1.04 + .055t)^{1/2}\right).
\end{aligned}
$$

We observe that the initial wealth W_0 will not affect the selection of the optimal value of t so we will ignore that leading factor. Setting the first derivative with respect to t of the expected utility function to zero yields:

$$\frac{1}{2}(-.053)(1.04 - .053t)^{-1/2} + \frac{1}{2}(.055)(1.04 + .055t)^{-1/2} = 0$$

$$
\begin{aligned}
\implies & (.055)(1.04 + .055t)^{-1/2} = (.053)(1.04 - .053t)^{-1/2} \\
\implies & (.055)(1.04 - .053t)^{1/2} = (.053)(1.04 + .055t)^{1/2} \\
\implies & (.055)^2(1.04 - .053t) = (.053)^2(1.04 + .055t) \\
\implies & \left((.055)^2 - (.053)^2\right)(1.04) = (.053)^2(.055t) + (.055)^2(.053t).
\end{aligned}
$$

The solution t to this linear equation can be found to be about $t = .714$, which is the proportion invested in the risky asset, leaving about $.286$ of the initial wealth in the non-risky asset. The plot in Figure 2.29 shows that the expected utility of final wealth is quite flat in a wide region including the maximum point. This means that the utility is not very sensitive to changes in the weight t.

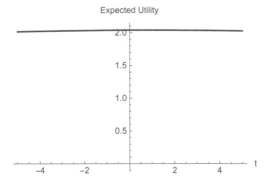

FIGURE 2.29 Expected utility of final wealth as a function of risky weight t.

If the utility function is changed to $U(x) = 3x^{1/3}$, then the solution works out as follows. The values w_1 and w_2 of the final wealth do not change, but we have:

$$
\begin{aligned}
E\left[U\left(W_1\right)\right] & = \tfrac{1}{2}\left(U\left(w_1\right)+U\left(w_2\right)\right) \\
& = \tfrac{1}{2}\cdot 3\cdot\left(w_1^{1/3}+w_2^{1/3}\right) \\
& = \tfrac{3}{2}\sqrt[3]{W_0}\left((1.04-.053t)^{1/3}+(1.04+.055t)^{1/3}\right).
\end{aligned}
$$

The constants in front may be ignored for the purpose of finding the optimal t, and so we set the following first derivative to zero:

$$
\frac{1}{3}(-.053)(1.04-.053t)^{-2/3}+\frac{1}{3}(.055)(1.04+.055t)^{-2/3}=0
$$

$$
\begin{aligned}
\Longrightarrow \quad & (.055)(1.04+.055t)^{-2/3}=(.053)(1.04-.053t)^{-2/3} \\
\Longrightarrow \quad & (.055)(1.04-.053t)^{2/3}=(.053)(1.04+.055t)^{2/3} \\
\Longrightarrow \quad & (.055)^{3/2}(1.04-.053t)=(.053)^{3/2}(1.04+.055t) \\
\Longrightarrow \quad & \left((.055)^{3/2}-(.053)^{3/2}\right)(1.04)=(.053)^{3/2}(.055t)+(.055)^{3/2}(.053t).
\end{aligned}
$$

The solution t to the equation changes to around $t = .535$ as the proportion invested in the risky asset, so that about $.465$ of the initial wealth is put into the non-risky asset. Notice that in making this change to the utility, the relative risk aversion has changed from $1-\gamma = 1-1/2 = 1/2$ to $1-\gamma = 1-1/3 = 2/3$. The investor has become more risk averse and holds less wealth in the risky asset. ∎

Example 10. Consider two risky assets for which the possible joint rate of return of outcomes at time 1 is in the table below. For an investor whose utility function is $U(x) = \log(x)$, find the optimal portfolio.

values of (R_1, R_2)	$(.021, 0)$	$(.044, .05)$	$(.06, .07)$
probability	$1/4$	$1/2$	$1/4$

Solution. Let $x_1, x_2 = 1 - x_1$ be the proportions of initial wealth invested in the two assets. If the initial wealth of the investor is W_0, then the final wealth is:

$$
\begin{aligned}
W_1 &= W_0 + x_1 W_0 R_1 + x_2 W_0 R_2 \\
&= W_0 \left(1 + x_1 R_1 + (1 - x_1) R_2\right) \\
&= W_0 \left(1 + R_2 + (R_1 - R_2) x_1\right).
\end{aligned}
$$

The utility of the final wealth is therefore:

$$
U(W_1) = \log(W_1) = \log(W_0) + \log(1 + R_2 + (R_1 - R_2) x_1).
$$

The additive constant $\log(W_0)$ will not change the makeup of the optimal portfolio, so we can dispose of that. The potential values that this utility can take on, using the rate of return outcomes from the table, are:

outcome 1 (prob 1/4) : $\log(1 + .021x_1)$
outcome 2 (prob 1/2) : $\log(1.05 - .006x_1)$
outcome 3 (prob 1/4) : $\log(1.07 - .01x_1)$.

The expected utility is the weighted sum of these values:

$$
E[U(W_1)] = \frac{1}{4}\log(1 + .021x_1) + \frac{1}{2}\log(1.05 - .006x_1) + \frac{1}{4}\log(1.07 - .01x_1).
$$

Setting the derivative of this function with respect to x_1 equal to zero gives the non-linear equation:

$$
\frac{1}{4} \cdot \frac{.021}{1 + .021x_1} - \frac{1}{2} \cdot \frac{.006}{1.05 - .006x_1} - \frac{1}{4} \cdot \frac{.01}{1.07 - .01x_1} = 0.
$$

Although this equation would bow to closed form analysis (by clearing the denominators via multiplication by the product of all denominators, producing a quadratic equation in x_1), numerical solution is easier. You can compute the critical point maximum as $x_1 = .382$; hence $x_2 = .618$. We plot the expected utility in Figure 2.30, and our answer is consistent with the picture. ∎

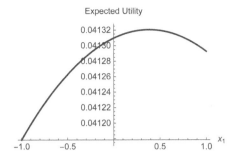

FIGURE 2.30 Expected log utility of final wealth as a function of weight w_1.

Exercises 2.4

1. Suppose that Ann chooses to devote $1/3$ of an initial $6000 investment to a non-risky asset with rate of return $r = .01$ and the remainder to a risky asset whose possible rates of return and probabilities are in the table below. Draw a tree diagram as in Figure 2.18 that displays Ann's possible final wealth values and corresponding probabilities for this portfolio security.

asset rates	−.02	0	.02	.05
asset probs	.4	.2	.2	.2

2. Referring to Exercise 1, assuming maximum and minimum returns of $M = \$6500$ and $m = \$5500$, for what probability h does the simple security Y_h have the same expectation as the portfolio? Does this mean that Y_h is equivalent to the portfolio?

3. For an investor with a linear indifference curve, what simple gamble Y_h is equivalent to sure wealth $x = 10$, if $m = 5, M = 20$?

4. Find the sure amount x that is equivalent to a simple gamble Y_h with $h = .6, M = \$900, m = 0$ for an investor with indifference curve $f(x) = k\sqrt{x} - c$. Check that $CE(Y_h) \leq E[Y_h]$.

5. Show that if Axiom 5 holds, then substitution of k of the outcomes x_i, where $k \leq n$, by equivalent simple securities results in a new security that is equivalent to the original one.

6. Argue that a risk-preferring individual must have a convex (i.e. concave up) indifference curve.

7. Suppose that two points $(36, .4)$ and $(101, .9)$ lie on the indifference curve of a risk-averse investor. Is the point $(62, .6)$ on the curve, above it, or beneath it? Why?

8. Suppose that an investor has an indifference curve of the form $h = f(x) = (x-a)^\gamma/\gamma$ for constants $a > 0$, $0 < \gamma < 1$. If the investor has the characteristic that $\mathrm{CE}\,(Y_{.3}) = 2 + \left(\frac{.21}{8}\right)^{8/7}$, find the minimum and maximum wealth earned on simple gambles, and compute $\mathrm{CE}\,(Y_{.5})$ and $E\,[Y_{.5}]$.

9. Consider the indifference curve given by $h = b\log(x-4)$.
   (a) If the maximum wealth received on a simple gamble is 10, find the certainty equivalents of the simple securities $Y_{.2}$ and $Y_{.8}$;
   (b) If $b = \frac{1}{2}$, find the maximum wealth on a simple gamble, and find the certainty equivalents of the simple securities $Y_{.3}$ and $Y_{.5}$.

10. If an investor has an indifference curve of the form $h = 3 - 3e^{-.75x}$, then what are $E\,[Y_{.5}]$ and $\mathrm{CE}\,(Y_{.5})$?

11. (Technology helpful) What indifference curve of the form $h = b\sqrt[3]{x} - c$ is such that the certainty equivalent of a simple security with probability .5 of achieving the high value is 6 and the certainty equivalent of a simple security with probability .6 of achieving the high value is 6.5?

12. If you have exponential utility function $U(x) = 3 - 2e^{-.5x}$, what is your certainty equivalent of a gamble that returns \$1 with probability .5 and \$3 with probability .5?

13. Use substitution to find the certainty equivalent of the security in Example 5, if the indifference curve is given by the function $h = f(x) = .5\log(x+1)$, $x \in [0, e^2 - 1]$. Recall that the distribution of the security is:

x	0	1	2	3
$p(x)$	1/4	1/4	1/4	1/4

14. Consider an investor with quadratic utility function $U(x) = -x^2 + 1000x$. Use the formula $U(\mathrm{CE}(X)) = E[U(X)]$ to find $\mathrm{CE}(X)$ for the security whose distribution is below.

wealth	\$100	\$200	\$300	\$400
probability	.4	.3	.2	.1

15. (Technology required) Suppose that you are indifferent between receiving \$1000 for certain, or either \$300 or \$2000 with equal probability. If your utility function is of power form $U(x) = x^\gamma/\gamma$, what is your γ? Based on this, which of the following investments do your prefer: one paying \$2000 with certainty, or another paying \$1000 with probability 1/4, \$1800 with probability 1/4, and \$3000 with probability 1/2?

16. Let $A(x)$ be the absolute risk aversion of a utility function $U(x)$. What is the absolute risk aversion of the utility function $a + bU(x)$, where a and b are constant? How about the relative risk aversion?

17. (a) For a simple security Y_h, show directly that $U\left(\text{CE}\left(Y_h\right)\right) = E\left[U\left(Y_h\right)\right]$, where $U(x) = a + bf(x)$ is the linear function of the indifference curve function in Theorem 1.

    (b) If X is a discrete security taking on values x_i with probabilities p_i, $i = 1, 2, ..., n$, write an expression for $E[U(X)]$.

    (c) By the substitution axiom, $\text{CE}(X) = \text{CE}\left(Y_h\right)$, where Y_h is the equivalent simple security obtained by carrying out substitutions in which each x_i is replaced by the simple security Y_{h_i} where $f\left(x_i\right) = h_i$ and hence $x_i = \text{CE}\left(Y_{h_i}\right)$. The formula for h is: $\sum_{i=1}^{n} p_i h_i$. Use this to write expressions for $\text{CE}(X)$ and $U(\text{CE}(X))$.

    (d) Verify that the expressions in parts (b) and (c) are the same, thereby proving formula (2.60).

18. If an investor's utility function is linear, what does that say about the investor? (Justify your answer.)

19. Derive the absolute and relative risk aversions of the log utility $U(x) = b \cdot \log(x) + c$.

20. (Technology helpful) Suppose that an investor has the power utility $U(x) = \frac{4}{3}x^{3/4}$. The market consists of two risky assets and two possible pairs of values for their rates of return as in the table below. Find the investor's optimal portfolio.

value of (R_1, R_2)	$(.03, .01)$	$(.04, .07)$
probability	$1/2$	$1/2$

21. Suppose that a market has a risk-free asset with rate of return r and a risky asset for which there are two possible outcomes at time 1: the rate of return on its value is either b with probability p or a with probability $1 - p$, where we assume that $a < r < b$. For an investor with utility function $U(x) = \log(x)$, show that the optimal proportion of wealth to put into the risky asset is:

$$x = \frac{(1+r)(p(b-a) - (r-a))}{(b-r)(r-a)}.$$

Show further that, denoting the risky asset prices at times 0 and 1 by A_0 and A_1, this x is the same as:

$$x = \left(\frac{(1+r)\left(E\left[A_1/A_0\right] - (1+r)\right)}{(b-r)(r-a)} \right).$$

2.5 Multiple Period Portfolio Problems

In the preceding section, we introduced the notion of the utility function as a way of capturing the degree of an investor's risk aversion, and we showed how to solve some simple one-period portfolio problems in which the investor's task was to maximize the utility of final wealth. Now we would like to extend that problem to multiple time periods, and also to introduce a variation in which the investor derives benefit from consuming some wealth from the system as time evolves. The goal is to maximize the total expected utility of all consumptions, plus the expected utility of the final remaining wealth.

We will not attempt full generality in the problem solution, but in the first subsection we will give the general statement of the portfolio-consumption problem for assets moving in discrete time and the description of the dynamic programming approach to solving these problems. The second subsection gives some illustrative examples in relatively simple cases. We close this section with a brief observation about the connection between the idea of martingale processes and the solution of portfolio problems.

2.5.1 Problem Description and Dynamic Programming Approach

In the general model, we have a non-risky asset S^0, normalized so that at time 0 its value is 1, and increasing at a deterministic rate r per period. Then,

$$S_0^0 = 1, S_1^0 = 1 + r, S_2^0 = (1+r)^2, ... \tag{2.75}$$

For convenience, write $R = 1 + r$ for the multiplier that is applied to the non-risky value in each period. This will make some expressions in the ensuing paragraphs just a little shorter. We also have n risky assets whose price processes are $S^1, S^2, S^3, ..., S^n$, and we denote by S_k^j the price of the jth asset at time k. In examples where there is a single risky asset under consideration, the superscript will be dropped.

For each individual risky asset price process, we will assume that the one period ratios of return defined by:

$$Z_k^j = \frac{S_{k+1}^j}{S_k^j} \tag{2.76}$$

are mutually independent for all k, although we do not necessarily assume that the kth ratio for one asset is independent of the kth ratio for another. Thus, $S_{k+1}^j = Z_k^j \cdot S_k^j$, and Z_k^j represents $1 +$ rate of return on asset j between times k and $k + 1$.

An investor is to choose, for each time k, a vector of portfolio weights $\boldsymbol{w}_k = \left(w_k^0, w_k^1, w_k^2, ..., w_k^n\right)$ for the assets in such a way that the sum of the entries is 1. The investor also chooses a consumption amount c_k, indicating wealth to be removed from the overall investment at time k. Each weight vector $\boldsymbol{w}_k$ and consumption value c_k is to depend only on the current overall wealth W_k in the system at that time, and the known, non-random parameters governing the motion of the assets, and not on any past information. We also assume for simplicity that there are no limits on borrowing and short-selling, and no transaction costs for trades.

There is a utility function U_1 for consumptions, and another utility function U_2, possibly different from U_1, for final wealth at the end of trading, which occurs at time T. These two functions may or may not depend explicitly on time as well as wealth, at least in the sense of incorporating a time discount factor. The investor's problem is to determine the optimal weight vectors $\boldsymbol{w}_k$ and consumptions c_k, for times $k = 0, 1, ..., T - 1$, to optimize:

$$V_0 = \max_{\substack{\boldsymbol{w}_0, \boldsymbol{w}_1, ..., \boldsymbol{w}_{T-1} \\ c_0, c_1, ..., c_{T-1}}} E\left[\sum_{k=0}^{T-1} U_1(k, c_k) + U_2(T, W_T)\right]. \qquad (2.77)$$

Next, we would like to find an equation describing the dynamic behavior of the overall wealth process $W_0, W_1, W_2, ..., W_T$. Here we are writing W_k for the beginning wealth in time interval $[k, k+1]$, prior to the consumption c_k.

If the investor begins with total wealth W_k at time k and then consumes c_k, the amount $(W_k - c_k)$ is left to allocate among the assets. If portfolio vector $\boldsymbol{w}_k = (w_k^0, w_k^1, ..., w_k^n)$ is chosen, then the amount invested in asset j is $(W_k - c_k) \cdot w_k^j$. Growth factor Z_k^j is applied to that amount, so that we can write the new total wealth over all assets as:

$$W_{k+1} = \sum_{j=0}^{n} (W_k - c_k) \cdot w_k^j \cdot Z_k^j. \qquad (2.78)$$

It will be helpful to eliminate the non-risky asset as a variable, and in so doing to eliminate the constraint that $\sum_{j=0}^{n} w_k^j = 1$. Therefore, in formula (2.78), factor out $(W_k - c_k)$ and split off the $j = 0$ term to obtain:

$$
\begin{aligned}
W_{k+1} &= (W_k - c_k) \cdot \left(w_k^0 \cdot R + \sum_{j=1}^{n} w_k^j \cdot Z_k^j\right) \\
&= (W_k - c_k) \cdot \left(\left(1 - \sum_{j=1}^{n} w_k^j\right) \cdot R + \sum_{j=1}^{n} w_k^j \cdot Z_k^j\right) \qquad (2.79) \\
&= (W_k - c_k) \cdot \left(R + \sum_{j=1}^{n} w_k^j \cdot \left(Z_k^j - R\right)\right).
\end{aligned}
$$

Since the consumption amounts c_k and the portfolio vectors $\boldsymbol{w}_k$ are to be functions of the wealth W_k at time k for each k, and the ratios Z_k^j are independent in time, formula (2.79) implies that the wealth process is a **Markov process**; that is, the next value W_{k+1} is conditionally independent of the past values of wealth given the present value W_k. The wealth process and the sequence of actions $a_k = (c_k, \boldsymbol{w}_k)$ taken by the investor at the transaction times fulfill the conditions of a problem known as the **Markov decision problem**. The general approach to the solution of such a problem (see for instance reference [9]), is via **dynamic programming**. In the method of dynamic programming, we define the sequence of partial value functions starting at each time l, similarly to V_0 in formula (2.77), by:

$$V_l(l, W_l) = \max_{\substack{w_l, w_{l+1}, \dots, w_{T-1} \\ c_l, c_{l+1}, \dots, c_{T-1}}} E\left[\sum_{k=l}^{T-1} U_1(k, c_k) + U_2(T, W_T)\right], l = 0, 1, 2, \dots, T-1.$$

$$(2.80)$$

Also, define $V_T = U_2(T, W_T)$. Then it is known that the value functions can be solved for using backwards recursion, starting with the known V_T, and using the so-called **Dynamic Programming (DP) equation**:

$$V_l(l, W_l) = \max_{w_l, c_l} E\left[U_1(l, c_l) + V_{l+1}(l+1, W_{l+1}) \,|\, W_l\right]. \qquad (2.81)$$

In this equation, the multiple time period problem is converted to a one period problem by maximizing the total of the expected utility of current consumption plus the optimal total utility value of all later consumptions and terminal wealth. The optimal investment-consumption strategy for the full problem consists of the sequence of all of the consumption-portfolio vector pairs $(c_l, w_l)_{l=0,1,2,\dots,T-1}$ that achieve the maximum values in formula (2.81).

2.5.2   Examples

In the examples below, for simplicity, we will restrict our attention to the case where each risky asset process in the market is of binomial branch type, in which the probabilistic law of motion is:

$$S_{k+1} = \begin{cases} (1+b)S_k & \text{with probability } p \\ (1+a)S_k & \text{with probability } 1-p. \end{cases} \qquad (2.82)$$

A 2-step model is displayed in Figure 2.31. The exercise set has a few problems in which the asset price motion is not of this type, and the dynamic programming approach encapsulated by formulas (2.79) and (2.81) does not require this simplification. But since the binomial branch model will lead to a standard continuous-time price model that we will describe later, results for the portfolio problem in this context may also lead to corresponding results in continuous time.

We illustrate four cases, with different initial assumptions.

Example 1. To begin, let us find the optimal portfolio allocation strategy for a two period problem without consumption, in the case of logarithmic utility of final wealth $U_2(W) = c\log(W) + d$. There is no explicit time dependence

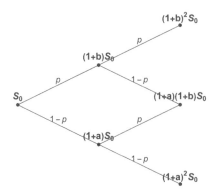

FIGURE 2.31 General 2-step binomial branch model.

in this utility. We suppose that there is a non-risky asset with rate r, and a single binomial branch asset as in Figure 2.31.

Solution. The first move is to step back from $V_2(W_2) = U_2(W_2) = c \log(W_2) + d$ to $V_1(W_1)$ and determine the optimal weight vector w_1 at time 1. Since there is no consumption allowed here, each $c_k = 0$. Also, there is only $n = 1$ risky asset, so we will simplify notation and write w_1 for the proportion of wealth devoted to the risky asset at time 1. The DP equation (2.81), combined with the recursive relation (2.79) for the wealth process, yields the following formula for V_1:

$$
\begin{aligned}
V_1(W_1) &= \max_{w_1} E\left[V_2(W_2)\,|W_1\right] \\
&= \max_{w_1} E\left[V_2(W_1(R + w_1 \cdot (Z_1 - R)))\,|W_1\right] \\
&= \max_{w_1} E\left[c \cdot \log(W_1 \cdot (R + w_1 \cdot (Z_1 - R))) + d\,|W_1\right] \\
&= c \cdot \max_{w_1} E\left[\log(W_1 \cdot (R + w_1 \cdot (Z_1 - R)))\,|W_1\right] + d \\
&= c \cdot (\log(W_1) + \max_{w_1} E\left[\log(R + w_1 \cdot (Z_1 - R))\,|W_1\right]) + d.
\end{aligned}
\tag{2.83}
$$

Now remember that $R = 1 + r$, and the growth random variable Z_1 takes on the possible values $1 + b$ with probability p and $1 + a$ with probability $1 - p$, independently of the wealth W_1. Therefore, the last line in formula (2.83) can be further rewritten as:

$$
\begin{aligned}
V_1(W_1) &= c \cdot (\log(W_1) + \max_{w_1} E\left[\log(R + w_1 \cdot (Z_1 - R))\right]) + d \\
&= c \cdot (\log(W_1) + \max_{w_1}(p \cdot \log(1 + r + w_1(b - r)) \\
&\quad + (1 - p) \cdot \log(1 + r + w_1(a - r)))) + d.
\end{aligned}
\tag{2.84}
$$

To maximize, set the derivative of the expression inside the maximum with respect to w_1 to 0, which produces, after some simplification:

$$
\frac{d}{dw_1}(p \cdot \log(1 + r + w_1(b - r)) + (1 - p) \cdot \log(1 + r + w_1(a - r)))
$$
$$
= \frac{p(b-r)}{1+r+w_1(b-r)} + \frac{(1-p)(a-r)}{1+r-w_1(r-a)} = 0
$$

$$\Longrightarrow \quad \frac{p(b-r)}{1+r+w_1(b-r)} = \frac{(1-p)(r-a)}{1+r-w_1(r-a)}$$

$$\Longrightarrow \quad p(b-r)\,(1+r-w_1(r-a)) = (1-p)(r-a)\,(1+r+w_1(b-r))$$

$$\Longrightarrow \quad p(b-r)(1+r) - (1-p)(r-a)(1+r)$$
$$= (p(b-r)(r-a) + (1-p)(r-a)(b-r))w_1$$

$$\Longrightarrow \quad (1+r)(p(b-a)-(r-a)) = (b-r)(r-a)w_1$$

$$\Longrightarrow \quad w_1 = \frac{(1+r)(p(b-a)-(r-a))}{(b-r)(r-a)}.$$

In order to use dynamic programming to find the optimal strategy at time 0, we must substitute the expression for the optimal w_1 into the time 1 value function in (2.84). Let us simplify each component of the expression separately.

$$
\begin{aligned}
p \cdot \log\left(1+r+w_1(b-r)\right) &= p \cdot \log\left(1+r+\tfrac{(1+r)(p(b-a)-(r-a))}{(b-r)(r-a)}(b-r)\right)\\
&= p \cdot \log\left((1+r)\left(1+\tfrac{p(b-a)-(r-a)}{(r-a)}\right)\right)\\
&= p \cdot \log\left((1+r)\tfrac{p(b-a)}{(r-a)}\right)\\
&= p \cdot (\log(1+r) + \log(p) + \log(b-a)\\
&\quad - \log(r-a));
\end{aligned}
$$

$$
\begin{aligned}
(1-p) \cdot \log(1+r+w_1(a-r)) &= (1-p) \cdot \log(1+r\\
&\quad + \tfrac{(1+r)(p(b-a)-(r-a))}{(b-r)(r-a)}(a-r))\\
&= (1-p) \cdot \log((1+r)(1-\tfrac{p(b-a)-(r-a)}{(b-r)}))\\
&= (1-p) \cdot \log((1+r)\tfrac{(1-p)(b-a)}{(b-r)})\\
&= (1-p) \cdot (\log(1+r) + \log(1-p)\\
&\quad + \log(b-a) - \log(b-r)).
\end{aligned}
$$

Thus,

$$
\begin{aligned}
V_1(W_1) &= c \cdot (\log(W_1) + p \cdot (\log(1+r) + \log(p) + \log(b-a) - \log(r-a))\\
&\quad + (1-p) \cdot (\log(1+r) + \log(1-p)\\
&\quad + \log(b-a) - \log(b-r))) + d\\
&= c \cdot (\log(W_1) + \log(1+r) + p \cdot (\log(p) + \log(b-a) - \log(r-a))\\
&\quad + (1-p) \cdot (\log(1-p) + \log(b-a) - \log(b-r))) + d.
\end{aligned}
$$

$$(2.85)$$

The details of formula (2.85) for V_1 are not important; what is important is that the form of the function is $V_1(W_1) = c \log(W_1) + d_1$, where d_1 is a constant determined by the problem parameters p, r, a, and b. This implies that when we back up to time 0 by solving the DP equation:

$$
\begin{aligned}
V_0(W_0) &= \max_{w_0} E\left[V_1(W_1)|W_0\right]\\
&= \max_{w_0} E\left[c\log(W_1) + d_1|W_0\right]\\
&= \max_{w_0} E\left[c \cdot \log(W_0 \cdot (R + w_0 \cdot (Z_0 - R))) + d_1|W_0\right],
\end{aligned}
$$

we will be doing the exact same computations as we did for the time 1 value function. Thus, the optimal risky portfolio weight at time 0 will still be $w_0 = \frac{(1+r)(p(b-a)-(r-a))}{(b-r)(r-a)}$. ∎

Example 2. Next, let us add two ingredients to the model in Example 1. The investor can consume money from the system at each time 0 and 1, and also the value of money is discounted at a rate δ per period, which may or may not coincide with the risk-free rate r. Find the optimal investment-consumption strategy $((c_0, w_0), (c_1, w_1))$ for the log utilities $U_1(k, x) = U_2(k, x) = (1 + \delta)^{-k}(c\log(x) + d)$.

Solution. As in the last example, we begin with the suitably modified DP equation, initialize $V_2 = U_2$, and substitute the expression for the wealth at time 2 in terms of W_1

$$
\begin{aligned}
V_1(1, W_1) &= \max_{w_1, c_1} E[U_1(1, c_1) + V_2(2, W_2)|W_1] \\
&= \max_{w_1, c_1} E[U_1(1, c_1) + V_2(2, (W_1 - c_1) \cdot (R \\
&\quad + w_1 \cdot (Z_1 - R)))|W_1] \\
&= \max_{w_1, c_1} E[(1 + \delta)^{-1}(c\log(c_1) + d) + (1 + \delta)^{-2} \\
&\quad (c \cdot \log((W_1 - c_1) \cdot (R + w_1 \cdot (Z_1 - R))) + d)|W_1] \\
&= (1 + \delta)^{-1} \cdot (\max_{w_1, c_1}(c\log(c_1) + d) + (1 + \delta)^{-1} \\
&\quad (c \cdot E[\log((W_1 - c_1) \cdot (R + w_1 \cdot (Z_1 - R)))|W_1] + d)) \\
&= (1 + \delta)^{-1} \cdot (\max_{w_1, c_1}(c\log(c_1) + d) + (1 + \delta)^{-1} \\
&\quad (c \cdot (\log(W_1 - c_1) + E[\log(R + w_1 \cdot (Z_1 - R))]) + d)).
\end{aligned}
$$
(2.86)

Once again, the random variable Z_1 takes on possible values $1 + b$ with probability p or $1 + a$ with probability $1 - p$, and $R = 1 + r$. Therefore, V_1 can be rewritten as:

$$
\begin{aligned}
V_1(1, W_1) &= (1 + \delta)^{-1}(\max_{w_1, c_1}(c\log(c_1) + d) \\
&\quad + (1 + \delta)^{-1}(c(\log(W_1 - c_1) + p\log((1 + r) + w_1(b - r)) \\
&\quad + (1 - p)\log((1 + r) + w_1(a - r))) + d)).
\end{aligned}
$$
(2.87)

To solve for the optimal consumption c_1 at time 1 and the optimal proportion w_1 of wealth to invest in the risky asset, we set the partial derivatives with respect to each of c_1 and w_1 of the function f within the max to zero. First, for c_1:

$$
\begin{aligned}
\frac{\partial f}{\partial c_1} = 0 &\implies \frac{c}{c_1} - (1 + \delta)^{-1}\frac{c}{W_1 - c_1} = 0 \\
&\implies \frac{1}{c_1} = \frac{(1+\delta)^{-1}}{W_1 - c_1} \\
&\implies (1 + \delta)^{-1}c_1 = W_1 - c_1 \\
&\implies c_1 = \frac{W_1}{(1+\delta)^{-1}+1}.
\end{aligned}
$$
(2.88)

Thus, a particular proportion of wealth is consumed at time 1; for instance if $\delta = .02$, then that proportion is:

$$
\frac{1}{(1 + \delta)^{-1} + 1} = \frac{1}{(1.02)^{-1} + 1} \approx .505.
$$

Turning to the partial derivative with respect to w_1, we get:

$$\frac{\partial f}{\partial w_1} = 0 \implies (1+\delta)^{-1}\left(\frac{p(b-r)}{1+r+w_1(b-r)} + \frac{(1-p)(a-r)}{1+r-w_1(r-a)}\right) = 0$$
$$\implies \frac{p(b-r)}{1+r+w_1(b-r)} + \frac{(1-p)(a-r)}{1+r-w_1(r-a)} = 0. \tag{2.89}$$

But this is the same equation for w_1 as in Example 1, which had the solution:

$$w_1 = \frac{(1+r)(p(b-a)-(r-a))}{(b-r)(r-a)}. \tag{2.90}$$

The inclusion of the consumption and the discount factor have not changed the optimal portfolio balance between the risky and the non-risky asset, only the amount held in the system. The coefficients c and d, as before, do not affect the solution.

To complete the problem we must step back to time 0, which requires us to know about the optimal time 1 value function V_1. Notice from formula (2.87) that two of the key terms are:

$$\begin{aligned}
c\log(c_1) &= c\log\left(\frac{W_1}{(1+\delta)^{-1}+1}\right) \\
&= c\log(W_1) - c\log\left((1+\delta)^{-1}+1\right);
\end{aligned}$$

$$\begin{aligned}
(1+\delta)^{-1}c\log(W_1-c_1) &= (1+\delta)^{-1}c\log\left((1+\delta)^{-1}c_1\right) \\
&= (1+\delta)^{-1}c\log\left((1+\delta)^{-1}\frac{W_1}{(1+\delta)^{-1}+1}\right) \\
&= (1+\delta)^{-1}\left(c\log(W_1)+c\log\left(\frac{(1+\delta)^{-1}}{(1+\delta)^{-1}+1}\right)\right).
\end{aligned} \tag{2.91}$$

After substituting in for all of the w_1 terms using the expression in (2.90), ignoring the details of the constants, we get that the form of V_1 is:

$$V_1(1,W_1) = (1+\delta)^{-1}\left(k_1\log(W_1)+k_2\right), \tag{2.92}$$

where $k_1 = c\left(1+(1+\delta)^{-1}\right)$ and k_2 are constants. This means that when we set up the DP equation for time 0, we obtain:

$$\begin{aligned}
V_0(0,W_0) &= \max_{w_0,c_0} E[U_0(0,c_0)+V_1(1,W_1)|W_0] \\
&= \max_{w_0,c_0} E[U_0(0,c_0) \\
&\quad + V_1(1,(W_0-c_0)\cdot(R+w_0\cdot(Z_0-R)))|W_0] \\
&= \max_{w_0,c_0} E[(c\log(c_0)+d)+(1+\delta)^{-1} \\
&\quad \cdot(k_1\log((W_0-c_0)(R+w_0(Z_0-R)))+k_2)|W_1].
\end{aligned} \tag{2.93}$$

As in the time 1 computation, the optimal portfolio weight w_0 will not depend on the coefficients k_1, k_2, c, d, or δ, and so the portfolio balance will remain the same:

$$w_0 = \frac{(1+r)(p(b-a)-(r-a))}{(b-r)(r-a)}. \tag{2.94}$$

The optimal consumption c_0 will change a bit. Setting the partial derivative with respect to c_0 inside the maximum to zero, we now get:

$$
\begin{aligned}
\frac{\partial f}{\partial c_0} = 0 \implies & \frac{c}{c_0} - (1+\delta)^{-1} \frac{k_1}{W_0 - c_0} = 0 \\
\implies & \frac{c}{c_0} - (1+\delta)^{-1} \frac{c\left(1+(1+\delta)^{-1}\right)}{W_0 - c_0} - 0 \\
\implies & \frac{1}{c_0} = \frac{(1+\delta)^{-1}\left(1+(1+\delta)^{-1}\right)}{W_0 - c_0} \\
\implies & (1+\delta)^{-1}\left(1 + (1+\delta)^{-1}\right) c_0 = W_0 - c_0 \\
\implies & c_0 = \frac{W_0}{1 + (1+\delta)^{-1} + (1+\delta)^{-2}}.
\end{aligned}
\tag{2.95}
$$

Then, taking for example $\delta = .02$ again, the proportion of wealth consumed at time zero would be:

$$
\frac{1}{1 + (1+\delta)^{-1} + (1+\delta)^{-2}} = \frac{1}{1 + (1.02)^{-1} + (1.02)^{-2}} \approx .340. \ \blacksquare
$$

Example 3. This time let us look at a numerical problem in which there are two independent risky assets undergoing binomial branch motion, with parameters $p_1 = .5, a_1 = -.02, b_1 = .04$, and $p_2 = .6, a_2 = -.05, b_2 = .05$. Investment and consumption occurs over two periods, there is no non-risky asset in this market, and the discount factor δ will be taken to be .01. The utility function for both consumption and final wealth is of square root form, $U(x) = 2\sqrt{x}$, discounted by the factor $(1+\delta) = 1.01$ raised to the appropriate power. We will find the optimal investment-consumption policy for arbitrary (but known) initial wealth W_0, using $w =$ proportion of wealth in asset 2 as one of our two control variables, and per period consumption as the other. Of course, each asset 1 portfolio weight is just $1-$ the asset 2 weight at that particular time. Specifically, we find the pairs $(c_0, w_0), (c_1, w_1)$ of consumptions and asset 2 weights at times 0 and 1 to optimize:

$$
V_0 = \max_{\substack{w_0, w_1 \\ c_0, c_1}} E\left[\sum_{k=0}^{1} (1.01)^{-k} \cdot 2 \cdot \sqrt{c_k} + (1.01)^{-2} \cdot 2 \cdot \sqrt{W_2} \right].
\tag{2.96}
$$

Solution. Backing up a little, we need a slightly revised recursive expression for the total wealth at each time $k = 1, 2$ in terms of the previous wealth. You are asked in Exercise 8 to check that:

$$
W_{k+1} = (W_k - c_k) \cdot \left(Z_k^1 + \left(Z_k^2 - Z_k^1 \right) w_k \right),
\tag{2.97}
$$

where Z_k^1 and Z_k^2 are the ratios of return for assets 1 and 2 at times $k = 0, 1$. All four of these random variables are being assumed to be independent in this example, and the distribution of both Z_0^1 and Z_1^1 is:

$$
P\left[Z^1 = z \right] = \begin{cases} .5 & \text{if } z = 1.04; \\ .5 & \text{if } z = .98, \end{cases}
$$

while the distribution of both Z_0^2 and Z_1^2 is:

$$P\left[Z^2 = z\right] = \begin{cases} .6 & \text{if } z = 1.05; \\ .4 & \text{if } z = .95. \end{cases}$$

To solve the problem, we initialize $V_2(2, W_2) = (1.01)^{-2} \cdot 2 \cdot \sqrt{W_2}$, and step backwards twice using the DP equation:

$$
\begin{aligned}
V_l(l, W_l) &= \max_{w_l, c_l} E\left[U_1(l, c_l) + V_{l+1}(l+1, W_{l+1}) | W_l\right] \\
&= \max_{w_l, c_l} E\left[(1.01)^{-l} \cdot 2 \cdot \sqrt{c_l} + V_{l+1}(l+1, W_{l+1}) | W_l\right], l = 0, 1.
\end{aligned}
\tag{2.98}
$$

For time $l = 1$, we have:

$$
\begin{aligned}
V_1(1, W_1) &= \max_{w_1, c_1} E[(1.01)^{-1} \cdot 2 \cdot \sqrt{c_1} + (1.01)^{-2} \cdot 2 \cdot \sqrt{W_2} | W_1] \\
&= \max_{w_1, c_1} (1.01)^{-1} (2 \cdot \sqrt{c_1} + (1.01)^{-1} \cdot E[2 \cdot \sqrt{W_2} | W_1]) \\
&= (1.01)^{-1} \max_{w_1, c_1} (2 \cdot \sqrt{c_1} + (1.01)^{-1} \\
&\quad \cdot E[2 \cdot \sqrt{(W_1 - c_1) \cdot (Z_1^1 + (Z_1^2 - Z_1^1)w_1)} | W_1]) \\
&= (1.01)^{-1} \max_{w_1, c_1} (2 \cdot \sqrt{c_1} + (1.01)^{-1} \\
&\quad \cdot 2 \cdot \sqrt{(W_1 - c_1)} \cdot E[\sqrt{(Z_1^1 + (Z_1^2 - Z_1^1)w_1)}]).
\end{aligned}
\tag{2.99}
$$

There are four cases for the joint value of the random variables Z_1^1 and Z_1^2; therefore the expectation in the fourth expression in formula (2.99) can be written:

$$
\begin{aligned}
E\left[\sqrt{(Z_1^1 + (Z_1^2 - Z_1^1)w_1)}\right] &= (.5)(.6)\sqrt{1.04 + (1.05 - 1.04)w_1} \\
&\quad + (.5)(.6)\sqrt{.98 + (1.05 - .98)w_1} \\
&\quad + (.5)(.4)\sqrt{1.04 + (.95 - 1.04)w_1} \\
&\quad + (.5)(.4)\sqrt{.98 + (.95 - .98)w_1} \\
&= .3\sqrt{1.04 + .01w_1} + .3\sqrt{.98 + .07w_1} \\
&\quad + .2\sqrt{1.04 - .09w_1} + .2\sqrt{.98 - .03w_1}.
\end{aligned}
$$

Define the preceding function as $f(w_1)$, and note that to optimize in formula (2.99) with respect to w_1, it is sufficient to optimize this f. A graph of f is shown in Figure 2.32, which suggests that the optimal weight is somewhere between .2 and .3. Differentiating f using the chain rule,

$$f'(w_1) = \frac{0.0015}{\sqrt{1.04 + .01w_1}} + \frac{0.0105}{\sqrt{0.98 + .07w_1}} - \frac{0.009}{\sqrt{1.04 - .09w_1}} - \frac{0.003}{\sqrt{0.98 - .03w_1}}.$$

The optimal w_1 solves the equation $f'(w_1) = 0$, which is not easy to do analytically. *Mathematica* yields the solution $w_1 = .271852$, and the optimal value of f is then 1.00491 after substituting this value of w_1 into f.

This means that to optimize with respect to c_1 we must maximize the function:

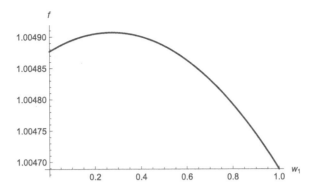

FIGURE 2.32 Optimization with respect to weight w_1 on asset 2 at time 1.

$$g(c_1) = 2 \cdot \sqrt{c_1} + (1.01)^{-1} \cdot 2 \cdot \sqrt{(W_1 - c_1)} \cdot 1.00491.$$

We have that:

$$g'(c_1) = \frac{1}{\sqrt{c_1}} - (1.01)^{-1} \frac{1.00491}{\sqrt{(W_1 - c_1)}} = 0 \implies \frac{1}{\sqrt{c_1}} = \frac{(1.01)^{-1} \cdot 1.00491}{\sqrt{(W_1 - c_1)}}.$$

Cross-multiplying and squaring yields the equation:

$$
\begin{aligned}
\sqrt{(W_1 - c_1)} &= (1.01)^{-1} \cdot 1.00491 \sqrt{c_1} \\
\implies W_1 - c_1 &= \left((1.01)^{-1} \cdot 1.00491\right)^2 c_1 \\
\implies c_1 &= \frac{W_1}{1 + ((1.01)^{-1} \cdot 1.00491)^2} = .502526 W_1.
\end{aligned}
$$

In this case it was important to have an explicit analytical expression for c_1, because its form implies something important about the value function V_1. Returning to formula (2.99) and substituting the optimal values of w_1 and c_1, we now have:

$$
\begin{aligned}
V_1(1, W_1) &= (1.01)^{-1} \max_{w_1, c_1} (2 \cdot \sqrt{c_1} + (1.01)^{-1} \\
&\quad \cdot 2 \cdot \sqrt{W_1 - c_1} \cdot E[\sqrt{(Z_1^1 + (Z_1^2 - Z_1^1) w_1)}]) \\
&= (1.01)^{-1} (2 \cdot \sqrt{.502526 W_1} + (1.01)^{-1} \cdot 2 \cdot \sqrt{W_1 - .502526 W_1} \\
&\quad \cdot 1.00491) \\
&= (1.01)^{-1} \cdot 2.82131 \sqrt{W_1}.
\end{aligned}
$$

$$(2.100)$$

Hence the time 1 value function is also of (discounted) square root form, with constant coefficient now equal to 2.82131 rather than 2.

This means that our computation of the time 0 optimal consumption c_0 and portfolio weight w_0 on asset 2 can proceed along very similar lines. For time $k = 0$ the DP equation becomes:

$$
\begin{aligned}
V_0(0, W_0) &= \max_{w_0, c_0} E[2\sqrt{c_0} + (1.01)^{-1} \cdot 2.82131\sqrt{W_1}|W_0] \\
&= \max_{w_0, c_0} (2\sqrt{c_0} + (1.01)^{-1} \\
&\quad \cdot E[2.82131 \cdot \sqrt{(W_0 - c_0) \cdot (Z_0^1 + (Z_0^2 - Z_0^1)w_0)}|W_0]) \\
&= \max_{w_0, c_0} (2 \cdot \sqrt{c_0} + (1.01)^{-1} \\
&\quad \cdot 2.82131\sqrt{(W_0 - c_0)} \cdot E[\sqrt{(Z_0^1 + (Z_0^2 - Z_0^1)w_0)}]).
\end{aligned}
$$
$$(2.101)$$

But since the ratios of return Z_0^1 and Z_0^2 have the same distribution at time 0 as at time 1, the expectation in this expression is again the same function $f(w_0)$ as above. Thus, the optimal weight for asset 2 is again $w_0 = .271852$, and the optimal value of f is 1.00491. For the optimal consumption at time 0, we repeat the steps that we did for time 1. We must maximize:

$$
g(c_0) = 2 \cdot \sqrt{c_0} + (1.01)^{-1} \cdot 2.82131 \cdot \sqrt{W_0 - c_0} \cdot 1.00491.
$$

To do this, we compute:

$$
\begin{aligned}
g'(c_0) &= \frac{1}{\sqrt{c_0}} - (1.01)^{-1} \frac{(2.82131)(1.00491)(.5)}{\sqrt{W_0 - c_0}} = 0 \\
&\implies \frac{1}{\sqrt{c_0}} = \frac{1.40355}{\sqrt{W_0 - c_0}} \\
&\implies \sqrt{W_0 - c_0} = 1.40355\sqrt{c_0} \\
&\implies W_0 - c_0 = (1.40355)^2 c_0 \\
&\implies c_0 = \frac{W_0}{1 + (1.40355)^2} = .336707 W_0.
\end{aligned}
$$

To summarize, the investor consumes about 33.7% of current wealth at time 0, and about 50.3% of wealth at time 1. At both times, it is optimal for the investor to allocate about 27.2% of wealth to asset 2, and therefore about 72.8% to asset 1. ∎

Example 4. As our last example, we turn the focus to a more general problem. Consider an n period model with a non-risky asset with rate of return r per period, and a single, binomial branch risky asset with general parameters a, b, and p. There will be a discount factor δ, and there will be no consumption prior to the final time. The utility of final wealth will be quadratic:

$$
U(n, W_n) = (1 + \delta)^{-n} \left(d \cdot W_n - e \cdot (W_n)^2 \right).
$$

(In order to have a domain in which U is increasing, the final wealth W_n must stay less than the x-coordinate of the vertex of the quadratic function, namely $d/2e$.) Show inductively that the intermediate value functions V_k are of discounted quadratic form $V_k(k, W_k) = (1 + \delta)^{-k} \left(c_k + d_k \cdot W_k - e_k \cdot (W_k)^2 \right)$, and derive a formula for the optimal weight w_k on the risky asset at time k in terms of the coefficients c_k, d_k, and e_k.

Solution. We do an inductive argument, backwards in k starting with $k = n$. At this last time, we are given $V_n = U_1$, which is of the desired form with

$c_n = 0, d_n = d$, and $e_n = e$. Assume that V_{k+1} is of discounted quadratic form. By the DP equation, the inductive hypothesis, and formula (2.79), we can write:

$$
\begin{aligned}
V_k(k, W_k) &= \max_{w_k} E[V_{k+1}(k+1, W_{k+1})|W_k] \\
&= \max_{w_k} E[(1+\delta)^{-(k+1)}(c_{k+1} + d_{k+1} \cdot W_{k+1} \\
&\quad - e_{k+1} \cdot (W_{k+1})^2)|W_k] \\
&= \max_{w_k} E[(1+\delta)^{-(k+1)}(c_{k+1} + d_{k+1} \cdot W_k \\
&\quad \cdot (R + w_k(Z_k - R)) - e_{k+1}(W_k(R + w_k \cdot (Z_k - R)))^2)|W_k] \\
&= (1+\delta)^{-k} \cdot \max_{w_k} (1+\delta)^{-1}(c_{k+1} + d_{k+1} \cdot W_k \\
&\quad \cdot (R + w_k \cdot E[(Z_k - R)]) - e_{k+1} \cdot (W_k)^2 \\
&\quad \cdot E[(R + w_k \cdot (Z_k - R))^2]).
\end{aligned}
\tag{2.102}
$$

It will be helpful to give labels to two expectations:

$$
\begin{aligned}
m &= E[Z_k - R] \\
&= p(1 + b - (1 + r)) + (1 - p)(1 + a - (1 + r)) \\
&= p(b - r) + (1 - p)(a - r),
\end{aligned}
\tag{2.103}
$$

$$
v = E\left[(Z_k - R)^2\right] = p(b - r)^2 + (1 - p)(a - r)^2.
\tag{2.104}
$$

These expectations do not depend on time. Then the last line of (2.102) becomes:

$$
\begin{aligned}
V_k(k, W_k) &= (1+\delta)^{-k} \max_{w_k} (1+\delta)^{-1}(c_{k+1} + d_{k+1} \cdot W_k \cdot ((1+r) \\
&\quad + w_k \cdot m) - e_{k+1}(W_k)^2((1+r)^2 + 2(1+r)mw_k + vw_k^2)).
\end{aligned}
\tag{2.105}
$$

Take the derivative with respect to w_k of the function inside the max and set that to zero to obtain:

$$
\begin{aligned}
d_{k+1}mW_k &= e_{k+1} \cdot (W_k)^2 (2(1+r)m + 2vw_k) \\
&\implies w_k = \frac{d_{k+1}m}{2ve_{k+1}W_k} - \frac{(1+r)m}{v},
\end{aligned}
\tag{2.106}
$$

after a little algebra. Formula (2.106) answers the question of how w_k depends on the constants, and notice that the optimal proportion of wealth in the risky asset does depend, inversely, on the current level of wealth W_k. Then w_k will be computable as we recursively determine the coefficients c_k, d_k, and e_k, starting with values $c_n = 0$, $d_n = d$, and $e_n = e$, respectively, at time n.

To finish the induction and to find the recursive relationships between d_k, e_k and d_{k+1}, e_{k+1}, we substitute formula (2.106) for w_k into formula (2.105) to get:

$$
\begin{aligned}
V_k(k, W_k) &= (1+\delta)^{-k}((1+\delta)^{-1}(d_{k+1}W_k((1+r) + (\frac{d_{k+1}m}{2ve_{k+1}W_k} \\
&\quad - \frac{(1+r)m}{v})m) - e_{k+1} \cdot (W_k)^2((1+r)^2 + 2(1+r)m(\frac{d_{k+1}m}{2ve_{k+1}W_k} \\
&\quad - \frac{(1+r)m}{v}) + v(\frac{d_{k+1}m}{2ve_{k+1}W_k} - \frac{(1+r)m}{v})^2))).
\end{aligned}
\tag{2.107}
$$

The function above is clearly a discounted quadratic function of wealth W_k of the proper form $c_k + d_k \cdot W_k - e_k \cdot (W_k)^2$, with rather complicated coefficients, which turn out to be:

$$c_k = (1+\delta)^{-1}\left(\frac{d_{k+1}^2 m^2}{4e_{k+1}v}\right);$$
(2.108)

$$d_k = -(1+\delta)^{-1}\frac{d_{k+1}(1+r)\left(m^2 - v\right)}{v};$$
(2.109)

$$e_k = -(1+\delta)^{-1}\frac{e_{k+1}(1+r)^2\left(m^2 - v\right)}{v}.$$
(2.110)

The constant coefficient c_k is not needed to find w_k, and the linear and quadratic coefficients d_k and e_k turn out to just be proportional to their own successors, which makes computations easy (see Exercise 12). ∎

2.5.3   Optimal Portfolios and Martingales

There is an interesting and unexpected connection between the problem of optimizing portfolios in discrete time and the idea of the risk neutral probability q introduced in Section 1.6. Recall that in the context of the binomial branch process, where a risk-free asset with rate r was present, derivative valuation was accomplished by converting the natural up probability p to probability q defined by:

$$q = \frac{r - a}{b - a}.$$
(2.111)

A consequence of the way that q is defined is that the price process satisfies the condition:

$$E_q\left[(1+r)^{-1}S_1\right] = S_0.$$
(2.112)

A similar statement can be made about the conditional expectation of a later price S_k of the asset given its predecessor S_{k-1}. We observed that the discounted process $\tilde{S}_k = (1+r)^{-k}S_k$ forms a martingale in this case.

To see this connection, consider a portfolio optimization problem with no consumption in which the investor is interested in maximizing expected utility at the final time. There is just the risk-free asset and a single risky binomial branch asset in the market. For simplicity we will just consider a one-step problem, although the ideas can extend to multiple step problems. Denote by $R = (S_1 - S_0)/S_0$ the random step 1 rate of return on the risky asset, and denote by w the proportion of wealth devoted to the risky asset. By previous work, we have that the total wealth at time 1 will be:

$$W_1 = W_0(1 + r + (R - r)w).$$
(2.113)

In the binomial case, given W_0 there are two possible values of W_1, which we can denote W^u and W^d, depending on whether the risky rate of return R is b or a, respectively.

$$W^u = W_0(1 + r + (b - r)w); \quad W^d = W_0(1 + r + (a - r)w). \qquad (2.114)$$

The investor wishes to maximize:

$$
\begin{aligned}
E\left[U\left(W_1\right)|W_0\right] &= E\left[U\left(W_0(1 + r + (R - r)w)\right)|W_0\right] \\
&= p \cdot U\left(W^u\right) + (1 - p) \cdot U\left(W^d\right).
\end{aligned}
\qquad (2.115)
$$

As always, the optimal weight w^* is found by formally differentiating the expression in formula (2.115) with respect to w and equating to zero. This would result in the equation:

$$E\left[U'\left(W_1\right)(R - r)\right] = 0 \implies E\left[U'\left(W_1\right) \cdot R\right] = r \cdot E\left[U'\left(W_1\right)\right]. \qquad (2.116)$$

Writing R in terms of the risky asset price we get that the optimal w^* satisfies:

$$
\begin{aligned}
E\left[U'\left(W_1\right) \cdot \tfrac{S_1 - S_0}{S_0}\right] &= r \cdot E\left[U'\left(W_1\right)\right] \\
\implies E\left[U'\left(W_1\right) \cdot (S_1 - S_0)\right] &= r \cdot S_0 E\left[U'\left(W_1\right)\right] \\
\implies E\left[U'\left(W_1\right) \cdot S_1\right] &= (1 + r) \cdot S_0 E\left[U'\left(W_1\right)\right] \\
\implies E\left[\tfrac{U'\left(W_1\right)}{E[U'(W_1)]} \cdot S_1\right] &= (1 + r) \cdot S_0.
\end{aligned}
$$

$$ (2.117) $$

The expectation in the last line of this derivation, when written out fully, is:

$$E\left[\frac{U'\left(W_1\right)}{E\left[U'\left(W_1\right)\right]} \cdot S_1\right] = p \cdot \frac{U'\left(W^u\right)}{E\left[U'\left(W_1\right)\right]} \cdot S^u + (1 - p) \cdot \frac{U'\left(W^d\right)}{E\left[U'\left(W_1\right)\right]} \cdot S^d, \qquad (2.118)$$

where $S^u = (1 + b)S_0$ and $S^d = (1 + a)S_0$ are the up and down values of S_1. Notice that since:

$$E\left[U'\left(W_1\right)\right] = p \cdot U'\left(W^u\right) + (1 - p) \cdot U'\left(W^d\right),$$

the sum of the coefficients of S^u and S^d in formula (2.118) is 1. So let:

$$q = p \cdot \frac{U'\left(W^u\right)}{E\left[U'\left(W_1\right)\right]} = p \cdot \frac{U'\left(W^u\right)}{p \cdot U'\left(W^u\right) + (1 - p) \cdot U'\left(W^d\right)}. \qquad (2.119)$$

Then formula (2.117) says that:

$$(1 + r) \cdot S_0 = E\left[\frac{U'\left(W_1\right)}{E\left[U'\left(W_1\right)\right]} \cdot S_1\right] = E_q\left[S_1\right],$$

which is the martingale condition.

There is only one probability q that makes the discounted price process a martingale, namely the risk-neutral probability $q = (r - a)/(b - a)$. The bottom line is that the optimal weight w^*, which gives values for W^u and

W^d as in formula (2.114), leads to a probability q under which the discounted risky asset process is a martingale. In theory, for a given utility function U, W^u and W^d can be substituted for in terms of w^*, giving an equation for w^* that can be algebraically solved. This is not necessarily the most efficient or straightforward way of going about the problem, however. For an interesting consequence regarding the wealth process, see Exercise 15.

Exercises 2.5

1. In reference to Example 1, explain why, in a three period version of the no consumption model, the optimal risky asset weights w_0, w_1, and w_2 are all the same.

2. In Example 1, with one non-risky asset with rate of return r and a general, binomial branch risky asset, suppose again that there is no consumption, but add the discount factor δ to the model. Does the optimal allocation strategy change? Explain in detail why or why not.

3. Redo Example 1 if the utility of final wealth is of cube root type, $U_2(x) = 3x^{1/3}$.

4. In Example 1, suppose that the ratio of return variables Z_0 and Z_1 are general independent and identically distributed random variables. Formally differentiate formula (2.83) in order to obtain an equation for the optimal portfolio weight w_1. Argue that at time 1 the value function is still of logarithmic form, and that consequently the optimal portfolio weight w_0 will be the same as w_1.

5. (Technology required) In Example 2, how large must the discounting constant δ be in order that at least 55% of wealth is consumed at time 1? At least 35% at time 0? Plot a graph of the proportion consumed at time 1 as a function of δ for δ values between 0 and .05.

6. Redo Example 2 if the utilities of consumption and final wealth are of discounted square root type: $U(k, x) = (1 + \delta)^{-k} \cdot 2\sqrt{x}$.

7. In Example 9 of Section 2.4 an investor's goal was to maximize the expected value of a power utility $U(x) = 2x^{1/2}$ evaluated at the final wealth. There was a risk-free asset with rate of return $r = .04$, and a risky asset following the binomial branch model with possible rates of return $-.013$ and $.095$, each with probability .5. In that problem, the time horizon was 1; now turn it into a 3-period model, with no discount factor and no consumption and find the optimal portfolio weights on the risky asset.

8. In Example 3, check formula (2.97) for W_{k+1}.

9. (Technology required) Redo Example 3 if the utilities of consumption and final wealth are $U(k,x) = (1.01)^{-k} \cdot 10 \log(x)$.

10. In Exercise 9 with log utility, let us give the ratios of return Z^1 and Z^2 a joint distribution that does not assume independence. Specifically, suppose that

$$P\left[Z^1 = 1.04, Z^2 = 1.05\right] = .3, P\left[Z^1 = .98, Z^2 = .95\right] = .3,$$
$$P\left[Z^1 = 1.04, Z^2 = .95\right] = .2, P\left[Z^1 = .98, Z^2 = 1.05\right] = .2.$$

From one time to the next, however, we still suppose that the ratios of return are independent. Notice that we are setting up a situation in which the two assets tend to go up or down together, but not in perfect lockstep. Recalculate the optimal portfolio weight for asset 2 and the optimal consumptions at times 0 and 1.

11. Referring to Example 3, argue by induction that, in a problem with n time periods, the optimal value function at each time l will always be of discounted square root form $V_l(l, W_l) = (1.01)^{-l} \cdot \left(2a_l \sqrt{W_l} + b_l\right)$, and the optimal weight devoted to the second risky asset will not change.

12. In the context of Example 4, derive the optimal portfolio weights, as functions of current wealth, in a 4-period problem for each of times 3, 2, 1, and 0, if $r = \delta = .02$, $b = .04$, $a = -.01$, and $p = .7$. Assume that the utility of final wealth is $U(4, W_4) = (1 + \delta)^{-4} \left(10 W_4 - (W_4)^2\right)$.

13. Prove that in a general portfolio-consumption problem of n periods with discounted logarithmic utility $U(k, x) = (1 + \delta)^k \log(x)$, one risky and one non-risky asset, and independent and identically distributed ratios of return Z_k, the optimal consumption at time $n - 1$ is a constant proportion of wealth, even when the asset price model is not the binomial branch. Then, step back to time $n - 2$ to find c_{n-2}, and finally guess at a general formula for the consumption c_k at time k (you need not prove the formula).

14. Solve the following portfolio problem in two periods. An investor has two independent risky assets between which to allocate wealth so as to maximize the expected value of the discounted utility value of terminal wealth at time 2, where the utility function is: $U(2, W_2) = (1.01)^{-2} \left(.1 W_2 - .005 W_2^2\right)$. There is to be no consumption of wealth at time 0 nor at time 1. The investor has $1000 to invest, and asset 1 has initial price $20 at time 0, and its rates of return R_0^1 and R_1^1 at times 0 and 1 are independent, normally distributed random variables with mean .1 and standard deviation .05. Similarly, risky asset

2 has rates of return R_0^2 and R_1^2 that are i.i.d. $N\left(.06, .03^2\right)$ random variables. Find the optimal asset 1 portfolio weights w_0 and w_1 at times 0 and 1.

15. In the context of the discussion in the subsection on Optimal Portfolios and Martingales, show that the discounted wealth process $\tilde{W}_0, \tilde{W}_1$ is also a one-step martingale under the probability q of formula (2.111) for the optimal allocation strategy.

16. Use Equations (2.114) and (2.119) to solve for the optimal portion of wealth w^* devoted to the risky asset for a one-period portfolio allocation problem with a single risk-free asset with per period rate of return $r = .01$, and a risky asset following a binomial branch process with parameters $p = .5$, $b = .04$, $a = -.02$. Use the utility function $U(x) = (1 + r)^{-1} \log(x)$, and show that initial wealth W_0 and the initial value S_0 of the risky asset are immaterial in the answer.

3

Discrete-Time Derivatives Valuation

We introduced the problem of finding initial values of derivatives that expire in a single time period in Chapter 1. Now we would like to generalize in a thorough fashion to discrete-time derivatives based on binomial branch assets, which expire several time periods in the future. Section 3.1 builds in a natural way on the one-period case and shows a recursive process to value derivatives at times that are intermediate between their issue and expiration times. In the process, we will introduce in the second and third sections some background discrete probability (a step toward measure-theoretic probability) using which we can give better justification of the intuitive computations. This study culminates in Section 3.4 where we show the so-called Fundamental Theorems of Asset Pricing. The theorems also give justification for generalizing the derivatives pricing procedure for such cases as non-binomial branch assets and non-vanilla derivatives. With the fundamental theorems in place, we look at exotic options in Section 3.5. When analytical results are difficult to obtain, the analyst can fall back on simulation methods to give approximate valuations of derivatives. These are covered in Section 3 6. Finally, in Section 3.7 we start to bridge the gap between the discrete-time problem and the continuous-time problem that is explored in more detail in the rest of the book.

3.1 Options Pricing for Multiple Time Periods

3.1.1 Introduction

As mentioned in the introduction, in Section 1.6 we saw how to price derivative assets in a single time period. A brief summary of the reasoning and the results is as follows. The underlying asset was assumed to have an initial value of S_0 which is observable at time 0. In the simple binomial branch model, the time 1 asset value S_1 takes on one of two possible values, S_u or S_d. There is a derivative contract defined on this asset, whose random value V_1 at time 1 can be either V_u or V_d, depending on whether the underlying asset went up or down. The problem was to find an arbitrage-free initial price V_0 at time 0 for the derivative. Figure 3.1 illustrates the situation, reproducing Figure 1.10 from Chapter 1.

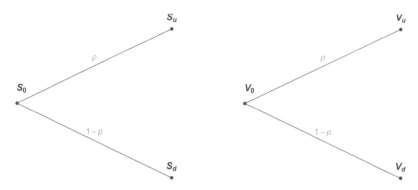

FIGURE 3.1 Single period binomial branch model for underlying asset and derivative.

To solve the problem, we formed a replicating portfolio for the derivative, which was a combination of Δ shares of the asset and c units of a risk-free asset whose per period rate of return is r. The Law of One Price, a consequence of the no-arbitrage assumption, allowed us to conclude that the initial price V_0 of the derivative must be the same as the initial value of the replicating portfolio, and so it remained only to find that portfolio. The parameters Δ and c satisfied the linear system:

$$\begin{cases} \Delta \cdot S_u + c(1+r) = V_u \\ \Delta \cdot S_d + c(1+r) = V_d. \end{cases} \tag{3.1}$$

This was easily solved to give:

$$\Delta = \frac{V_u - V_d}{S_u - S_d}, \quad c = (1+r)^{-1}\left(V_d - \frac{V_u - V_d}{S_u - S_d} \cdot S_d\right). \tag{3.2}$$

Substitution of these expressions into the formula $V_0 = \Delta S_0 + c$ yielded a formula for V_0:

$$V_0 = \frac{V_u - V_d}{S_u - S_d} \cdot S_0 + (1+r)^{-1} \cdot \frac{S_u \cdot V_d - S_d \cdot V_u}{S_u - S_d}. \tag{3.3}$$

(See Theorem 2 of Section 1.6.)

But we were not done. After some manipulation we discovered the alternative equivalent expression:

$$V_0 = (1+r)^{-1}\left(q \cdot V_u + (1-q)V_d\right) = E_q\left[(1+r)^{-1}V_1\right], \tag{3.4}$$

where:

$$q = \frac{S_0(1+r) - S_d}{S_u - S_d}. \tag{3.5}$$

By formula (3.4), the solution of the valuation problem is such that the arbitrage-free initial value of the derivative is just the expected present value of the claim random variable V_1 under q. If the problem is reparameterized so that $S_u = (1+b)S_0$ and $S_d = (1+a)S_0$, then the expression for this so-called *risk-neutral probability* q reduces to the simple form:

$$q = \frac{r-a}{b-a}. \qquad (3.6)$$

As an interesting side note, it turned out that if q is assumed to be the up probability for the underlying asset, then:

$$E_q\left[(1+r)^{-1}S_1\right] = S_0. \qquad (3.7)$$

It is the goal of this section to extend these ideas to the case of multiple discrete time periods. In Figure 3.2(a) we show a binomial branch model in which there are three time periods after the initial time 0. There are therefore three price random variables S_1, S_2, S_3 in addition to the known initial price S_0. The natural transition probabilities p and $1-p$ from parent to child nodes are shown in the figure, although we found that in the single-period case they have no bearing on the valuation of the derivative.

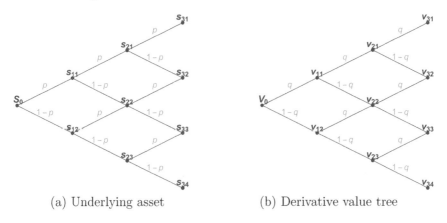

(a) Underlying asset   (b) Derivative value tree

FIGURE 3.2 Three period binomial branch model.

Nodes in the underlying asset tree are labeled with two subscripts, the first indicating the level of the tree that they reside in, and the second indicating the position of the node in that level, starting from 1 at the top and extending to $l+1$, where l is the level (beginning with level 0). The same notational scheme is used for the corresponding derivative tree in part (b) of the figure. Following this convention, the known initial price S_0 can also be written s_{01} and the initial derivative value V_0 can be written v_{01}.

At each node of the asset tree prior to the final time, there are two branches leading to child nodes. We will be assuming for simplicity that at each node s_{lj} at level l, the two possibilities for the asset price S_{l+1} at the next level are:

$s_{l+1,j} = (1 + b)s_{lj}$ or $s_{l+1,j+1} = (1 + a)s_{lj}$, where b and a remain constant throughout time. (We use a comma between subscripts when necessary for clarity.) This makes the asset price tree a **recombining tree**. For example, there are two paths which lead from S_0 to node s_{22}: an up followed by a down and a down followed by an up. In both cases, the price S_2 has the value $s_{22} = (1 + b)(1 + a)S_0$. If we allowed b and a to change as time progressed (which we could do for greater generality) then the two paths would not necessarily bring the price process to the same node.

Both recursive and non-recursive formulas for the price values at nodes will be useful. From the discussion above, it is clear that:

$$s_{l,j} = (1 + b)s_{l-1,j} \text{ if } j \leq l; s_{l,l+1} = (1 + a)s_{l-1,l}, \tag{3.8}$$

because $s_{l,j}$ is the upper child node of level $l-1$ node $s_{l-1,j}$, and the bottom-most node $s_{l,l+1}$ at level l is the lower child of node $s_{l-1,l}$. Also, a closed form for the price value at node $s_{l,j}$ for an n-period tree is:

$$s_{l,j} = (1 + b)^{l+1-j}(1 + a)^{j-1}S_0, l = 1, ..., n, j = 1, ..., l+1. \tag{3.9}$$

(See Exercise 3.)

Formulas (3.4), (3.6), and (3.7) play key roles in developing two valuation techniques: (i) valuation by **chaining**, a recursive strategy in which one moves backward through the derivative tree starting at the right side, computing values at parent nodes level by level; and (ii) valuation by **martingales**, in which the value of the derivative V_0 at time 0 is computed all at once as an expected time l discounted value of the claim value V_l at the final time l. Although there is ample intuitive evidence that both of these strategies are correct and give the one and only arbitrage-free derivative value V_0, the full justification relies on the two Fundamental Theorems of Options Pricing, which will be discussed in Section 3.4.

3.1.2   Valuation by Chaining

The key observation that justifies the chaining idea is this: in multiple time periods, arbitrage opportunities must be eliminated not just at time 0, but at intermediate times as well. To take advantage of this fact, extract from a general binomial branch tree any one particular parent-child configuration as in Figure 3.3. Do the same for the derivative tree. The current price S_{l-1} at time $l - 1$ happens to be $s_{l-1,j}$ for this particular outcome. A replicating portfolio for the next time step can be created just as in formula (3.1), with coefficients $\Delta_{l-1,j}$ and $c_{l-1,j}$ satisfying:

$$\begin{cases} \Delta_{l-1,j} \cdot s_{l,j} + c_{l-1,j}(1 + r) = v_{l,j} \\ \Delta_{l-1,j} \cdot s_{l,j+1} + c_{l-1,j}(1 + r) = v_{l,j+1}. \end{cases} \tag{3.10}$$

We have already done the algebra in formula (3.2), which leads to the solution:

$$\Delta_{l-1,j} = \frac{v_{l,j} - v_{l,j+1}}{s_{l,j} - s_{l,j+1}}$$

$$c_{l-1,j} = (1+r)^{-1} \left(v_{l,j+1} - \Delta_{l-1,j} \cdot s_{l,j+1} \right). \tag{3.11}$$

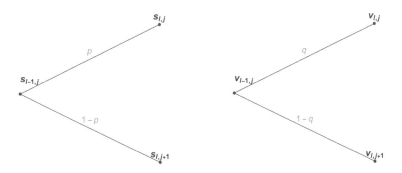

FIGURE 3.3 Chaining from child nodes to parent node.

What this means is that at any node at level $l-1$ and position j within the level, we have a single-period valuation problem of the same kind as in Section 1.6, for which we have already done all of the work. The risk-neutral probability q is computed again as:

$$q = \frac{s_{l-1,j}(1+r) - s_{l,j+1}}{s_{l,j} - s_{l,j+1}} = \frac{r-a}{b-a}, \tag{3.12}$$

and the value of the derivative at node $v_{l-1,j}$ is computed recursively from the child node values by:

$$
\begin{aligned}
v_{l-1,j} \quad &- \quad (1+r)^{-1} \left(q \cdot v_{l,j} + (1-q)v_{l,j\,|\,1} \right) \\
&= \quad E_q \left[(1+r)^{-1} V_l | V_{l-1} = v_{l-1,j} \right].
\end{aligned}
\tag{3.13}
$$

Notice that this is not an ordinary expectation, but a conditional expectation given that the derivative assumes the value $v_{l-1,j}$ at level $l-1$, which is the same as saying that the underlying asset has value $s_{l-1,j}$ at level $l-1$. To initialize this recursive approach, we must know the possible final values $v_{n,j}$ at the level (that is the time) n at which the derivative is to be executed, which are typically available as simple functions of the corresponding final values of the underlying asset.

In summary, here is how chaining is carried out for a binomial branch model terminating at time n:

Chaining Algorithm

1. Compute $q = (r-a)/(b-a)$;
2. Determine the possible final claim values $v_{n,1}, v_{n,2}, ..., v_{n,n+1}$;
3. For each level $l = n, n-1, ... , 2, 1$ find all derivative values $v_{l-1,j}$ at the

next lower level by:

$$v_{l-1,j} = (1+r)^{-1}\left(q \cdot v_{l,j} + (1-q)v_{l,j+1}\right). \blacksquare$$

Then $v_{0,1}$ is the initial value V_0 of the derivative. As a side benefit, we get the values of the derivative at each possible interior node. We can also, if we choose, find the replicating portfolio for the derivative at each node using formulas (3.11).

Example 1. Example 5 of Section 1.6 described a put option on an underlying asset following the binomial branch model such that $b = .07$, $a = -.02$, and $S_0 = \$200$. The exercise price of the option was $E = \$198$, and the risk-free interest rate was $r = .04$. Let us now suppose that the option is of European type, and it expires two time periods after being issued. Let us find the option price at each node, including the initial price, and also compute the replicating portfolios at each parent node.

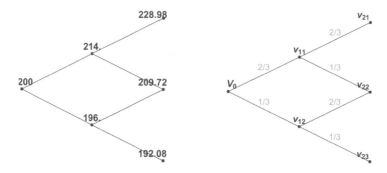

FIGURE 3.4 Two-period asset and derivative trees, $S_0 = \$200$, $b = .07$, $a = -.02$.

Solution. As in the earlier example, we can compute the risk-neutral probability as:

$$q = \frac{r-a}{b-a} = \frac{.04-(-.02)}{.07-(-.02)} = \frac{.06}{.09} = \frac{2}{3}.$$

The possible price values of the underlying asset are:

$S_0 = \$200,$
$s_{11} = \$200(1.07) = \$214,$ $s_{12} = \$200(.98) = \$196,$
$s_{21} = \$200(1.07)^2 = \$228.98,$ $s_{22} = \$200(1.07)(.98) = \$209.72,$
$s_{23} = \$200(.98)^2 = \$192.08.$

These are shown in Figure 3.4. Since the derivative is a put option with exercise price \$198, the final derivative values at level 2 are:

$$v_{21} = \max\{198 - s_{21}, 0\} = \max\{198 - 228.98, 0\} = 0;$$
$$v_{22} = \max\{198 - s_{22}, 0\} = \max\{198 - 209.72, 0\} = 0;$$
$$v_{23} = \max\{198 - s_{23}, 0\} = \max\{198 - 192.08, 0\} = \$5.92.$$

The value of the option at node v_{11} can be found as follows:

$$
\begin{aligned}
v_{11} &= (1+r)^{-1}(q \cdot v_{21} + (1-q)v_{22}) \\
&= (1.04)^{-1}\left(\tfrac{2}{3} \cdot 0 + \tfrac{1}{3} \cdot 0\right) = 0.
\end{aligned}
$$

Similarly, at the second node in level 1,

$$
\begin{aligned}
v_{12} &= (1+r)^{-1}(q \cdot v_{22} + (1-q)v_{23}) \\
&= (1.04)^{-1}\left(\tfrac{2}{3} \cdot 0 + \tfrac{1}{3} \cdot 5.92\right) = 1.89744.
\end{aligned}
$$

Therefore, the initial value of this option is:

$$
\begin{aligned}
V_0 = v_{01} &= (1+r)^{-1}(q \cdot v_{11} + (1-q)v_{12}) \\
&= (1.04)^{-1}\left(\tfrac{2}{3} \cdot 0 + \tfrac{1}{3} \cdot 1.89744\right) = .608154.
\end{aligned}
$$

We are also asked to compute at each of the nodes v_{11}, v_{12}, v_{01} the portfolio of cash and underlying asset that replicates completely the possible values of the option at each of the children of these respective nodes. By equation (3.11), we have at the (1,1) node:

$$\Delta_{11} = \frac{v_{21} - v_{22}}{s_{21} - s_{22}} = \frac{0 - 0}{228.98 - 209.72} = 0;$$

$$
\begin{aligned}
c_{11} &= (1+r)^{-1}\left(v_{22} - \tfrac{v_{21}-v_{22}}{s_{21}-s_{22}} \cdot s_{22}\right) \\
&= (1.04)^{-1}(0 - 0 \cdot 209.72) = 0.
\end{aligned}
$$

Similarly, at the (1,2) node:

$$\Delta_{12} = \frac{v_{22} - v_{23}}{s_{22} - s_{23}} = \frac{0 - 5.92}{209.72 - 192.08} = -.335601;$$

$$
\begin{aligned}
c_{12} &= (1+r)^{-1}\left(v_{23} - \tfrac{v_{22}-v_{23}}{s_{22}-s_{23}} \cdot s_{23}\right) \\
&= (1.04)^{-1}(5.92 - (-.335601) \cdot 192.08) \\
&= 67.6752.
\end{aligned}
$$

And at the initial node we obtain:

$$\Delta_{01} = \frac{v_{11} - v_{12}}{s_{11} - s_{12}} = \frac{0 - 1.89744}{214 - 196} = -.105413;$$

$$
\begin{aligned}
c_{01} &= (1+r)^{-1}\left(v_{12} - \tfrac{v_{11}-v_{12}}{s_{11}-s_{12}} \cdot s_{12}\right) \\
&= (1.04)^{-1}(1.89744 - (-.105413)196) \\
&= 21.6908.
\end{aligned}
$$

Let us check the replication property at node $(1, 2)$. First, we found the value of the derivative to be $v_{12} = 1.89744$ at that node. Also, $\Delta_{12} = -.335601$ and $c_{12} = 67.6752$, so the value of the portfolio of risky and non-risky asset at node $(1, 2)$ is:

$$(-.335601)(196) + 67.6752 = 1.8974.$$

Hence there is agreement between the portfolio and the derivative directly at that node. But more importantly, the portfolio should agree with the derivative for both of the possible transitions, to node $(2, 2)$ or $(2, 3)$. If the underlying asset moves up to node $(2, 2)$, then since the non-risky asset gains value at rate .04, the total value of the replicating portfolio is:

$$(-.335601)(209.72) + 67.6752(1 + .04) = 0,$$

and if the asset moves down to node $(2, 3)$ the portfolio value is:

$$(-.335601)(192.08) + 67.6752(1 + .04) = 5.92.$$

(Both of these calculations are subject to small rounding errors.) So in each case the portfolio matches the behavior of the derivative, which is what it is designed to do. In Exercise 4 you are asked to verify the replication property at node $(1, 1)$ (which is very easy) and the initial node $(0, 1)$. ■

If you are experienced with spreadsheets you may have noticed that the chaining method is particularly well-suited for spreadsheet solution. Good practice suggests that a portion of the top of the sheet should be reserved for the underlying parameters S_0, b, a, and r, and the risk-neutral probability q can be computed from these in another cell. These cells can be given names as well for use in formulas described below. The asset prices $s_{l,j}$ can be generated iteratively from left to right; for instance a rectangular block of cells in the spreadsheet can be set aside and relevant cells can be computed as:

		$s_{21} = s_{11}(1 + b)$	
	$s_{11} = S_0(1 + b)$		
value of S_0		$s_{22} = s_{12}(1 + b)$	$\cdots$
	$s_{12} = S_0(1 + a)$		
		$s_{23} = s_{12}(1 + a)$	

The value of the leftmost cell is just set equal to the value of the home cell where S_0 is held. As the columns go by, each child is $(1 + b)$ times the entry one cell to the left and one below, except for the bottommost in the column, which is $(1 + a)$ times the entry one cell to the left and one above. Copying and pasting judiciously makes the task of computing these asset prices very simple, even for large trees. To compute the option values, you would need to set aside an identically sized block of cells, start from the right and work left. The final claim values will be functions f (such as $\max(s_{n,j} - E, 0)$ or

$\max(E - s_{n,j}, 0)$ for call and put options, respectively) of the corresponding rightmost cells of the asset tree. Move one-by-one to the left until the initial cell is reached, using the chaining condition. Each parent cell is $(1 + r)^{-1}$ times the quantity $(q \cdot \text{upper child cell} + (1 - q) \cdot \text{lower child cell})$. For a small two-period option for instance, the display below indicates how to proceed.

			$v_{21} = f(s_{21})$
		$v_{11} = \frac{q \cdot v_{21} + (1-q)v_{22}}{1+r}$	
$V_0 = \frac{q \cdot v_{11} + (1-q)v_{12}}{1+r}$			$v_{22} = f(s_{22})$
		$v_{12} = \frac{q \cdot v_{22} + (1-q)v_{23}}{1+r}$	
			$v_{23} = f(s_{23})$

Example 2. Example 2 of Section 1.6 described an underlying asset initially priced at $20 per share with up rate $b = .04$ and down rate $a = -.04$. The assumed risk-free rate was $r = .01$. Suppose that a European call option exists on this asset with strike price $21 and expiration time $n = 3$. Use chaining to find all values of the option at all nodes.

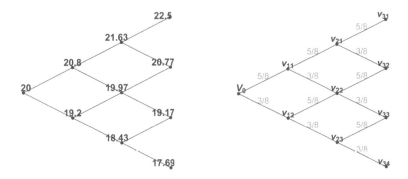

FIGURE 3.5 Three-period asset and derivative trees, $S_0 = \$20$, $b = .04$, $a = -0.04$.

Solution. The situation is shown in Figure 3.5. We calculate the risk-neutral probability as:

$$q = \frac{r - a}{b - a} = \frac{.01 - (-.04)}{.04 - (-.04)} = \frac{.05}{.08} = \frac{5}{8}.$$

The price values for the risky asset nodes are:

$$S_0 = 20,$$
$$s_{11} = 20(1.04) = 20.8, \qquad s_{12} = 20(.96) = 19.2,$$
$$s_{21} = 20(1.04)^2 = 21.632, \qquad s_{22} = 20(1.04)(.96) = 19.968,$$
$$s_{23} = 20(.96)^2 = 18.432,$$
$$s_{31} = 20(1.04)^3 = 22.4973, \qquad s_{32} = 20(1.04)^2(.96) = 20.7667,$$
$$s_{33} = 20(1.04)(.96)^2 = 19.1693, \quad s_{34} = 20(.96)^3 = 17.6947.$$

This time the option is a call, valued at execution time by $V_3 = \max\{S_3 - 21, 0\}$. Since the three lowest asset values at time 3 all fall below the strike price of \$21, the final claim values at those nodes on the derivative tree are zero. So for the initialization of the chaining process, we have:

$$v_{31} = \max\{22.4973 - 21, 0\} = 1.49728; \; v_{32} = v_{33} = v_{34} = 0.$$

Stepping back to time 2, the derivative values are:

$$
\begin{aligned}
v_{21} &= (1+r)^{-1}\left(q \cdot v_{31} + (1-q)v_{32}\right) \\
&= (1.01)^{-1}\left(\tfrac{5}{8} \cdot 1.49728 + \tfrac{3}{8} \cdot 0\right) \\
&= .926535; \\
v_{22} &= v_{23} = 0.
\end{aligned}
$$

The lower two nodes at this level have value 0 because their child nodes at level 3 all have value 0. Continuing back to time 1, we compute:

$$
\begin{aligned}
v_{11} &= (1+r)^{-1}\left(q \cdot v_{21} + (1-q)v_{22}\right) \\
&= (1.01)^{-1}\left(\tfrac{5}{8} \cdot .926535 + \tfrac{3}{8} \cdot 0\right) \\
&= .573351 \\
v_{12} &= (1+r)^{-1}\left(q \cdot v_{22} + (1-q)v_{23}\right) \\
&= (1.01)^{-1}\left(\tfrac{5}{8} \cdot 0 + \tfrac{3}{8} \cdot 0\right) \\
&= 0.
\end{aligned}
$$

At level 0, the value of the derivative is:

$$
\begin{aligned}
V_0 = v_{01} &= (1+r)^{-1}\left(q \cdot v_{11} + (1-q)v_{12}\right) \\
&= (1.01)^{-1}\left(\tfrac{5}{8} \cdot .573351 + \tfrac{3}{8} \cdot 0\right) \quad \blacksquare \\
&= .354796.
\end{aligned}
$$

Something interesting came out in the computation in Example 2. Because only one path of the underlying asset led to the final state in which the derivative had value, in the end the recursive calculations gave us:

$$
\begin{aligned}
v_{01} = V_0 &= (1.01)^{-1}\left(\tfrac{5}{8}\right) v_{11} \\
&= (1.01)^{-1}\left(\tfrac{5}{8}\right) \cdot (1.01)^{-1}\left(\tfrac{5}{8}\right) v_{21} \\
&= (1.01)^{-1}\left(\tfrac{5}{8}\right) \cdot (1.01)^{-1}\left(\tfrac{5}{8}\right) \cdot (1.01)^{-1}\left(\tfrac{5}{8}\right) v_{31} \qquad (3.14) \\
&= (1.01)^{-3}\left(\tfrac{5}{8}\right)^3 v_{31}.
\end{aligned}
$$

But this is exactly the expected present value:

$$V_0 = E_q \left[(1+r)^{-3} V_3 \right], \tag{3.15}$$

where $q = 5/8$ and $r = .01$, because of the fact that the claim value V_3 was only non-zero for one node, which is reached with probability $(5/8)^3$. It turns out that this result is not just an artifact of the simple situation in the example, but instead is a general way to compute the initial value of a derivative, as the next subsection discusses.

3.1.3 Valuation by Martingales

In Section 1.1 we defined a **martingale process** as a stochastic process $(X_t)_{t \geq 0}$ satisfying the condition:

$$E[X_{t+s} | X_t] = X_t, \tag{3.16}$$

for all times $t, s \geq 0$. Thus, a martingale is constant in (conditional) expectation. The idea has relevance now, because if q is the risk-neutral probability in a binomial branch situation, then:

$$E_q \left[(1+r)^{-1} S_1 \right] = S_0.$$

This leads us to guess that if we define the **discounted asset price process** $(\bar{S}_j)_{j \geq 0}$ by:

$$
\begin{aligned}
\bar{S}_0 &= (1+r)^0 S_0 = S_0, \\
\bar{S}_1 &= (1+r)^{-1} S_1, \\
\bar{S}_2 &= (1+r)^{-2} S_2, \\
&\vdots \\
\bar{S}_n &= (1+r)^{-n} S_n, ...,
\end{aligned}
\tag{3.17}
$$

then the discounted process may be a martingale under the measure q. The same result should hold for the derivative value process, since we know that

$$V_0 = (1+r)^{-1} \left(q \cdot V_u + (1-q) V_d \right) = E_q \left[(1+r)^{-1} V_1 \right]. \tag{3.18}$$

So analogously, let us also introduce the **discounted derivative value process** $(\bar{V}_j)_{j=0,1,2,...}$, defined by:

$$
\begin{aligned}
\bar{V}_0 &= (1+r)^0 V_0 = V_0, \\
\bar{V}_1 &= (1+r)^{-1} V_1, \\
\bar{V}_2 &= (1+r)^{-2} V_2, \\
&\vdots \\
\bar{V}_n &= (1+r)^{-n} V_n, ...
\end{aligned}
\tag{3.19}
$$

The random variables $V_1, V_2, ..., V_n$ mimic the underlying asset process, in the sense that $S_l = s_{l,j}$ if and only if $V_l = v_{l,j}$.

We will verify the martingale property for both the discounted underlying asset process and the derivative process in Theorem 1 below, but first consider the consequences. If the discounted derivative process is a martingale, then by the defining condition with $s = n$ and $t = 0$ applied at the derivative execution time n:

$$E_q\left[\bar{V}_n | \bar{V}_0\right] = \bar{V}_0 = V_0. \tag{3.20}$$

Taking expectation of both sides in (3.20), and noting that V_0 is actually non-random, we obtain:

$$E_q\left[\bar{V}_n\right] = E_q\left[\bar{V}_0\right] = V_0 \implies V_0 = E_q\left[(1+r)^{-n}V_n\right]. \tag{3.21}$$

Thus, in one expected value calculation we can compute the initial value of the derivative as the expected present value of the claim value of the derivative at its execution time. This method of finding the initial value of the derivative will be called the **martingale method**. The martingale method eliminates the need to chain back from time n just to find V_0; however it doesn't tell you the values of the derivative at intermediate nodes. (But, see Exercise 8 for a way of using the same idea for such intermediate valuations.)

Example 3. Consider the asset process of Example 1, in which the initial state was $S_0 = \$200$, the up and down rates were $b = .07$, $a = -.02$, and the risk-free rate was $r = .04$. These combined to yield the risk-neutral probability $q = 2/3$. Check that the first three random discounted values $\bar{S}_1, \bar{S}_2, \bar{S}_3$ satisfy the condition $E_q\left[\bar{S}_i | \bar{S}_0\right] = \bar{S}_0$.

Solution. We had already computed the possible values of S_1 and S_2, which are repeated below:

$$s_{11} = 214, s_{12} = 196,$$
$$s_{21} = 228.98, s_{22} = 209.72, s_{23} = 192.08.$$

For level 3,

$$s_{31} = 200(1.07)^3 = 245.009, \qquad s_{32} = 200(1.07)^2(.98) = 224.4,$$
$$s_{33} = 200(1.07)(.98)^2 = 205.526, \qquad s_{34} = 200(.98)^3 = 188.238.$$

Recall that $\bar{S}_0 = S_0 = \$200$, which is deterministic. At level 1 we have:

$$
\begin{aligned}
E_q\left[\bar{S}_1 | \bar{S}_0\right] &= E_q\left[\bar{S}_1\right] \\
&= E_q\left[(1.04)^{-1}S_1\right] \\
&= (1.04)^{-1}\left(\tfrac{2}{3} \cdot 214 + \tfrac{1}{3} \cdot 196\right) \\
&= 200 = S_0.
\end{aligned}
$$

At level 2, price s_{21} is reached if both moves are up; hence its probability is $(2/3)^2 = 4/9$. Price s_{22} has two equally likely paths ending there, whose total probability is $2(2/3)(1/3) = 4/9$. Finally, price s_{23} comes about in the case of two downs, with probability $(1/3)^2 = 1/9$. Therefore the expectation we want is:

$$
\begin{aligned}
E_q\left[\bar{S}_2|\bar{S}_0\right] &= E_q\left[\bar{S}_2\right] \\
&= E_q\left[(1.04)^{-2}S_2\right] \\
&= (1.04)^{-2}\left(\tfrac{4}{9}\cdot 228.98 + \tfrac{4}{9}\cdot 209.72 + \tfrac{1}{9}\cdot 192.08\right) \\
&= 200 = S_0.
\end{aligned}
$$

At level 3, price s_{31} occurs if there are three consecutive up moves, with probability $(2/3)^3 = 8/27$. The price reaches s_{32} if there are two up moves and one down move. There are three cases of this kind, each with probability $(2/3)^2(1/3) = 4/27$, for a total probability of $12/27$. Similarly state s_{33} is reached in three different cases of one up and two down moves. The probability associated with this price is therefore $3(2/3)(1/3)^2 = 6/27$. The lowest price s_{34} has probability $(1/3)^3 = 1/27$. Thus, the expected discounted value at level 3 is:

$$
\begin{aligned}
E_q\left[\bar{S}_3|\bar{S}_0\right] &= E_q\left[\bar{S}_3\right] \\
&= E_q\left[(1.04)^{-3}S_3\right] \\
&= (1.04)^{-3}\big(\tfrac{8}{27}\cdot 245.009 + \tfrac{12}{27}\cdot 224.4 + \tfrac{6}{27}\cdot 205.526 \\
&\quad + \tfrac{1}{27}\cdot 188.238.\big) \\
&= 200 = S_0. \ \blacksquare
\end{aligned}
$$

The main theorem that establishes the correctness of formula (3.21) is next. The proof is possible to do with elementary methods, although it would be lengthy, and we omit it for the time being. We will be able to do it more efficiently in Section 3.3 after we have a better handle on the concept of conditional expectation.

Theorem 1. Suppose that an underlying asset follows the binomial branch model with up and down rates b and a, and there is a risk-free asset with rate of return r. Let q be the risk-neutral probability $q = (r - a)/(b - a)$. Suppose that a derivative on this asset has expiration time n. Then both the discounted asset price $(\bar{S}_l)$ and the discounted derivative value process $(\bar{V}_l)$ are martingales under q; hence the initial value of the derivative is:

$$
V_0 = E_q\left[(1+r)^{-n}V_n\right]. \ \blacksquare \tag{3.22}
$$

Remark. To use the formula $V_0 = E_q\left[(1+r)^{-n}V_n\right]$ conveniently for valuation, we need an easy way to find the probabilities associated with the final claim values $v_{n,j}$ of the derivative. But these are just binomial probabilities for an experiment with n trials, with a success event defined as an up move, and success probability q. The only challenge is to relate the number of successes to the second subscript j. To reach node $v_{n,1}$, all transitions must be ups, which occurs with probability q^n. To reach node $v_{n,2}$, all but one transition is an up move; to reach $v_{n,3}$ all but two transitions are up, etc. So we have:

$$P\left[V_n = v_{n,1}\right] = \binom{n}{n} q^n (1-q)^0;$$

$$P\left[V_n = v_{n,2}\right] = \binom{n}{n-1} q^{n-1} (1-q);$$

$$P\left[V_n = v_{n,3}\right] = \binom{n}{n-2} q^{n-2} (1-q)^2; \qquad (3.23)$$

$$\vdots$$

$$P\left[V_n = v_{n,n+1}\right] = \binom{n}{0} q^0 (1-q)^n.$$

The pattern that emerges is that if j is the second subscript, then $j-1$ is subtracted from n in both the lower index of the binomial coefficient and in the exponent of q, and $j-1$ is the exponent of $1-q$. In summary,

$$P\left[V_n = v_{n,j}\right] = \binom{n}{n-(j-1)} q^{n-(j-1)} (1-q)^{j-1}. \qquad (3.24)$$

Example 4. Example 2 dealt with an underlying asset with $S_0 = \$20$, $b = .04$, $a = -.04$, and risk-free rate $r = .01$. Find the initial value of a European call option expiring at time 5 with strike price \$22.

Solution. Since only the initial value of the derivative is desired, and since the relatively distant expiration time makes it tedious to use chaining to find the value, this is an ideal situation for martingale valuation. To make things even easier, we will also only need the possible asset values at time 5, as opposed to computing the asset value at all intermediate times. These are:

$$
\begin{aligned}
s_{5,1} &= (1+b)^5 (1+a)^0 (20) = (1.04)^5 (.96)^0 (20) = 24.3331; \\
s_{5,2} &= (1+b)^4 (1+a)^1 (20) = (1.04)^4 (.96)^1 (20) = 22.4613; \\
s_{5,3} &= (1+b)^3 (1+a)^2 (20) = (1.04)^3 (.96)^2 (20) = 20.7335; \\
s_{5,4} &= (1+b)^2 (1+a)^3 (20) = (1.04)^2 (.96)^3 (20) = 19.1386; \\
s_{5,5} &= (1+b)^1 (1+a)^4 (20) = (1.04)^1 (.96)^4 (20) = 17.6664; \\
s_{5,6} &= (1+b)^0 (1+a)^5 (20) = (1.04)^0 (.96)^5 (20) = 16.3075.
\end{aligned}
$$

Since the strike price is $E = \$22$, the call option has no claim value for the lowest four of these final prices. For the other two price nodes:

$$v_{5,1} = 24.3331 - 22 = 2.3331; \quad v_{5,2} = 22.4613 - 22 = .4613.$$

The risk-neutral probability was computed as $q = 5/8$. The initial call option value is therefore:

$$
\begin{aligned}
V_0 &= E_q\left[(1.01)^{-5}V_5\right] \\
&= (1.01)^{-5}E_q\left[V_5\right] \\
&= (1.01)^{-5}\left(v_{5,1}\cdot P\left[V_5 = v_{5,1}\right] + v_{5,2}\cdot P\left[V_5 = v_{5,2}\right]\right) \\
&= (1.01)^{-5}\left((2.3331)\cdot P\left[V_5 = v_{5,1}\right] + (.4613)\cdot P\left[V_5 = v_{5,2}\right]\right) \\
&= (1.01)^{-5}\left((2.3331)\cdot \binom{5}{5}\left(\frac{5}{8}\right)^5 + (.4613)\cdot \binom{5}{4}\left(\frac{5}{8}\right)^4\left(\frac{3}{8}\right)\right) \\
&= .337276. \ \blacksquare
\end{aligned}
$$

Because of formulas (3.22) and (3.24), we can get open forms for the values of certain specific options, which can be numerically evaluated and plotted as functions of the parameters of the valuation problem. By doing this we can conduct qualitative analyses of dependence of the derivative value on parameters. Consider, for instance, a European call option with strike price E and exercise time n. The general valuation formula would be:

$$
E_q\left[(1+r)^{-n}\max\left(S_n - E, 0\right)\right],
$$

where, as usual, $q = \frac{r-a}{b-a}$. It is simpler to depart for a moment from our double subscript notation $s_{n,j}$ for the possible asset values at level n. The possible values of the underlying asset at time n have the form $S_0(1+b)^k(1+a)^{n-k}$, for $k = 0, 1, 2, ...n$, where k represents the number of up moves. The probabilities associated with those states are binomial with parameters n and q. Therefore the initial value of the call is given in open form by:

$$
V_0 = (1+r)^{-n}\sum_{k=0}^{n}\binom{n}{k}q^k(1-q)^{n-k}\max\left(S_0(1+b)^k(1+a)^{n-k} - E, 0\right).
$$

(3.25)

For instance, using the data of Example 1: $b = .07$, $a = -.02$, $S_0 = \$200$, $r = .04$, we would have the following form for V_0 as a function of n and E:

$$
V_0 = (1.04)^{-n}\sum_{k=0}^{n}\binom{n}{k}\left(\frac{2}{3}\right)^k\left(\frac{1}{3}\right)^{n-k}\max\left(200(1.07)^k(.98)^{n-k} - E, 0\right).
$$

(3.26)

Specifically, for $n = 5$ and $E = \$202$, it can be computed that

$$
\begin{aligned}
V_0 &= (1.04)^{-5}\sum_{k=0}^{5}\binom{5}{k}\left(\frac{2}{3}\right)^k\left(\frac{1}{3}\right)^{5-k}\max\left(200(1.07)^k(.98)^{5-k} - \$202, 0\right) \\
&= \$34.1985.
\end{aligned}
$$

More usefully, formula (3.25) allows us to study the dependence of the option value on the exercise time and strike price.

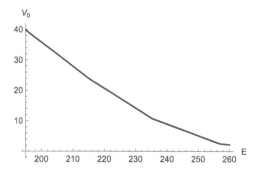

FIGURE 3.6 Call option value V_0 as a function of strike price E, $n = 5$ periods.

Figure 3.6 displays the relationship between the initial value of the call option on this asset and the strike price E, assuming a 5-period time horizon. The decreasing behavior is to be expected, since it is more difficult for the call option to be in the money (that is, for the asset to exceed the strike price) as the strike price grows. The apparent corners are more subtle. But if you consider the possible final asset prices $\$200(1.07)^k(.98)^{5-k}$, the maximum expression in (3.25) will drop to 0 at a node as the exercise price grows, thereby suddenly excluding the node from the total sum. Although you would expect the value function to be differentiable as a function of E strictly between states, as E passes a state and a term is lost from the sum, non-differentiability results.

In Figure 3.7, we see the dependence of the call value for this asset on the expiration time n, for values of n ranging from 3 through 10, assuming that the strike price is $\$202$. For options of longer duration, the underlying asset has a better chance of rising up past the strike price, which increases the desirability of the option.

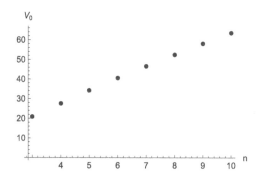

FIGURE 3.7 Call option value V_0 as a function of number of periods n, $E = \$202$.

Although we will not try to look at the issue here, intuitively you might expect that the broader the range of values for the underlying asset, that is, the wider the separation between b and a in the binomial branch model, the better the chance that the underlying asset will end at a high price that exceeds the strike price by a great deal. This should favorably affect the initial value of the option. Therefore, variance is in fact a good thing in the context of valuing call options. When we introduce the continuous-time option pricing model we will pay more attention to this phenomenon.

Exercises 3.1

1. Consider a risky asset following the binomial branch assumptions with initial price $S_0 = \$50$, and parameters $b = .03$, $a = -.02$. Compute the prices: (a) s_{32}; (b) s_{41}; (c) s_{64}.

2. Suppose that a risky asset satisfies the binomial branch model with $b = .03$ and $a = -.02$. Its initial price is $S_0 = \$50$, and its natural up probability is $p = 1/2$. Find the probability distribution and the mean value of S_3 under p, and also under the risk neutral probability q, assuming a risk-free rate of .01.

3. Verify the closed formula (3.9) for the value s_{lj} of a binomial branch process at level l in position j.

4. In Example 1, verify the replication properties at the other two nodes $(1, 1)$ and $(0, 1)$.

5. In Example 2, find the replicating portfolios at each of the nodes $(2, 1)$, $(1, 1)$, and $(0, 1)$.

6. In Exercise 4 of Section 1.6 we had a stock with initial price $40 following a binomial branch process with parameters $b = .03$ and $a = -.01$, and the risk-free rate was assumed to be $r = .01$. Find the current value of a three-period European put option on this asset with strike price $40 using the chaining approach.

7. Suppose that an asset of initial price $50 per share follows a binomial branch process, with natural up probability $p = .6$, and up and down parameters $b = .05$, $a = -.02$. Suppose that the risk-free rate is $r = .02$.
   (a) Use chaining to find the initial value, and the possible values at time 1, of a European call option with exercise time 2 and strike price $54;
   (b) Compare the initial value found in (a) to the expected present value of the claim value under the natural probability p: $E_p\left[1.02^{-2}V_2\right]$.
   (c) Find the replicating portfolios at each of the nodes $(0, 1)$, $(1, 1)$, and $(1, 2)$.

8. Give an informal argument that the value of a derivative at node (l, j) may be computed by the martingale method as:

$$v_{l,j} = E_q \left[(1+r)^{-(n-l)} V_n | V_l = v_{l,j} \right].$$

9. Use the result of Exercise 8 to compute values at node $(3, 2)$ and $(3, 3)$ of a 5-period put option on an asset whose initial price is $80, with up and down rates $b = .05$, $a = .01$. The risk-free rate is $r = .02$, and the exercise price of the put is $88.

10. In Example 3, verify that for each price value at time 1, $E_q \left[\bar{S}_3 | \bar{S}_1 = \bar{s}_{1,j} \right] = \bar{s}_{1,j}$.

11. Exercise 7 of Section 1.6 assumed that there was a risky asset following the binomial branch model with initial value $S_0 = 200, up rate $b = .08$, and down rate $a = -.04$. Consider a derivative on this asset that pays $10 if the asset reaches a value of at least $210 at time 4, and $-$5$ otherwise. Assume as in that exercise that the risk-free rate is $r = .02$. Use martingale valuation to give an initial value to this derivative.

12. A risky asset follows a binomial branch model with parameters $a = .01$, $b = .10$, $S_0 = 350. Assume that the risk-free rate is $r = .03$. Use the martingale valuation technique to compute the initial value of a three-period European put option with exercise price $E = 380.

13. In the scenario of Exercise 6 involving a three-period put option, use the values of the derivative at the intermediate nodes to show that:
   (a) $E_q \left[\bar{V}_3 | \bar{V}_2 = \bar{v}_{23} \right] = \bar{v}_{23}$;
   (b) $E_q \left[\bar{V}_3 | \bar{V}_1 = \bar{v}_{12} \right] = \bar{v}_{12}$.

14. In the setting of Example 4, use martingale valuation to find the initial value of a put option with exercise time 6 and strike price $18.

15. (Technology required) Produce a graph showing the dependence of the initial value of the derivative described here on the exercise time. The underlying asset begins at price $S_0 = 40 and follows a binomial branch process with up rate $b = .05$ and down rate $a = -.01$, and there is a risk-free asset with rate of return $r = .02$. The derivative expires at time n, and has strike price $E = 43, but the claim value is the fixed amount of $10 if the asset exceeds $43 at the expiration time; otherwise the derivative expires worthless.

16. At the end of the section, there was a brief discussion about the effect of increasing variance on the initial value of a call option. Do you think that

as the separation $b - a$ between the up and down rates in a binomial branch model increases, this also increases the value of a put option? Why or why not?

17. Throughout the section we used the binomial branch process to model the motion of an underlying asset. For this process, transitions are multiplicative, that is the possible next states are constant multiples of the previous state, not depending on time. Suppose for this problem that we assume constant differences instead, specifically an asset that starts with value $100 either gains value by $5 or loses value by $3 in each time period. Let us also assume that martingale valuation still holds in this new model. Given a risk-free rate of $r = .02$, find probabilities on the branches of a two-step binary tree that make the discounted asset process a martingale, and compute the initial value of a two-period call option with strike price $101.

18. (Technology required) For a European call option, the initial value should increase as the initial price of the underlying asset increases, all other things being equal, since it becomes more likely for the asset to exceed the strike price. Similarly, the initial value of the call should decrease as the strike price increases, because the asset is less likely to be in the money. With this in mind, consider a binomial branch asset with initial price S_0, and parameters $b = .01$ and $a = -.01$, and suppose that the risk-free rate is $r = .005$. A call option with expiration $n = 6$ is available with strike price E.

    (a) If $S_0 = \$60$, how small must the strike price be so that the initial option value is at least $1?

    (b) Now suppose that the strike price is $60. How large must the initial asset price S_0 be so that the value of the option is at least $2?

3.2 Key Ideas of Discrete Probability, Part I

In valuing derivative assets, the martingale condition below plays a key role:

$$E\left[X_{t+s}|X_t\right] = X_t. \qquad (3.27)$$

But what does this equation really mean? What is meant, particularly, by conditioning on "knowledge" of a random variable? It is the task of this section and the next to put the concepts of knowledge and conditional expectation on firm footing. Not only will this help to prove that the martingale method is correct, but it will also set up the important Fundamental Theorems of Asset Pricing in Section 3.4.

3.2.1   Algebras and Measurability

We must start from first principles. There is a sample space Ω of outcomes of a random experiment. We usually think of this as a finite set, although it does not have to be. And there are events, which are subsets of this sample space, whose probability we try to compute. We also often need compound event probabilities, such as unions $A \cup B$, intersections $A \cap B$, and complements A^c. So if A and B are events, then such new sets should also be events. This leads to the following definition.

Definition 1. A collection of events $\mathcal{A}$ in a sample space Ω is called an **algebra** if it includes Ω itself and is closed under complementation and finite union, that is:

(a) $\Omega \in \mathcal{A}$;
(b) if $A \in \mathcal{A}$, then $A^c \in \mathcal{A}$;
(c) if $A_1, A_2, ..., A_n \in \mathcal{A}$, then $A_1 \cup A_2 \cup \cdots \cup A_n \in \mathcal{A}$. ∎

Remark. Because DeMorgan's Laws imply that:

$$A_1 \cap A_2 \cap \cdots \cap A_n = (A_1^c \cup A_2^c \cup \cdots \cup A_n^c)^c, \tag{3.28}$$

properties (b) and (c) of the definition of algebra yield that an algebra is also closed under finite intersection. (See Exercise 2) So we have not unfairly privileged union over intersection. Additionally, since the sample space Ω is in every algebra $\mathcal{A}$, and $\mathcal{A}$ is closed under complementation, the empty set $\emptyset$ is also in every algebra.

Next is the idea of a basic building block of an algebra.

Definition 2. An **atom** A of an algebra $\mathcal{A}$ is an event that is in the algebra and does not strictly contain any other event in $\mathcal{A}$. The **algebra generated by a collection of atoms** $A_1, A_2, ..., A_n$ is the smallest algebra containing them all. We write $\mathcal{A}(A_1, A_2, ..., A_n)$ for this algebra. ∎

Example 1. Let sample space Ω be partitioned into four disjoint subsets B_1, B_2, B_3, and B_4, as in Figure 3.8. Treat these sets as atoms, and list all of the members of the algebra generated by them.

Solution. The algebra generated by the atoms B_1, B_2, B_3, and B_4 must be closed under finite union. Also, Ω and $\emptyset$ must be in the collection of sets. So we at least need all of the following to be in the algebra:

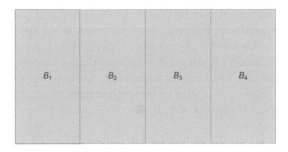

FIGURE 3.8 Atoms of an algebra.

$$\{\emptyset,$$
$$B_1, B_2, B_3, B_4,$$
$$B_1 \cup B_2, B_1 \cup B_3, B_1 \cup B_4, B_2 \cup B_3, B_2 \cup B_4, B_3 \cup B_4, \qquad (3.29)$$
$$B_1 \cup B_2 \cup B_3, B_1 \cup B_2 \cup B_4, B_1 \cup B_3 \cup B_4, B_2 \cup B_3 \cup B_4,$$
$$\Omega\}.$$

The strategy in display (3.29) was to list first all subsets of Ω with no atoms, then all subsets with one atom, then all with unions of two, three, and four atoms. We can form no further unions that are not redundant; for instance, $(B_1 \cup B_3 \cup B_4) \cup B_1$ is just $B_1 \cup B_3 \cup B_4$ again. Also, each set in the list has a complement that is also in the list, for example $(B_2 \cup B_3)^c = B_1 \cup B_4$. Therefore this list of sets is an algebra and is the smallest possible algebra that contains all of the B_i's. Notice that there are $1 + 4 + 6 + 4 + 1 = 16$ events in the algebra, which we can see by identifying sets in the algebra uniquely with subsets of $\{1, 2, 3, 4\}$. A set in the algebra is associated with the subset that gives the atom subscripts that are present in the set; for example $\{1, 2, 4\}$ is associated with $B_1 \cup B_2 \cup B_4$. There are $2^4 = 16$ such subsets of $\{1, 2, 3, 4\}$. ∎

Remark. Notice that atoms of an algebra cannot intersect each other. To see this, suppose two distinct atoms A_1 and A_2 had outcomes in common. Certainly neither atom can be a subset of the other, by definition of an atom. Then the set $A_1 \cap A_2$ is non-empty, and a proper subset of both A_1 and A_2. But by closure of an algebra under intersection, this would mean that one of A_1 or A_2 has a proper subset in the algebra, violating the defining condition for an atom. It is also an important property that arbitrary events in an algebra are just unions of zero (in case of $\emptyset$) or more atoms (see Exercise 5). Also, the union of all of the atoms of an algebra must exhaust the entire sample space Ω; otherwise if it did not, then the complement of that union would be a non-empty event in the algebra which does not contain any other event in the algebra, that is, it would be another atom that was unaccounted for. These arguments establish that the set of all atoms of an algebra forms a partition of the sample space.

The idea of algebra is our first pass at characterizing partial knowledge in a random phenomenon. Knowing that an event in an algebra occurred in an experiment narrows down the possible outcomes to those that are in that event, although it doesn't distinguish which particular outcome happened. To get at the concept of a hierarchy of knowledge, we have the following idea.

Definition 3. An algebra $\mathcal{B}$ is said to be a **refinement** of an algebra $\mathcal{A}$ if every event in $\mathcal{A}$ is also in $\mathcal{B}$, that is $\mathcal{A} \subset \mathcal{B}$. ∎

For example, if set B_1 in Figure 3.8 were subdivided into two disjoint subsets, $B_1 = B_{11} \cup B_{12}$ as in Figure 3.9, and we considered the algebra $\mathcal{B}$ generated by the larger collection of atoms, it would have the same sets as in display (3.29), but also many more, such as the individual atoms B_{11} and B_{12} and the union $B_{11} \cup B_2$. Then $\mathcal{B}$ would be a refinement of the original algebra $\mathcal{A}$ in the example. From a knowledge perspective, knowing that a member of $\mathcal{B}$ occurred can narrow down some possible cases: we can distinguish those outcomes in B_{11} from those in B_{12}, which we could not have done before.

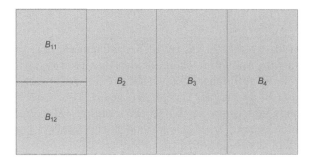

FIGURE 3.9 Refining an algebra by subdividing an atom.

Example 2. Consider the three-period risky asset process $S_0 = 20, S_1, S_2, S_3$ from Example 2 of Section 3.1, reproduced here as Figure 3.10. If the sample space Ω is defined as the set of 8 possible paths from the root of the price tree to its leaves, find the atoms of algebras $\mathcal{A}_0, \mathcal{A}_1, \mathcal{A}_2$, and $\mathcal{A}_3$ that characterize full path information available at times 0, 1, 2, and 3. Also, define algebras $\mathcal{F}_i$ at each time that contain only the information of the values of the current prices, that is, $\mathcal{F}_0$ contains information about S_0, $\mathcal{F}_1$ contains information about S_1, etc. Explore the containment relationships among all these algebras.

Solution. First, let us list and name all of the outcomes, that is, paths through the tree. We characterize them by lists of prices, but annotate the lists to indicate which up and down transitions were made.

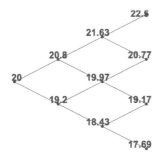

FIGURE 3.10 History of a binomial branch process.

$$
\begin{aligned}
\omega_1 &= (20, 20.8, 21.63, 22.50), &&(\text{up, up, up}) \\
\omega_2 &= (20, 20.8, 21.63, 20.77), &&(\text{up, up, down}) \\
\omega_3 &= (20, 20.8, 19.97, 20.77), &&(\text{up, down, up}) \\
\omega_4 &= (20, 20.8, 19.97, 19.17), &&(\text{up, down, down}) \\
\omega_5 &= (20, 19.2, 19.97, 20.77), &&(\text{down, up, up}) \\
\omega_6 &= (20, 19.2, 19.97, 19.17), &&(\text{down, up, down}) \\
\omega_7 &= (20, 19.2, 18.43, 19.17), &&(\text{down, down, up}) \\
\omega_8 &= (20, 19.2, 18.43, 17.69) &&(\text{down, down, down})
\end{aligned}
$$

At time 0, only the deterministic initial price is known; any of the 8 outcomes are still possible. The only atom is the whole sample space, hence:

$$\mathcal{A}_0 = \mathcal{A}(\Omega) = \{\emptyset, \{\omega_1, \omega_2, ..., \omega_8\}\}).$$

The algebra $\mathcal{A}_1$ follows the path through time 1, and so it distinguishes whether the first move is up or down. Therefore it has two atoms: $\{\omega_1, \omega_2, \omega_3, \omega_4\}$ and $\{\omega_5, \omega_6, \omega_7, \omega_8\}$. The full algebra consists of all possible unions and complements, but this is simply the atoms together with Ω and $\emptyset$:

$$
\begin{aligned}
\mathcal{A}_1 &= \mathcal{A}(\{\omega_1, \omega_2, \omega_3, \omega_4\}, \{\omega_5, \omega_6, \omega_7, \omega_8\}) \\
&= \{\emptyset, \Omega, \{\omega_1, \omega_2, \omega_3, \omega_4\}, \{\omega_5, \omega_6, \omega_7, \omega_8\}\}.
\end{aligned}
$$

Clearly, $\mathcal{A}_1$ is a refinement of $\mathcal{A}_0$.

At time 2 we know the second transition that occurred as well as the first. We cannot yet distinguish ω_1 from ω_2, in which the first two transitions are ups, but we can distinguish these from ω_3 and ω_4 for instance, because for the latter pair the second move was down rather than up. This reasoning shows that the atoms are:

$$\mathcal{A}_2 = \mathcal{A}(\{\omega_1, \omega_2\}, \{\omega_3, \omega_4\}, \{\omega_5, \omega_6\}, \{\omega_7, \omega_8\}).$$

There are many more sets in this full algebra, such as $\{\omega_1, \omega_2\}^c = \{\omega_3, \omega_4, \omega_5, \omega_6, \omega_7, \omega_8\}$, but in the interest of space we will not list them all. But notice that since $\{\omega_1, \omega_2, \omega_3, \omega_4\} = \{\omega_1, \omega_2\} \cup \{\omega_3, \omega_4\}$ and $\{\omega_5, \omega_6, \omega_7, \omega_8\} =$

$\{\omega_5, \omega_6\} \cup \{\omega_7, \omega_8\}$, the atoms of $\mathcal{A}_2$ can generate all of the sets in $\mathcal{A}_1$; thus $\mathcal{A}_2$ is a refinement of $\mathcal{A}_1$.

Time 3 gives us the final information about all transitions. So algebra $\mathcal{A}_3$ has all of the singleton sets as its atoms:

$$\mathcal{A}_3 = \mathcal{A}\left(\{\omega_1\}, \{\omega_2\}, \{\omega_3\}, \{\omega_4\}, \{\omega_5\}, \{\omega_6\}, \{\omega_7\}, \{\omega_8\}\right).$$

The algebra must contain all unions of these, which means that $\mathcal{A}_3$ is a refinement of each of the previous algebras, and in fact, $\mathcal{A}_3$ is the power set of Ω. So we have:

$$\mathcal{A}_0 \subset \mathcal{A}_1 \subset \mathcal{A}_2 \subset \mathcal{A}_3.$$

It is in this sense that the increasing sequence of algebras represents the accumulation of new information as time passes. Such a sequence of algebras ordered by refinement is called a *history*.

Now the algebras generated by only the price values S_i do not have information as to what path led to the price, and so they generally contain less information. For example, if S_2 is \$19.97 we do not know which of $\omega_3, \omega_4, \omega_5$, or ω_6 occurred. At time 0, as in the previous analysis, S_0 is all that is known, so that:

$$\mathcal{F}_0 = \mathcal{A}_0 = \mathcal{A}(\Omega) = \{\emptyset, \{\omega_1, \omega_2, ..., \omega_8\}\}.$$

At time 1, knowing the price is the same as knowing whether the first step was up or down; thus:

$$\mathcal{F}_1 = \mathcal{A}_1 = \mathcal{A}\left(\{\omega_1, \omega_2, \omega_3, \omega_4\}, \{\omega_5, \omega_6, \omega_7, \omega_8\}\right).$$

We begin to see differences at time 2. In addition to what we already noted, knowing S_2 does not help to distinguish between outcomes ω_1 and ω_2, nor between outcomes ω_7 and ω_8. Hence,

$$\mathcal{F}_2 = \mathcal{A}\left(\{\omega_1, \omega_2\}, \{\omega_3, \omega_4, \omega_5, \omega_6\}, \{\omega_7, \omega_8\}\right).$$

Algebra $\mathcal{A}_2$ already has the atoms $\{\omega_1, \omega_2\}$ and $\{\omega_7, \omega_8\}$, and it can generate the other by the union of its atoms $\{\omega_3, \omega_4,\} \cup \{\omega_5, \omega_6\}$. Therefore $\mathcal{A}_2$ refines $\mathcal{F}_2$. But this time $\mathcal{F}_2$ does not refine $\mathcal{F}_1$, because for example $\mathcal{F}_2$ has no way of generating the atom $\{\omega_1, \omega_2, \omega_3, \omega_4\}$ of $\mathcal{F}_1$ via unions, intersections, or complements. In Exercise 7 you are asked to verify that the algebra generated by S_3 is as below, that it does not refine $\mathcal{F}_2$, but that $\mathcal{A}_3$ refines $\mathcal{F}_3$.

$$\mathcal{F}_3 = \mathcal{A}\left(\{\omega_1\}, \{\omega_2, \omega_3, \omega_5\}, \{\omega_4, \omega_6, \omega_7\}, \{\omega_8\}\right). \blacksquare$$

Example 2 introduced a concept that is worth defining in general.

Definition 4. The *algebra generated by a discrete random variable* X, denoted by $\mathcal{A}(X)$, is the smallest algebra containing all of the atoms $A = \{X = x\}$, ranging over all possible values x of X. $\blacksquare$

The algebras $\mathcal{F}_i$ in Example 2 were exactly the algebras $\mathcal{A}(S_i)$. For example, the atom $\{\omega_4, \omega_6, \omega_7\}$ of $\mathcal{F}_3$ was the set of outcomes such that $S_3 = \$19.17$.

Notice from this definition that because of the property of closure under finite union, for any finite subset B of the state space of X:

$$\{\omega : X(\omega) \in B\} = \{X \in B\} = \underset{x \in B}{\cup} \{\omega : X(\omega) = x\} \in \mathcal{A}(X). \qquad (3.30)$$

Thus $\mathcal{A}(X)$ contains the collection $\mathcal{B}$ of all events $\{X \in B\}$ for finite sets B in the state space of X.

Conversely, notice that this collection $\mathcal{B}$ is an algebra, since:

$$\{X \in B\}^c = \{X \in B^c\} \in \mathcal{B}, \qquad (3.31)$$

and for any finite subcollection of sets in $\mathcal{B}$, $\{X \in B_1\}$, $\{X \in B_2\}$, ... $\{X \in B_n\}$, we have:

$$\{X \in B_1\} \cup \{X \in B_2\} \cup \cdots \cup \{X \in B_n\} = \left\{ X \in \overset{n}{\underset{i=1}{\cup}} B_i \right\} \in \mathcal{B}. \qquad (3.32)$$

(See Exercise 8.) The collection $\mathcal{B}$ also contains all atoms $\{X = x\}$, which are the same as $\{X \in \{x\}\}$. Since $\mathcal{A}(X)$ is the smallest algebra containing the atoms, $\mathcal{A}(X) \subset \mathcal{B}$. We have proved the following characterization of $\mathcal{A}(X)$.

Theorem 1. The algebra $\mathcal{A}(X)$ generated by a discrete random variable X is the collection of all events $\{X \in B\}$ for B ranging through all finite subsets of the state space of X. The atoms of the algebra are events of the form $A = \{X = x\}$. ∎

Example 3. Let a sample space be given by $\Omega = \{-3, -2, -1, 0, 1, 2, 3\}$, and let a random variable be defined as $X(\omega) = |\omega|$. Find the atoms of $\mathcal{A}(X)$, the algebra generated by X.

Solution. The atoms are subsets of Ω on which X has constant value. Thus, they are:

$$\begin{aligned} A_0 &= \{X = 0\} = \{0\}, \\ A_1 &= \{X = 1\} = \{-1, 1\}, \\ A_2 &= \{X = 2\} = \{-2, 2\}, \\ A_3 &= \{X = 3\} = \{-3, 3\}. \end{aligned}$$

Typical elements of $\mathcal{A}(X)$ are sets $\{X \in B\}$, such as:

$$\{X \in \{1, 2\}\} = A_1 \cup A_2 = \{-2, -1, 1, 2\}. \blacksquare$$

As in Example 2, we sometimes have several algebras of events under consideration, and we would like to extend the concept of information carried by an algebra to random variables. Does one algebra "know" what value a random variable X takes on, while another does not? The next definition points the way.

Definition 5. A discrete random variable X is called **measurable with respect to an algebra** $\mathcal{A}$ if it is constant on the atoms of $\mathcal{A}$. In particular, X is measurable with respect to $\mathcal{A}(X)$. ∎

Example 4. In the setting of Example 3, show that X is measurable with respect to the algebra $\mathcal{A}_1$ whose atoms are the singletons $\{-3\}$, $\{-2\}$, $\{-1\}$, $\{0\}$, $\{1\}$, $\{2\}$, $\{3\}$. But X is not measurable with respect to the algebra $\mathcal{A}_2$ whose atoms are $\{-3, -2, -1\}$, $\{0\}$, $\{1, 2, 3\}$.

Solution. Clearly X has a single value on all of the atoms of $\mathcal{A}_1$, for example $X(-3) = 3$. But X doesn't have constant value on two of the atoms of $\mathcal{A}_2$, for example $X(-3) = 3$ but $X(-2) = 2$. ∎

So if X is measurable with respect to an algebra $\mathcal{A}$, and one of the atoms of the algebra occurs, then X can have only one value, which is the sense in which we mean that an algebra "knows" the value of X. If an event A in $\mathcal{A}$ has occurred, then it is a union of atoms, on which X takes on particular single values, so that there is partial knowledge about the value that X has taken on.

There is an alternative characterization of measurability in the next theorem, which is often taken as the defining condition in other texts, and is more generalizable to the continuous case.

Theorem 2. A random variable X is measurable with respect to an algebra $\mathcal{A}$ if and only if all events $\{X \in B\}$ are in $\mathcal{A}$, for all B ranging through finite subsets of the state space of X.

Proof. We first prove that if all events $\{X \in B\}$ are in $\mathcal{A}$, then X must be constant on the atoms of $\mathcal{A}$; hence X is $\mathcal{A}$-measurable. In particular, for sets $B = \{x\}$ of single values, we are supposing that $\{X = x\} \in \mathcal{A}$ for all states x. But these events are the atoms of $\mathcal{A}(X)$. Since $\mathcal{A}$ is therefore an algebra containing the atoms of $\mathcal{A}(X)$, and $\mathcal{A}(X)$ is the smallest such algebra, $\mathcal{A}(X) \subset \mathcal{A}$. Consider an atom A of $\mathcal{A}$. This atom A cannot strictly contain an atom $\{X = x\}$ of $\mathcal{A}(X)$, which is in $\mathcal{A}$, or else A would not be an atom. Nor can A intersect an atom $\{X = x\}$ of $\mathcal{A}(X)$ in a proper subset, which would again violate the atomic property of A. Thus A is either equal to an atom of $\mathcal{A}(X)$, or else A is a proper subset of an atom of $\mathcal{A}(X)$. In either case, since atoms of $\mathcal{A}(X)$ are of the form $\{X = x\}$, X must have constant value on A; thus X is $\mathcal{A}$-measurable.

Conversely, suppose that X is $\mathcal{A}$-measurable. We must show that $\{X \in B\} \in \mathcal{A}$ for all finite subsets B. For concreteness, consider a specific set of states $B = \{x_1, x_2, ..., x_n\}$. Then,

$$\{X \in B\} = \bigcup_{i=1}^{n} \{X = x_i\} = \bigcup_{i=1}^{n} C_i,$$

where $C_i = \{X = x_i\}$. Since X is constant on atoms of $\mathcal{A}$, for each state x_i under consideration, there is some individual atom, or perhaps a union of several atoms, that exhausts all outcomes on which X takes the value x_i. That is, $C_i = A_{i1} \cup A_{i2} \cup \cdots \cup A_{ik_i}$, for some atoms A_{ij} of $\mathcal{A}$. But this means that:

$$\{X \in B\} = \bigcup_{i=1}^{n} C_i = \bigcup_{i=1}^{n} (A_{i1} \cup A_{i2} \cup \cdots \cup A_{ik_i}).$$

By the closure property under finite union for algebra $\mathcal{A}$, it follows that $\{X \in B\} \in \mathcal{A}$, which completes the proof. ∎

Remark. In the same way as we define the algebra generated by a single random variable, we can define the algebra generated by several random variables. For example, if $S_1, S_2,$ and S_3 are the price variables of Example 2, we would define the algebra $\mathcal{A}(S_1, S_2, S_3)$ as the smallest algebra containing the atoms $\{S_1 = s_1, S_2 = s_2, S_3 = s_3\}$ as s_1, s_2, s_3 range through all price combinations at the three times. This algebra would consist of all sets $\{S_1 \in B_1, S_2 \in B_2, S_3 \in B_3\}$, ranging over subsets B_i of possible states at the three times. In fact, since the process only runs for three time periods, algebra $\mathcal{A}(S_1, S_2, S_3)$ would be the same as the algebra $\mathcal{A}_3$ corresponding to the complete history.

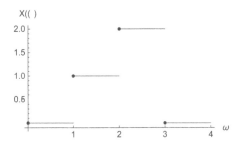

FIGURE 3.11 A discrete random variable on a continuous sample space.

Example 5. As long as the random variable under consideration has a finite state space of possible values, it is not necessary in any of the definitions and reasoning above for the sample space Ω itself to be finite. Suppose that a random variable X is defined on sample space $\Omega = [0, 4]$ and has value 0 for $\omega \in [0, 1)$, 1 for $\omega \in [1, 2)$, 2 for $\omega \in [2, 3)$, and again 0 for $\omega \in [3, 4]$, as shown in Figure 3.11. Then there are three atoms of $\mathcal{A}(X)$:

$$[0, 1) \cup [3, 4], \quad [1, 2), \quad [2, 3),$$

since these are the events over which X has constant value 0, 1, and 2 respectively. Also, X would be measurable with respect to the refinement $\mathcal{A}$ of $\mathcal{A}(X)$, whose atoms are:

$$\mathcal{Q} = \{[0,1), [1,1.5), [1.5,2), [2,3), [3,3.25), [3.25,4]\},$$

since X has the constant values 0, 1, 1, 2, 0, and 0, respectively, over these six atoms. The function $Y = 2X$ is $\mathcal{A}(X)$-measurable; it has the constant values $2 \cdot 0 = 0$ on $[0,1) \cup [3,4]$, $2 \cdot 1 = 2$ on $[1,2)$, and $2 \cdot 2 = 4$ on $[2,3)$. In fact, since any $\mathcal{A}(X)$-measurable random variable Y must be constant on the atoms of $\mathcal{A}(X)$, it is easy to see that such a random variable must take the form:

$$Y(\omega) = \begin{cases} y_1 & \text{if } \omega \in [0,1) \cup [3,4] \\ y_2 & \text{if } \omega \in [1,2) \\ y_3 & \text{if } \omega \in [2,3). \end{cases}$$

Defining $f(0) = y_1$, $f(1) = y_2$, and $f(2) = y_3$, we see that an $\mathcal{A}(X)$-measurable random variable Y can only be $Y = f(x)$ for some function f. $\blacksquare$

The observation at the end of Example 5 is very general. The situation is illustrated for a random variable X with three possible values in Figure 3.12, and is stated as a theorem below. The idea is just that for each atom $\{X = x_i\}$ of $\mathcal{A}(X)$, Y has a constant value y_i, and the function f is defined by $f(x_i) = y_i$.

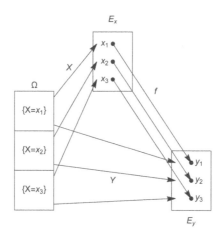

FIGURE 3.12 A random variable Y that is $\mathcal{A}(X)$-measurable.

Theorem 3. A random variable Y is $\mathcal{A}(X)$-measurable iff there is a function f such that $Y = f(X)$. ∎

The intuition is that the value of Y is known if the value of X is known.

3.2.2 Independence

We now provide a few reminders about the probabilistic notion of independence, and we extend the idea to independence of algebras.

From elementary probability, two events A and B in a sample space are **independent** of each other if:

$$P[A \cap B] = P[A] \cdot P[B]. \tag{3.33}$$

For events A and B of non-zero probability, this criterion for independence implies both of:

$$P[A|B] = P[A] \text{ and } P[B|A] = P[B]. \tag{3.34}$$

For more than two events, we say that events $A_1, A_2, ..., A_n$ are **mutually independent** if any subcollection of j among the n of them obeys the factorization law:

$$P\left[A_{k_1} \cap A_{k_2} \cap \cdots \cap A_{k_j}\right] = P\left[A_{k_1}\right] \cdot P\left[A_{k_2}\right] \cdots P\left[A_{k_j}\right]. \tag{3.35}$$

We can ratchet up the independence idea one more level by defining mutual independence of collections of events (such as algebras).

Definition 6. Collections of events $\mathcal{C}_1, \mathcal{C}_2, ..., \mathcal{C}_k$ are called **mutually independent** if for any selection of events $A_1 \in \mathcal{C}_1, A_2 \in \mathcal{C}_2, ..., A_k \in \mathcal{C}_k$ the A_i's are mutually independent. ∎

Example 6. For a binomial branch process with three transitions, define for each $i = 1, 2, 3$ the collection of four sets:

$$\mathcal{C}_i = \{\emptyset, \{U \text{ or } D \text{ on transition } i\}, \{U \text{ on transition } i\}, \{D \text{ on transition } i\}\}.$$

If, as usual, the conditions of the motion are conducive to independence assumptions, then $\mathcal{C}_1, \mathcal{C}_2$, and $\mathcal{C}_3$ are mutually independent collections of sets. You are asked to check some details in Exercise 14. ∎

Remark. To show that algebras $\mathcal{A}_1, \mathcal{A}_2, \ldots, \mathcal{A}_n$ are mutually independent, it suffices to show that their collections of atoms are mutually independent. This is straightforward but cumbersome to show in general. To get the idea, consider two algebras $\mathcal{A}, \mathcal{B}$ and suppose that the atoms of $\mathcal{A}$ are $\{A_1, A_2, A_3\}$ and the atoms of $\mathcal{B}$ are $\{B_1, B_2, B_3, B_4\}$. Assume that the two collections of atoms are independent. Then, for example, for event $C_1 = A_1 \cup A_2 \in \mathcal{A}$ and event $C_2 = B_1 \cup B_3 \cup B_4 \in \mathcal{B}$, we have:

$$
\begin{aligned}
P[C_1 \cap C_2] &= P[(A_1 \cup A_2) \cap (B_1 \cup B_3 \cup B_4)] \\
&= P[(A_1 \cap B_1) \cup (A_1 \cap B_3) \cup (A_1 \cap B_4) \\
&\quad \cup (A_2 \cap B_1) \cup (A_2 \cap B_3) \cup (A_2 \cap B_4)] \\
&= P[A_1 \cap B_1] + P[A_1 \cap B_3] + P[A_1 \cap B_4] \\
&\quad + P[A_2 \cap B_1] + P[A_2 \cap B_3] + P[A_2 \cap B_4] \\
&= P[A_1]P[B_1] + P[A_1]P[B_3] + P[A_1]P[B_4] \\
&\quad + P[A_2]P[B_1] + P[A_2]P[B_3] + P[A_2]P[B_4] \\
&= (P[A_1] + P[A_2])(P[B_1] + P[B_3] + P[B_4]) \\
&= P[C_1] \cdot P[C_2].
\end{aligned}
$$

The critical third line of the derivation is true because different sets such as $A_1 \cap B_1$ and $A_1 \cap B_4$ must be disjoint; otherwise atoms B_1 and B_4 would have outcomes in common, which cannot be. The assumption that the atoms are independent is used in line four.

In the setting of random variables, the elementary definition of *independence of discrete random variables* is:

$$
P[X_1 = x_1, X_2 = x_2, \ldots, X_n = x_n] = \begin{aligned}P[X_1 = x_1] \cdot P[X_2 = x_2] \\ \cdots P[X_n = x_n]\end{aligned} \tag{3.36}
$$

(similarly for subcollections of these X_i's). This can be re-expressed as the factorization of the joint probability mass function of the X_i's into the product of the marginal mass functions. But since the events $\{X_i = x_i\}$ are the atoms of their respective algebras $\mathcal{A}(X_i)$, and independence of the atoms implies independence of all other events in the algebras that are built from these atoms, and vice versa, this elementary definition is the same as the independence of the algebras $\mathcal{A}(X_i)$. Therefore we have the following theorem.

Theorem 4. Random variables $X_1, X_2, ..., X_n$ are are mutually independent in the sense of (3.36) if and only if the algebras $\mathcal{A}(X_1), \mathcal{A}(X_2), ..., \mathcal{A}(X_n)$ that they generate are mutually independent. Therefore, $X_1, X_2, ..., X_n$ are are mutually independent if and only if for all choices of subsets $B_1, B_2, ..., B_n$ of their respective state spaces:

$$
P[X_1 \in B_1, X_2 \in B_2, \ldots, X_n \in B_n] = \begin{aligned}P[X_1 \in B_1] \cdot P[X_2 \in B_2] \\ \cdots P[X_n \in B_n].\end{aligned} \blacksquare
$$

$$
\tag{3.37}
$$

Exercises 3.2

1. For a sample space defined by $\Omega = \{1, 2, 3, ..., 10\}$, give two different examples of algebras in Ω and give an example of a refinement of one of the algebras.

2. Prove the set identity in equation (3.28).

3. Suppose that a sample space Ω is partitioned into seven sets as in the diagram below. If these sets C_i are taken as the atoms of an algebra $\mathcal{C}$, list four different sets in that algebra other than the atoms themselves. How many sets are in $\mathcal{C}$?

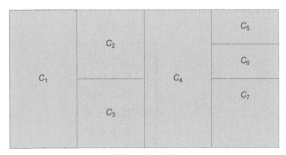

Exercise 3

4. Is the union of two algebras an algebra? Why or why not? What about the intersection of two algebras?

5. Show that any event in an algebra $\mathcal{A}$ is a disjoint union of zero or more atoms.

6. Consider the following algebra on the sample space $\Omega = \{0, 1, 2, 3, 4, 5, 6, 7, 8\}$. Find the atoms of $\mathcal{A}$.

$$\mathcal{A} = \{\{0, 1, 2, 3, 5, 6, 7, 8\}, \{0, 1, 2, 3, 4\}, \{4, 5, 6, 7, 8\}, \{0, 1, 2, 3\},$$
$$\{5, 6, 7, 8\}, \{4\}, \emptyset, \Omega\}$$

7. In reference to Example 2, argue that:

$$\mathcal{F}_3 = \mathcal{A}\left(\{\omega_1\}, \{\omega_2, \omega_3, \omega_5\}, \{\omega_4, \omega_6, \omega_7\}, \{\omega_8\}\right)$$

and that $\mathcal{F}_3 \subset \mathcal{A}_3$, and give an example to show that $\mathcal{F}_2$ is not contained in $\mathcal{F}_3$.

8. Verify set identities (3.31) and (3.32).

9. For the random variable X in Example 5, is the random variable below $\mathcal{A}(X)$-measurable?

$$Z(\omega) = \begin{cases} 1 & \text{if } \omega \in [0,2) \\ 2 & \text{if } \omega \in [2,4] \end{cases}$$

Find $\mathcal{A}(Z)$, the algebra generated by Z.

10. A random variable X is defined by:

$$X(\omega) = 2\omega^2 + 1 \text{ for outcomes } \omega \in \Omega = \{-3, -2, -1, 0, 1, 2, 3\}.$$

Find the atoms of the algebra generated by X.

11. A random variable X on sample space $\Omega = \{1, 2, 3, 4, 5, 6\}$ is defined as follows:

$$X(\omega) = \begin{cases} 0 & \text{if } \omega = 1; \\ 1 & \text{if } \omega = 2; \\ 2 & \text{if } \omega = 3; \\ 0 & \text{if } \omega = 4; \\ 1 & \text{if } \omega = 5; \\ 2 & \text{if } \omega = 6. \end{cases}$$

Find the algebra generated by X. Is X measurable with respect to the algebra generated by the atoms $A_1 = \{1, 3\}$, $A_2 = \{2, 4, 5\}$, $A_3 = \{6\}$?

12. Create a random variable that is measurable with respect to the algebra $\mathcal{A}(X)$, where X is the random variable in Exercise 11.

13. In Example 2, show that the price variable S_1 is measurable with respect to the algebra $\mathcal{A}(S_1, S_2)$ generated by S_1 and S_2.

14. A random asset moves through three price transitions at times 1, 2, and 3, and the transition at each time is independent of the transition at the other times. We just keep track of whether the asset moves up (U) or down (D) at each transition. In the first transition, up occurs with probability $2/3$, in the second $1/2$, and in the third $1/3$. Explicitly list the outcomes in the sample space, and define probabilities on outcomes consistent with independence. If C_i is the event of "up" on transition i for $i = 1, 2, 3$, and $\mathcal{A}_i$ is the algebra generated by C_i and its complement for each i, show that the algebras are independent collections of sets.

15. Consider a sample space $\Omega = \{1, 2, 3, 4, 5, 6\}$ and a probability measure P defined as: $P[1] = 1/12$, $P[2] = 2/12$, $P[3] = 2/12$, $P[4] = 4/12$, $P[5] = 1/12$, and $P[6] = 2/12$. Let two algebras on Ω be determined by the following atoms:

$$\mathcal{A}_1 = \mathcal{A}(\{1,2\},\{3,4\},\{5,6\}); \mathcal{A}_2 = \mathcal{A}(\{1,3,5\},\{2,4,6\}).$$

Show that $\mathcal{A}_1$ and $\mathcal{A}_2$ are independent algebras by showing that the atoms of $\mathcal{A}_1$ are independent of those of $\mathcal{A}_2$.

3.3   Key Ideas of Discrete Probability, Part II

The concepts of algebra, measurability of random variables with respect to algebras, and independence of random variables and algebras that were discussed in the last section play important roles as we now move to clarify the meaning and properties of conditional expectation. At the end of the section, we will show how these subtle ideas pertain to derivatives valuation.

3.3.1   Conditional Expectation

We would like to build up from the elementary definition of conditional probability of one event given another, $P[A|B] = P[A \cap B]/P[B]$, to a definition of conditional expectation of a random variable given an algebra of information $E[X|\mathcal{A}]$. This requires a sequence of steps, beginning with the notion of indicator random variable.

Definition 1. The ***indicator random variable*** $X = I_A$ of an event $A \subset \Omega$ is:

$$I_A(\omega) = \begin{cases} 1 & \text{if } \omega \in A; \\ 0 & \text{otherwise.} \end{cases} \quad \blacksquare \qquad (3.38)$$

Notice that the ordinary expectation of I_A is just the probability of A:

$$E[I_A] = 1 \cdot P[\omega \in A] + 0 \cdot P[\omega \in A^c] = 1 \cdot P[A] + 0 \cdot P[A^c] = P[A]. \quad (3.39)$$

Along the same lines we can define the conditional expectation of an indicator given an event. Notice, in formula (3.40) below, that the product of indicators $I_B I_A$ is the same as the indicator $I_{A \cap B}$, since both of these equal 1 for outcomes ω that are in both A and B, and they equal zero otherwise.

Step 1: Expectation of indicator random variable given an event:

$$E[I_A|B] = \frac{E[I_B I_A]}{P[B]} = \frac{P[A \cap B]}{P[B]} = P[A|B]. \qquad (3.40)$$

Indicator random variables are especially useful as building blocks for general discrete random variables. If random variable X takes on possible values

$x_1, x_2, ..., x_n$ on the atomic sets $A_i = \{\omega : X(\omega) = x_i\} = \{X = x_i\}$, then X can be written as a finite linear combination of indicator random variables:

$$X(\omega) = \sum_{i=1}^{n} x_i \cdot I_{A_i}(\omega), \tag{3.41}$$

because both sides have the value x_i for exactly the outcomes $\omega \in A_i$.

This leads us to the second step in the construction.

Step 2: Expectation of general discrete r.v. given an event:

If $X = \sum_{i=1}^{n} x_i I_{A_i}$, then requiring the linearity property to hold for conditional expectation, we must have:

$$
\begin{aligned}
E[X|B] &= E\left[\sum_{i=1}^{n} x_i I_{A_i} \,\middle|\, B \right] \\
&= \sum_{i=1}^{n} x_i E\left[I_{A_i} | B \right] \\
&= \sum_{i=1}^{n} x_i \frac{E[I_{A_i} I_B]}{P[B]} \\
&= \sum_{i=1}^{n} \frac{E[x_i I_{A_i} I_B]}{P[B]} \\
&= \frac{E\left[I_B \sum_{i=1}^{n} x_i I_{A_i} \right]}{P[B]} \\
&= \frac{E[I_B X]}{P[B]}.
\end{aligned}
\tag{3.42}
$$

When $X = I_A$, this agrees with the definition in Step 1.

Step 3: Expectation of a general discrete r.v. given an algebra:

If X is a random variable and $\mathcal{A}$ is an algebra whose atoms are $A_1, A_2, ..., A_n$, then define the ***conditional expectation of X given $\mathcal{A}$*** to be the random variable:

$$E[X|\mathcal{A}](\omega) = E\left[X | A_i \right] = \frac{E\left[I_{A_i} X \right]}{P\left[A_i \right]} \text{ if } \omega \in A_i. \tag{3.43}$$

Since $E[X|\mathcal{A}]$ is therefore constant on the atoms of $\mathcal{A}$, we immediately have the following result.

Theorem 1. If X is a random variable defined on a sample space Ω and $\mathcal{A}$ is an algebra of events on Ω, then $E[X|\mathcal{A}]$ defined by (3.43) is an $\mathcal{A}$-measurable random variable. ∎

What other properties does $E[X|\mathcal{A}]$ satisfy? First, it has the linearity property that you would expect of something that is called an expectation, as stated in our second theorem.

Theorem 2. If X and Y are random variables defined on a sample space Ω, if c and d are constants, and if $\mathcal{A}$ is an algebra of events on Ω, then:

$$E[cX + dY|\mathcal{A}] = c \cdot E[X|\mathcal{A}] + d \cdot E[Y|\mathcal{A}]. \tag{3.44}$$

Proof. Let $A_1, A_2, ..., A_n$ be the atoms of $\mathcal{A}$ and pick an arbitrary outcome ω, which belongs to some atom A_i. Then by formula (3.43),

$$E[cX + dY|\mathcal{A}](\omega) = E\left[cX + dY|A_i\right] = \frac{E\left[I_{A_i}(cX + dY)\right]}{P\left[A_i\right]}.$$

Using linearity of ordinary expectation, we obtain:

$$
\begin{aligned}
E[cX + dY|\mathcal{A}](\omega) &= \frac{E\left[I_{A_i}(cX + dY)\right]}{P[A_i]} \\
&= \frac{c \cdot E\left[I_{A_i} \cdot X\right] + d \cdot E\left[I_{A_i} \cdot Y\right]}{P[A_i]} \\
&= c \cdot \frac{E\left[I_{A_i} \cdot X\right]}{P[A_i]} + d \cdot \frac{E\left[I_{A_i} \cdot Y\right]}{P[A_i]} \\
&= c \cdot E[X|\mathcal{A}](\omega) + d \cdot E[Y|\mathcal{A}](\omega).
\end{aligned}
$$

Since ω was arbitrary, the linearity formula is proved. ∎

We can specialize the definition of conditional expectation to the case where the known information is the value of a random variable.

Definition 2. The ***conditional expectation of a random variable X given a random variable Y*** is the $\mathcal{A}(Y)$-measurable random variable:

$$E[X|Y] = E[X|\mathcal{A}(Y)]. \quad\blacksquare \tag{3.45}$$

Then since $\mathcal{A}(Y)$ is generated by the atoms $\{Y = y_i\}$, we have the concrete expression:

$$E[X|\mathcal{A}(Y)](\omega) = E\left[X|Y = y_i\right] = \frac{E\left[I_{\{Y = y_i\}} \cdot X\right]}{P\left[Y = y_i\right]} \text{ if } \omega \in \{Y = y_i\}. \tag{3.46}$$

Example 1. To see how all this plays out in our phenomenon of primary interest, consider a binomial branch process with initial value $S_0 = \$50$, up rate $b = .1$, down rate $a = -.1$, in a market with non-risky rate $r = .02$, under the risk-neutral measure q. We can compute q as:

$$q = \frac{r - a}{b - a} = \frac{.02 - (-.1)}{.1 - (-.1)} = \frac{.12}{.2} = .6.$$

Leave q general for the time being. We will find the random variable $E_q[S_3|S_1]$.

Solution. Since S_1 can only take on the values $50(1.1) = \$55$ and $50(.9) = \$45$, there are just two atoms, so that the algebra generated by S_1 is:

$$A(S_1) = \{\Omega, \emptyset, \{S_1 = 55\}, \{S_1 = 45\}\}.$$

For the first atom, formula (3.46) yields that, for $\omega \in \{S_1 = 55\}$:

$$
\begin{aligned}
E_q\left[S_3|S_1\right](\omega) &= E_q\left[S_3|A(S_1)\right](\omega) \\
&= E_q\left[S_3|S_1 = 55\right] \\
&= \frac{E_q\left[I_{\{S_1=55\}} \cdot S_3\right]}{P_q[S_1=55]}.
\end{aligned}
$$

The denominator of the last expression is q. The possible S_3 values are:

$$
\begin{aligned}
&\$50(1.1)^3 = \$66.55, \quad \$50(1.1)^2(.9) = \$54.45, \\
&\$50(1.1)(.9)^2 = \$44.55, \quad \$50(.9)^3 = \$36.45.
\end{aligned}
$$

These occur with probabilities $q^3, 3q^2(1-q), 3q(1-q)^2$, and $(1-q)^3$, respectively. The random variable $I_{\{S_1=55\}} \cdot S_3$ whose expectation is being taken in the numerator of our expression is equal to S_3 on paths for which the initial move was up, and otherwise is equal to 0. The paths that contribute are therefore (up, up, up), (up, up, down), (up, down, up), and (up, down, down). So the expectation $E_q\left[I_{\{S_1=55\}} \cdot S_3\right]$ can be written:

$$
\begin{aligned}
E_q\left[I_{\{S_1=55\}} \cdot S_3\right] &= q^3 \cdot 66.55 + 2q^2(1-q) \cdot 54.45 + q(1-q)^2 \cdot 44.55 \\
&= q\left(q^2 \cdot 66.55 + 2q(1-q) \cdot 54.45 + (1-q)^2 \cdot 44.55\right)
\end{aligned}
$$

Substituting into the expression for $E_q\left[S_3|S_1\right](\omega)$ gives, for $\omega \in \{S_1 = 55\}$:

$$
\begin{aligned}
E_q\left[S_3|S_1\right](\omega)] &= \frac{E_q\left[I_{\{S_1=55\}} \cdot S_3\right]}{P_q[S_1=55]} \\
&= \frac{q\left(q^2 \cdot 66.55 + 2q(1-q) \cdot 54.45 + (1-q)^2 \cdot 44.55\right)}{q} \\
&= q^2 \cdot 66.55 + 2q(1-q) \cdot 54.45 + (1-q)^2 \cdot 44.55.
\end{aligned}
$$

The last formula on the right matches exactly the elementary method of computing $E_q\left[S_3|S_1 = 55\right]$, since, starting from price $\$55$ at time 1, two up moves result in a price of $\$66.55$, one up and one down (two cases thereof) give a price of $\$54.45$, and two down moves result in a final price of $\$44.55$ at time 3.

Working similarly for the other atom $\{S_1 = 45\}$ of $A(S_1)$, we get, for $\omega \in \{S_1 = 45\}$:

$$
\begin{aligned}
E_q\left[S_3|S_1\right](\omega)] &= \frac{E_q\left[I_{\{S_1=45\}} \cdot S_3\right]}{P_q[S_1=45]} \\
&= \frac{(1-q)\left(q^2 \cdot 54.45 + 2q(1-q) \cdot 44.55 + (1-q)^2 \cdot 36.45\right)}{1-q} \\
&= q^2 \cdot 54.45 + 2q(1-q) \cdot 44.55 + (1-q)^2 \cdot 36.45.
\end{aligned}
$$

Again, this is the elementary expression for $E_q[S_3|S_1 = 45]$ from probability theory. Plugging in $q = .6$ and $1 - q = .4$, we get the numbers:

$$E_q[S_3|S_1 = 55] = .6^2 \cdot 66.55 + 2(.6)(.4) \cdot 54.45 + (.4)^2 \cdot 44.55 = 57.22;$$
$$E_q[S_3|S_1 = 45] = .6^2 \cdot 54.45 + 2(.6)(.4) \cdot 44.55 + (.4)^2 \cdot 36.45 = 46.818.$$

Note that the conditional expectation $E_q[S_3|S_1]$ is a random variable, taking on the value 57.22 for outcomes ω such that $S_1(\omega) = 55$, and 46.818 for outcomes with $S_1(\omega) = 45$. In both cases, you can check that $E_q[S_3|S_1] = (1 + r)^2 S_1$, verifying the martingale condition at times 1 and 3 for the discounted price process under q: $E_q\left[(1 + r)^{-3} S_3|S_1\right] = (1 + r)^{-1} S_1$. ∎

Conditional expectation has another important property relative to averaging. The random variable $E[X|\mathcal{A}]$, when averaged over sets in $\mathcal{A}$, has the same average value as X itself does. This makes $E[X|\mathcal{A}]$ a kind of best estimate of the behavior of X relative to events in $\mathcal{A}$. This is stated and proved in the next theorem.

Theorem 3. Let X be a random variable on a sample space Ω, let $\mathcal{A}$ be an algebra of events on Ω, and let $A \in \mathcal{A}$. Then:

$$E[I_A \cdot E[X|\mathcal{A}]] = E[I_A \cdot X] \tag{3.47}$$

Proof. The random variable $I_A \cdot E[X|\mathcal{A}]$ is zero for outcomes in A^c, and it equals $E[X|\mathcal{A}](\omega)$ for outcomes ω in A. Since A is in the algebra $\mathcal{A}$, it is a union of (disjoint) atoms $A = A_1 \cup A_2 \cup \cdots \cup A_m$. This implies that the indicator variable I_A splits apart:

$$I_A = I_{A_1} + I_{A_2} + \cdots + I_{A_m}.$$

For outcomes ω in atom A_i, $E[X|\mathcal{A}](\omega) = E[X|A_i]$. Therefore:

$$
\begin{aligned}
E[I_A \cdot E[X|\mathcal{A}]] &= E[(I_{A_1} + I_{A_2} + \cdots + I_{A_m}) \cdot E[X|\mathcal{A}]] \\
&= E[I_{A_1} \cdot E[X|A_1] + I_{A_2} \cdot E[X|A_2] \\
&\quad + \cdots + I_{A_m} \cdot E[X|A_m]] \\
&= E\left[I_{A_1} \cdot \frac{E[I_{A_1}X]}{P[A_1]} + I_{A_2} \cdot \frac{E[I_{A_2}X]}{P[A_2]} + \cdots + I_{A_m} \cdot \frac{E[I_{A_m}X]}{P[A_m]}\right] \\
&= \frac{E[I_{A_1}X]}{P[A_1]} E[I_{A_1}] + \frac{E[I_{A_2}X]}{P[A_2]} E[I_{A_2}] + \cdots \frac{E[I_{A_1}X]}{P[A_1]} E[I_{A_1}] \\
&= E[I_{A_1}X] + E[I_{A_2}X] + \cdots + E[I_{A_m}X] \\
&= E[(I_{A_1} + I_{A_2} + \cdots + I_{A_m}) \cdot X] \\
&= E[I_A \cdot X].
\end{aligned}
$$

In the second line we distributed the sum of indicators and used the fact that $I_{A_i} \cdot E[X|\mathcal{A}] = I_{A_i} E[X|A_i]$. After using the definition of $E[X|A_i]$ in line 3, in the fourth line we noted that each $\frac{E[I_{A_i}X]}{P[A_i]}$ is a constant and removed them from expectation, splitting the sum as well. But recall that $E[I_{A_i}] = P[A_i]$, so cancellation occurred in line 5, enabling a reuniting of terms in lines 6 and 7. This completes the proof. ■

Remark. In Exercise 4 you will show conversely that if Y is an $\mathcal{A}$-measurable random variable satisfying the property:

$$E[I_A \cdot Y] = E[I_A \cdot X], \tag{3.48}$$

for all events in the algebra $\mathcal{A}$, then $Y = E[X|\mathcal{A}]$. In other sources, the conditional expectation is taken to be the unique (up to sets of probability zero) $\mathcal{A}$-measurable random variable Y that satisfies (3.48), and we will use this fact without further proof from now on.

We will risk the insertion of a continuous example, though a simple one, to help illustrate the idea of averaging.

Example 2. Let $\Omega = [0, 1]$, and let $A_1 = [0, 1/3)$, $A_2 = [1/3, 2/3)$, and $A_3 = [2/3, 1]$ be the atoms of an algebra $\mathcal{A}$ on Ω. Let $X(\omega) = 2\omega$ for $\omega \in \Omega$, and let P be the uniform measure on $[0, 1]$. The random variable X is graphed as a line with slope 2 in Figure 3.13. Then the values of $E[X|\mathcal{A}]$ on the atoms are:

$$E[X|A_1] = \frac{E[X \cdot I_{A_1}]}{P[A_1]} = \frac{\int_0^{1/3} 2\omega \cdot 1 d\omega}{1/3} = \frac{\omega^2}{1/3}\Big|_0^{1/3} = 1/3;$$

$$E[X|A_2] = \frac{E[X \cdot I_{A_2}]}{P[A_2]} = \frac{\int_{1/3}^{2/3} 2\omega \cdot 1 d\omega}{1/3} = \frac{\omega^2}{1/3}\Big|_{1/3}^{2/3} = 1;$$

$$E[X|A_3] = \frac{E[X \cdot I_{A_3}]}{P[A_3]} = \frac{\int_{2/3}^{1} 2\omega \cdot 1 d\omega}{1/3} = \frac{\omega^2}{1/3}\Big|_{2/3}^{1} = 5/3.$$

The conditional expectation is then a step function:

$$E[X|\mathcal{A}] = \begin{cases} 1/3 & \text{if } \omega \in A_1; \\ 1 & \text{if } \omega \in A_2; \\ 5/3 & \text{if } \omega \in A_3. \end{cases}$$

The conditional expectation variable is shown in the picture. Its three constant values $1/3$, 1, and $5/3$ are clearly the average values of the random variable X on the atoms. ■

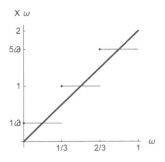

FIGURE 3.13 $Y = E[X|\mathcal{A}]$ is an average value of X over subinterval atoms.

Another important property of conditional expectation is below. It is useful as a computational device to find expectations by conditioning and unconditioning on information that becomes available as a process evolves.

Theorem 4. (Law of Total Probability for Conditional Expectation) Let B_1, B_2, ..., B_n be a partition of a sample space Ω (that is a collection of pairwise disjoint sets whose union is Ω), and let X be a random variable. Then:

$$E[X] = \sum_{i=1}^{n} E[X|B_i] \cdot P[B_i]. \tag{3.49}$$

Proof. First notice that since the B_i's form a partition, for all outcomes $\omega \in \Omega$:

$$\sum_{i=1}^{n} I_{B_i}(\omega) \cdot X(\omega) = X(\omega).$$

This is because each ω is in exactly one of the events B_i, so the sum reverts to a single term $I_{B_i}(\omega) \cdot X(\omega) = 1 \cdot X(\omega)$. Thus, by the definition of conditional expectation given an event, the right side of the desired equation is:

$$
\begin{aligned}
\sum_{i=1}^{n} E[X|B_i] \cdot P[B_i] &= \sum_{i=1}^{n} \frac{E[I_{B_i} X]}{P[B_i]} P[B_i] \\
&= \sum_{i=1}^{n} E[I_{B_i} \cdot X] \\
&= E\left[\sum_{i=1}^{n} I_{B_i} \cdot X\right] = E[X],
\end{aligned}
$$

which proves formula (3.49). ∎

Example 3. Consider again the binomial branch process that we have used in several examples, displayed in two periods in FIgure 3.14. The parameters were $S_0 = 20$, $b = .04$, and $a = -.04$, and the risk-free rate was $r = .01$. Find $E_q[S_2]$ using the law of total probability.
Solution. Let $B_1 = \{S_1 = 20.8\}$ and $B_2 = \{S_1 = 19.2\}$. Since exactly one of these two prices must occur at time 1, B_1 and B_2 partition the sample space of all paths. We had earlier found that the risk-neutral probability was $q = 5/8$.

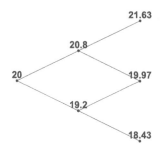

FIGURE 3.14 A binomial branch process.

To further simplify, recall that Example 1 established that conditional expectations given events can also be found in the way that elementary probability suggests. Then the law of total probability gives us that:

$$
\begin{aligned}
E_q\left[S_2\right] &= E_q\left[S_2|S_1 = 20.8\right] P\left[S_1 = 20.8\right] + E_q\left[S_2|S_1 = 19.2\right] P\left[S_1 = 19.2\right] \\
&= (q \cdot 21.63 + (1 - q)19.97)P\left[S_1 = 20.8\right] \\
&\quad + (q \cdot 19.97 + (1 - q)18.43)P\left[S_1 = 19.2\right] \\
&= \left(\tfrac{5}{8} \cdot 21.63 + \tfrac{3}{8} \cdot 19.97\right) \cdot \tfrac{5}{8} + \left(\tfrac{5}{8} \cdot 19.97 + \tfrac{3}{8} \cdot 18.43\right) \cdot \tfrac{3}{8} \\
&= 20.4019. \; \blacksquare
\end{aligned}
$$

We have a few more results before completing our study of the definitions and the algebra of conditional expectation. Next is a theorem that you would intuitively expect to be true regarding independence. If a random variable is independent of an algebra of information, then that algebra gives no additional information about the expected value of the random variable.

Theorem 5. If X is a random variable that is independent of an algebra $\mathcal{B}$ then:

$$
E[X|\mathcal{B}] = E[X]. \tag{3.50}
$$

Proof. By definition, the algebra $\mathcal{A}(X)$ is independent of $\mathcal{B}$, from which it follows that the atoms of $\mathcal{A}(X)$, i.e. events of the form $\{X = x\}$ are independent of the atoms $B_1, B_2, ..., B_n$ of $\mathcal{B}$. Consider an arbitrary particular atom B_i, and an arbitrary outcome ω in that atom. By definition of conditional expectation given an algebra:

$$
E[X|\mathcal{B}](\omega) = E\left[X|B_i\right] = \frac{E\left[I_{B_i}X\right]}{P\left[B_i\right]}.
$$

But the indicator random variable I_{B_i} is independent of random variable X (see Exercise 5). Therefore,

$$E[X|\mathcal{B}](\omega) = \frac{E\left[I_{B_i} X\right]}{P\left[B_i\right]} = \frac{E\left[I_{B_i}\right] \cdot E[X]}{P\left[B_i\right]} = \frac{P\left[B_i\right] \cdot E[X]}{P\left[B_i\right]} = E[X].$$

Since the atom B_i and the element ω of B_i were chosen arbitrarily, this shows that $E[X|\mathcal{B}] = E[X]$. ∎

Theorem 6. If X is $\mathcal{A}$-measurable, then $E[X|\mathcal{A}] = X$.

Proof. Let $A_1, A_2, ..., A_n$ be the atoms of $\mathcal{A}$, and let ω be a particular (but arbitrary) outcome in one of these A_i. Then,

$$E[X|\mathcal{A}](\omega) = \frac{E\left[I_{A_i} \cdot X\right]}{P\left[A_i\right]}.$$

Since X is $\mathcal{A}$-measurable, it is constant on the atoms; suppose that on A_i, X has the value x_i. Then $I_{A_i} \cdot X = 1 \cdot x_i$ for outcomes in A_i and is 0 otherwise. Hence $E\left[I_{A_i} \cdot X\right] = x_i \cdot P\left[A_i\right]$. This means that:

$$E[X|\mathcal{A}](\omega) = \frac{E\left[I_{A_i} \cdot X\right]}{P\left[A_i\right]} = \frac{x_i \cdot P\left[A_i\right]}{P\left[A_i\right]} = x_i = X(\omega).$$

Since atom A_i and outcome ω are arbitrary, $E[X|\mathcal{A}]$ agrees with X for all outcomes. ∎

Remark. The proof of Theorem 6 could have been done very easily using the alternative definition of conditional expectation. Random variable $Y = X$ is both $\mathcal{A}$-measurable and satisfies the condition $E\left[Y \cdot I_A\right] = E\left[X \cdot I_A\right]$ for all events $A \in \mathcal{A}$. Hence $Y = X = E[X|\mathcal{A}]$.

The intuition behind Theorem 6 is that if X is in the information base that algebra $\mathcal{A}$ carries, then it acts as a known constant as far as conditional expectation is concerned; the expected value of a constant is that constant. The next theorem extends this idea to the case where X multiplies another random variable Z inside expectation. Then if X is known in $\mathcal{A}$, it can be taken outside expectation as if it were constant.

Theorem 7. If X is measurable with respect to an algebra $\mathcal{A}$, and Z is another random variable, then $E[X \cdot Z|\mathcal{A}] = X \cdot E[Z|\mathcal{A}]$.

Proof. Since X is $\mathcal{A}$-measurable, it is constant on the atoms $A_1, A_2, ..., A_n$ of $\mathcal{A}$. Denote the value of X on A_i by x_i. Then, $I_{A_i} \cdot X \cdot Z = 1_{A_i} \cdot x_i \cdot Z$ for all outcomes, and so if an outcome ω is chosen in atom A_i, we have:

$$\begin{aligned}
E[X \cdot Z | \mathcal{A}](\omega) &= \frac{E[I_{A_i} \cdot X \cdot Z]}{P[A_i]} \\
&= \frac{x_i \cdot E[I_{A_i} \cdot Z]}{P[A_i]} \\
&= x_i \cdot E[Z | A_i](\omega) \\
&= X(\omega) \cdot E[Z | \mathcal{A}](\omega).
\end{aligned}$$

Because A_i and $\omega \in A_i$ were arbitrary, the proof is complete. ∎

Remark. An alternative proof of Theorem 7 is also possible using the equivalent definition of conditional expectation. If we define $Y = E[Z | \mathcal{A}]$, then Y is $\mathcal{A}$-measurable, and the product $X \cdot Y$ is also $\mathcal{A}$-measurable, and so it is a candidate for $E[X \cdot Z | \mathcal{A}]$. To show that the product is the conditional expectation, we must show that if $A \in \mathcal{A}$ then $E[X \cdot Z \cdot I_A] = E[X \cdot Y \cdot I_A]$. Here we can write $X = \sum_{i=1}^{n} x_i I_{A_i}$, so we have:

$$\begin{aligned}
E[X \cdot Z \cdot I_A] &= E[\sum_{i=1}^{n} x_i I_{A_i} \cdot Z \cdot I_A] \\
&= E[\sum_{i=1}^{n} x_i I_{A_i \cap A} \cdot Z] \\
&= \sum_{i=1}^{n} x_i E[I_{A_i \cap A} \cdot Z].
\end{aligned}$$

Since $A_i \cap A \in \mathcal{A}$, $E[I_{A_i \cap A} \cdot Z] = E[I_{A_i \cap A} \cdot Y]$ by definition of $Y = E[Z | \mathcal{A}]$. Thus, substituting into the last line on the right of the derivation above, and working backwards, we obtain:

$$\begin{aligned}
E[X \cdot Z \cdot I_A] &= \sum_{i=1}^{n} x_i E[I_{A_i \cap A} \cdot Z] \\
&= \sum_{i=1}^{n} x_i E[I_{A_i \cap A} \cdot Y] \\
&= E[\sum_{i=1}^{n} x_i I_{A_i \cap A} \cdot Y] \\
&= E[\sum_{i=1}^{n} x_i I_{A_i} \cdot Y \cdot I_A] \\
&= E[X \cdot Y \cdot I_A].
\end{aligned}$$

The next theorem contains what are referred to as the **Tower Laws** for conditional expectation, stated in formula (3.51). The meaning of these laws is that since a conditional expectation given an algebra is a random variable itself, its conditional expectation may be taken given a different algebra. This double conditional expectation reduces to the conditional expectation given the algebra that is the smaller of the two.

Theorem 8. If $\mathcal{A}$ and $\mathcal{B}$ are two algebras on a sample space Ω such that $\mathcal{A} \subset \mathcal{B}$, and if X is a random variable on Ω, then:

$$E[E[X | \mathcal{A}] | \mathcal{B}] = E[X | \mathcal{A}] \quad \text{and} \quad E[E[X | \mathcal{B}] | \mathcal{A}] = E[X | \mathcal{A}]. \tag{3.51}$$

Proof. Regarding the first equation, the random variable $Y = E[X | \mathcal{A}]$ is $\mathcal{A}$-measurable. Thus, all events of the form $\{Y \in C\}$ belong to $\mathcal{A}$ for sets C in the state space of Y. But, since $\mathcal{A} \subset \mathcal{B}$, all of these events also belong to $\mathcal{B}$. This means that $Y = E[X | \mathcal{A}]$ is $\mathcal{B}$-measurable also. By Theorem 6, $E[Y | \mathcal{B}] = Y$, which is the same as the equation we are to prove: $E[E[X | \mathcal{A}] | \mathcal{B}] = E[X | \mathcal{A}]$.

Now for the second equation, consider $E[X|\mathcal{B}]$. By formula (3.48), for all events $A \in \mathcal{B}$:

$$E\left[I_A \cdot E[X|\mathcal{B}]\right] = E\left[I_A \cdot X\right].$$

But again since $\mathcal{A} \subset \mathcal{B}$, the same holds in particular for all events $A \in \mathcal{A}$. Also, by formula (3.48) applied to $E[X|\mathcal{A}]$, for all events $A \in \mathcal{A}$:

$$E\left[I_A \cdot E[X|\mathcal{A}]\right] = E\left[I_A \cdot X\right].$$

Equating the two expressions gives:

$$E\left[I_A \cdot E[X|\mathcal{B}]\right] = E\left[I_A \cdot E[X|\mathcal{A}]\right] \text{ for all } A \in \mathcal{A},$$

so $E[X|\mathcal{A}]$ satisfies the criterion for being the conditional expectation $E[E[X|\mathcal{B}]|\mathcal{A}]$ and the second formula is proved. ∎

This section has been so packed with definitions and results about conditional expectation of discrete random variables that it will be helpful to bring them together in one place, and state them in brief.

Conditional expectation of an indicator given an event:

$$E\left[I_A|\, B\right] = \frac{E\left[I_B I_A\right]}{P[B]} = P[A|B].$$

Conditional expectation of a random variable given an event:

$$E[X|B] = \frac{E\left[I_B X\right]}{P[B]}.$$

Conditional expectation of a random variable given an algebra:

$$E[X|\mathcal{A}](\omega) = E\left[X\,|A_i\right] = \frac{E\left[I_{A_i} X\right]}{P\left[A_i\right]} \text{ if } \omega \in A_i.$$

Linearity:

$$E[cX + dY|\mathcal{A}] = c \cdot E[X|\mathcal{A}] + d \cdot E[Y|\mathcal{A}].$$

Conditional expectation of one random variable given another:

$$E[X|Y] = E[X|\mathcal{A}(Y)](\omega) = E\left[X|Y = y_i\right] = \frac{E\left[I_{\{Y=y_i\}} \cdot X\right]}{P\left[Y = y_i\right]}$$

for outcomes $\omega \in \{Y = y_i\}$.

Alternative definition:

$$Y = E[X|\mathcal{A}] \iff Y \text{is } \mathcal{A} - \text{measurable and} \\ E\left[I_A \cdot Y\right] = E\left[I_A \cdot X\right]$$

for all $A \in \mathcal{A}$.

Law of Total Probability: If sets B_i partition a sample space, then:

$$E[X] = \sum_{i=1}^{n} E[X | B_i] \cdot P[B_i].$$

Independence: If X is independent of an algebra $\mathcal{B}$, then

$$E[X|\mathcal{B}] = E[X].$$

Measurability $\Longrightarrow$ constancy in conditional expectation:

If X is $\mathcal{A} -$ measurable, then $E[X|\mathcal{A}] = X$.

Factoring out the known: If X is measurable with respect to $\mathcal{A}$ and Z is another random variable, then

$$E[X \cdot Z|\mathcal{A}] = X \cdot E[Z|\mathcal{A}].$$

Tower Laws: If algebras $\mathcal{A}$ and $\mathcal{B}$ satisfy $\mathcal{A} \subset \mathcal{B}$, then

$$E[E[X|\mathcal{A}]|\mathcal{B}] = E[X|\mathcal{A}] \text{ and } E[E[X|\mathcal{B}]|\mathcal{A}] = E[X|\mathcal{A}].$$

3.3.2   Application to Pricing Models

In this subsection we show how the theoretical substructure of algebras, measurability, independence, and conditional expectation applies to asset price models and derivative valuation. We do so via an example that invites a re-thinking of what the binomial branch process is, and secondly by giving a straightforward proof of the main theorem of Section 3.1 that asserts the correctness of the martingale derivative valuation method.

Example 4. To see the application of these ideas, consider a risky asset whose starting value is a fixed number $S_0 = s_0$ and which moves through a sequence of random states $S_1, S_2, S_3, \ldots$ as time progresses in such a way that each new price value is a random multiple R of the previous price, in other words:

$$S_{n+1} = R_{n+1} S_n, \quad n = 0, 1, 2, \ldots$$

We assume that the random multipliers R_i are independent of each other and of all previous S values in the price series. We will also assume that the common probability distribution of each R is:

$$f(r) = \begin{cases} p & \text{if } r = 1 + b; \\ 1 - p & \text{if } r = 1 + a, \end{cases}$$

where $a < b$. Notice that we are just giving a careful construction of a typical binomial branch process. Let Ω be the set of all possible paths $(s_0, S_1, S_2, ..)$, whose path length we leave unspecified but finite. Define $\mathcal{A}_1$ to be the algebra on Ω generated by S_1, define $\mathcal{A}_2$ to be the algebra generated by S_1 and S_2 together, whose atoms are of the form $\{\omega : S_1(\omega) = s_1, S_2(\omega) = s_2\}$, let $\mathcal{A}_3$ be the algebra generated by S_1, S_2, and S_3 together, whose atoms are of the form $\{\omega : S_1(\omega) = s_1, S_2(\omega) = s_2, S_3(\omega) = s_3\}$, etc. Then $\mathcal{A}_1, \mathcal{A}_2, \mathcal{A}_3, \ldots$ form the history of the price process, and they are successive refinements:

$$\mathcal{A}_1 \subset \mathcal{A}_2 \subset \mathcal{A}_3, \ldots.$$

Let us consider first the prediction of the next S value given the previous history. We have, because of the measurability of S_n with respect to $\mathcal{A}_n$ and the independence of R_{n+1} from $\mathcal{A}_n$,

$$
\begin{aligned}
E\left[S_{n+1}|\mathcal{A}_n\right] &= E\left[R_{n+1}S_n|\mathcal{A}_n\right] \\
&= S_n \cdot E\left[R_{n+1}|\mathcal{A}_n\right] \\
&= S_n \cdot E\left[R_{n+1}\right] \\
&= S_n \cdot (p(1+b) + (1-p)(1+a)).
\end{aligned}
\tag{3.52}
$$

We have used Theorems 7 and 5 in lines 2 and 3 of this derivation.

Next, suppose that at time n the investor selects a number of shares θ_{n+1} which will be held until time $n+1$. This θ_{n+1} may be a random variable, but to signify that we make such a choice of shares with reference only to past history, we assume that θ_{n+1} is $\mathcal{A}_n$-measurable for each $n = 0, 1, 2\ldots$. Then by Theorem 7 we can write the following formula for the expected amount of wealth in the asset at time $n+1$ given the history until time n:

$$
\begin{aligned}
E\left[\theta_{n+1}S_{n+1}|\mathcal{A}_n\right] &= \theta_{n+1} \cdot E\left[S_{n+1}|\mathcal{A}_n\right] \\
&= \theta_{n+1} \cdot E\left[R_{n+1}S_n|\mathcal{A}_n\right] \\
&= \theta_{n+1} \cdot S_n \cdot E\left[R_{n+1}\right] \\
&= \theta_{n+1} \cdot S_n \cdot (p(1+b) + (1-p)(1+a)).
\end{aligned}
\tag{3.53}
$$

This kind of reasoning will be a key step in our development of the Fundamental Theorems of Asset Pricing in the next section. Finally, by the Tower Law, we can write a formula for the expected price two time steps later given past history through time n, by conditioning and unconditioning on the history through time $n+1$, as follows:

$$
\begin{aligned}
E\left[S_{n+2}|\mathcal{A}_n\right] &= E\left[E\left[S_{n+2}|\mathcal{A}_{n+1}\right]|\mathcal{A}_n\right] \\
&= E\left[S_{n+1} \cdot (p(1+b) + (1-p)(1+a))|\mathcal{A}_n\right] \\
&= (p(1+b) + (1-p)(1+a))E\left[S_{n+1}|\mathcal{A}_n\right] \\
&= S_n \cdot (p(1+b) + (1-p)(1+a))^2.
\end{aligned}
\tag{3.54}
$$

Formula (3.52) was applied in the second step and again in the fourth. ∎

Let us close this section by filling in the gaps in the theoretical development of the techniques of Section 3.1 for valuing derivative assets. Previously, we have defined a martingale process as a stochastic process $(X_t)_{t \geq 0}$ satisfying the condition:

$$E[X_{t+s}|X_t] = X_t, \tag{3.55}$$

for all times $t, s \geq 0$. We did not have a firm foundation to stand on before, but now we know that the expression $E[X_{t+s}|X_t]$ means the conditional expectation $E[X_{t+s}|\mathcal{A}(X_t)]$ given the algebra generated by X_t. Actually, the more general definition of martingale that we need is below.

Definition 3. Let $\mathcal{A} = (\mathcal{A}_t)_{t \geq 0}$ be a history, that is, an increasing collection of algebras on a sample space Ω, so that $\mathcal{A}_t \subset \mathcal{A}_{t+s}$ for all $t, s \geq 0$. A stochastic process $(X_t)_{t \geq 0}$ is a ***martingale relative to history*** $\mathcal{A}$ if each X_t is $\mathcal{A}_t$-measurable, and:

$$E[X_{t+s}|\mathcal{A}_t] = X_t, \tag{3.56}$$

for all times $t, s \geq 0$. ∎

The history might well be the one generated by the process itself, that is, $\mathcal{A}_t = \mathcal{A}(X_r; r \leq t)$, but it could contain more information.

It would seem from formula (3.56) that to verify that a process is a martingale, we must check this powerful condition not only for all initial times t, but for all time increments s. The next result establishes that in the discrete time case, we only need to check the condition for the time increment $s = 1$.

Theorem 9. If $\mathcal{A} = (\mathcal{A}_t)_{t=0,1,2,...}$ is an increasing family of algebras, and $X = (X_t)_{t=0,1,2,...}$ is a process such that X_t is $\mathcal{A}_t$-measurable for each t, and $E[X_{t+1}|\mathcal{A}_t] = X_t$ for all $t \geq 0$, then X is a martingale relative to $\mathcal{A}$.

Proof. We prove this by induction on the time increment s. Let the initial time t be fixed but arbitrary. The anchor step $s = 1$ follows from the hypothesis of the theorem. Suppose that the defining equation for martingale $E[X_{t+r}|\mathcal{A}_t] = X_t$ is true for all r up to a particular time increment s. Then for increment $s + 1$, by the Tower Law,

$$E[X_{t+s+1}|\mathcal{A}_t] = E[E[X_{t+1+s}|\mathcal{A}_{t+1}]|\mathcal{A}_t] = E[X_{t+1}|\mathcal{A}_t] = X_t. \ ∎$$

Theorem 9 simplifies the proof of our main valuation theorem a great deal.

Theorem 10. Suppose that an underlying asset with price sequence $S_0 = s_0, S_1, S_2, ...$ follows the binomial branch model with up and down rates b and a, and there is a risk-free asset with rate of return r. Let $\mathcal{A} = (\mathcal{A}_k)_{k=0,1,2,...}$ be the history of algebras generated by the price process, and let q be the risk-neutral probability $q = (r - a)/(b - a)$. Suppose that a derivative on this asset has expiration time n. Then both the discounted asset price process $(\bar{S}_l)$

and the discounted derivative value process $(\bar{V}_l)$ are martingales relative to $\mathcal{A}$ under q; hence the initial value of the derivative is:

$$V_0 = E_q\left[(1+r)^{-n}V_n\right].$$

Proof. To verify the martingale condition, by Theorem 9 it is enough to show for each initial level $l = 0, 1, 2, \ldots$ $E_q\left[\bar{S}_{l+1}|\mathcal{A}_l\right] = \bar{S}_l$. By formula (3.52) in Example 4,

$$
\begin{aligned}
E_q\left[\bar{S}_{l+1}|\mathcal{A}_l\right] &= E_q\left[(1+r)^{-(l+1)}S_{l+1}|\mathcal{A}_l\right] \\
&= (1+r)^{-(l+1)}E_q\left[S_{l+1}|\mathcal{A}_l\right] \\
&= (1+r)^{-(l+1)}S_l \cdot (q(1+b) + (1-q)(1+a)) \\
&= (1+r)^{-l}S_l \cdot (1+r)^{-1}(q(1+b) + (1-q)(1+a)) \\
&= \bar{S}_l \cdot (1+r)^{-1}(q(1+b) + (1-q)(1+a)).
\end{aligned}
$$

Thus, it suffices to show algebraically that:

$$(1+r)^{-1}(q(1+b) + (1-q)(1+a)) = 1.$$

This equation is routine to verify. Manipulating the second factor, we have:

$$
\begin{aligned}
q(1+b) + (1-q)(1+a) &= \tfrac{r-a}{b-a}(1+b) + \tfrac{b-r}{b-a}(1+a) \\
&= \tfrac{r+br-a-ba+b+ab-r-ar}{b-a} \\
&= \tfrac{br-a+b-ar}{b-a} \\
&= \tfrac{(b-a)(1+r)}{b-a} \\
&= 1+r.
\end{aligned}
$$

Therefore $(1+r)^{-1}(q(1+b)+(1-q)(1+a)) = 1$ as desired, which shows that $E_q\left[\bar{S}_{l+1}|\mathcal{A}_l\right] = \bar{S}_l$, and hence the process $(\bar{S}_l)$ is an $\mathcal{A}$-martingale.

We also claim that from this, it follows that the discounted derivative process $(\bar{V}_l)$ is a martingale under measure q. Recall from Section 3.1 that there is a portfolio of the underlying asset and cash that can be constructed and adapted at each time l. The number of shares Δ_l of the asset to hold during time interval $[l, l+1)$ and the amount in the risk-free asset C_l are determined by the asset price S_l at the beginning of the interval according to the formula:

$$\Delta_l = \Delta_{l,j} \text{ and } C_l = c_{l,j} \text{ if } S_l = s_{l,j}, \tag{3.57}$$

where

$$\Delta_{l,j} = \frac{v_{l+1,j} - v_{l+1,j+1}}{s_{l+1,j} - s_{l+1,j+1}}, \quad c_{l,j} = (1+r)^{-1}\left(v_{l+1,j+1} - \Delta_{l,j} \cdot s_{l+1,j+1}\right). \tag{3.58}$$

The important part is that the variables Δ_l and C_l are functions of S_l and are therefore $\mathcal{A}_l$-measurable. The portfolio is so chosen that the level $l+1$ values match: $V_{l+1} = \Delta_l S_{l+1} + C_l(1+r)$, regardless of whether an up or down move of the underlying asset occurred. The level l values also match; that is, $V_l = \Delta_l S_l + C_l$. So we can derive:

$$
\begin{aligned}
E_q\left[\bar{V}_{l+1}|\mathcal{A}_l\right] &= E_q\left[(1+r)^{-(l+1)}V_{l+1}|\mathcal{A}_l\right] \\
&= (1+r)^{-(l+1)}E_q\left[V_{l+1}|\mathcal{A}_l\right] \\
&= (1+r)^{-(l+1)}E_q\left[\Delta_l S_{l+1} + C_l(1+r)|\mathcal{A}_l\right] \\
&= (1+r)^{-(l+1)}\left(\Delta_l E_q\left[S_{l+1}|\mathcal{A}_l\right] + C_l(1+r)\right) \\
&= (1+r)^{-(l+1)}(\Delta_l(S_l \cdot (q(1+b) + (1-q)(1+a)) \quad (3.59) \\
&\quad + C_l(1+r)) \\
&= (1+r)^{-(l+1)}\left(\Delta_l\left(S_l \cdot (1+r) + C_l(1+r)\right)\right) \\
&= (1+r)^{-l}\left(\Delta_l S_l + C_l\right) \\
&= (1+r)^{-l}V_l = \bar{V}_l.
\end{aligned}
$$

Line 4 used Theorems 6 and 7. In line 6 we use the identity established above that $q(1+b) + (1-q)(1+a) = 1 + r$. This establishes that $\left(\bar{V}_l\right)$ is a martingale and finishes the proof of the theorem. ∎

Exercises 3.3

1. If X is the random variable in Exercise 11 of Section 3.2, repeated below, if all outcomes are equally likely, and if $\mathcal{A}$ is the algebra whose atoms are $A_1 = \{1, 3\}$, $A_2 = \{2, 4, 5\}$, $A_3 = \{6\}$ compute $E[X|\mathcal{A}]$.

$$
X(\omega) = \begin{cases}
0 & \text{if } \omega = 1; \\
1 & \text{if } \omega = 2; \\
2 & \text{if } \omega = 3; \\
0 & \text{if } \omega = 4; \\
1 & \text{if } \omega = 5; \\
2 & \text{if } \omega = 6.
\end{cases}
$$

2. In the context of the discussion around formula (3.41), use that formula to compute $E[X]$, and verify that the result is the same as in the elementary definition of the expectation of a discrete random variable.

3. Let a sample space be defined by the consecutive integers $\Omega = \{1, ..., 9\}$ and suppose that probability measure P puts equal weight on all outcomes. Let X be a random variable on Ω defined by $X(\omega) = 2\omega$. For the algebra $\mathcal{A}$ of Ω whose atoms are the three events $B_1 = \{1, 2\}$, $B_2 = \{3, 4, 5, 6\}$, and $B_3 = \{7, 8, 9\}$, find $E[X|\mathcal{A}]$.

4. Prove the claim after Theorem 3 that if Y is an $\mathcal{A}$-measurable random variable satisfying:

$$
E\left[I_A \cdot Y\right] = E\left[I_A \cdot X\right]
$$

for all events $A \in \mathcal{A}$, then $Y = E[X|\mathcal{A}]$.

5. In the proof of Theorem 5, show that if B_i is an atom of the algebra $\mathcal{B}$ of which random variable X is independent, then the indicator random variable I_{B_i} is independent of X.

6. Suppose that X is a random variable on a sample space Ω that takes on non-negative values, and $\mathcal{A}$ is an algebra on Ω. Then show that $E[X|\mathcal{A}] \geq 0$.

7. Show that for any algebra $\mathcal{A}$ on a sample space Ω, if X is a constant random variable such that $X(\omega) = c$ for all outcomes, then $E[X|\mathcal{A}] = c$.

8. Argue that, if $\mathcal{B}$ is the algebra whose atoms are all outcomes in the sample space (i.e. the algebra of complete information), then $E[X|\mathcal{B}] = X$ for any random variable X. Also, show that if $\mathcal{A}$ is the trivial algebra consisting only of $\emptyset$ and Ω, then $E[X|\mathcal{A}] = E[X]$.

9. Prove that for all algebras $\mathcal{B}$ and random variables X defined on the same sample space:

$$E[E[X|\mathcal{B}]] = E[X].$$

(Hint: See Exercise 8.)

10. In Example 4, show that for each n, S_n is $\mathcal{A}_n$ − measurable.

11. In Example 4, compute (a) $E[S_{n+3}|\mathcal{A}_n]$; (b) $E[S_{n+k}|\mathcal{A}_n]$.

12. Suppose that a random variable X is measurable with respect to an algebra $\mathcal{A}$, Y is independent of $\mathcal{A}$, Z is measurable with respect to an algebra $\mathcal{B}$, and $\mathcal{A} \subset \mathcal{B}$. Simplify as far as possible:

$$E[2X + X \cdot Y - X \cdot Z|\mathcal{B}].$$

What is the expectation of this conditional expectation random variable? (Hint: See Exercise 9.)

Exercises 13-15 refer to the state tree below for a three-period process $X = (X_0, X_1, X_2, X_3)$, whose initial state x_0 is known, whose possible states for time 1 are labeled A and B, for time 2 are labeled C, D, E, and for time 3 are labeled F, G.

13. Find the history $\mathcal{A} = (\mathcal{A}_0, \mathcal{A}_1, \mathcal{A}_2, \mathcal{A}_3)$ generated by the process.

14. Continuing Exercise 13, let rewards be earned according to the rules:

$$Y_0 = 0,$$

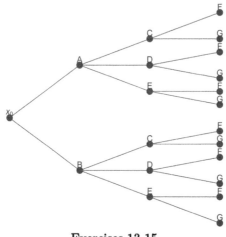

Exercises 13-15

$Y_1 = 1$ for state A and -1 for state B,
$Y_2 = 1$ for state C, 0 for state D, and-1 for state E, and
$Y_3 = 1$ for state F, and -1 for state G.

Define also the stochastic process Z of cumulative reward variables $Z_0 = Y_0$, $Z_1 = Y_0 + Y_1$, $Z_2 = Y_0 + Y_1 + Y_2$, and $Z_3 = Y_0 + Y_1 + Y_2 + Y_3$. Show that for each $i = 1, 2, 3$, Y_i and Z_i are $\mathcal{A}_i$-measurable.

15. Continuing Exercises 13 and 14, assume that the states at each time are equally likely, and that transitions are independent. Show that the Z process in Exercise 14 is an $\mathcal{A}$-martingale process.

16. True or false: A binomial branch process with up probability $p = 1/2$ must be a martingale. Explain.

3.4 Fundamental Theorems of Options Pricing

So far in this chapter we have considered derivative models in which there are only three assets: the risk-free asset with constant per-period rate of return r, a single, underlying risky asset whose initial price $S_0 = s_0$ is known and which moves through a sequence $S_1, S_2, \ldots$ of random states (usually following the simple binomial branch model), and a derivative asset based on the price process S of the underlying asset (usually dependent on the final value of the price process as opposed to the full path that it takes). We found in the binomial branch case that the absence of arbitrage opportunities implied that there was a unique risk-neutral probability q under which the discounted asset price process was a martingale, which led as well to the martingale property for the discounted derivative value process. It follows that the initial value V_0 of the derivative is $E_q\left[(1+r)^{-n}V_n\right]$, where n is the integer time at which the derivative is executed.

This leads us to wonder how far we can push the approach. Can the transition times be non-integer? Can the underlying asset process have a more general structure than the binomial branch? Can the final claim value of the derivative depend on features of the path taken by the underlying asset rather than just its final value? And can the market contain more than just one risky asset? The last two questions are particularly important, because there are common derivatives in real markets that are path dependent, such as American options and Asian options, and there are two-asset derivatives, whose claim value depends in some way on what two different risky assets are doing. Later in the text, as well, we will consider derivative valuation when time moves continuously rather than discretely, and the underlying asset moves through a continuous state space, but for now we will stick to the case of discrete time and state. The assumption of no arbitrage in the market, and the ideas of replicating portfolios and martingale measures on the market will again occupy the main stage, and we will see that under very general conditions the martingale valuation technique still applies. This will allow us to make an initial foray into the world of "non-vanilla" options in the following section.

The presentation in this section owes a great deal to the fine book by Roman [17], particularly the proofs of the two Fundamental Theorems of Asset Pricing.

3.4.1   The Market Model

To set up the model that we will use, first consider the time dimension. We suppose that the time at which we begin monitoring asset prices is designated as time $t_0 = 0$. The terminal time at which the derivative (or derivatives) of interest are to be executed is time T. For simplicity, we will restrict to the case where there are n times of price transitions, which are equally spaced $t_1 < t_2 < t_3 < \cdots < t_{n-1} < t_n = T$ as shown in Figure 3.15, where the common spacing is Δt, and so, explicitly:

$$t_k = k \cdot \Delta t, k = 0, 1, 2, ..., n. \tag{3.60}$$

In particular, at the terminal time, $T = n \cdot \Delta t$. This even spacing is not crucial to our development. Our special setup will not only simplify the presentation here but will also facilitate the passage to continuous time that will come later, in which we let Δt approach 0.

Our market will have some number a of risky assets, all of which make price transitions at the times t_k. We will use superscripts j to distinguish

FIGURE 3.15 Time axis for asset motion.

which asset we refer to, and subscripts as usual for the times, but to simplify a little, instead of indicating the real clock time t_k, we will just use the index k. Initial prices are assumed to be known constants. So our price process values are:

$$\text{asset1}: \quad S_0^1 = s_0^1, S_1^1, S_2^1, ..., S_n^1 \text{ (resp., at times } t_0 = 0, t_1, t_2, ...t_n = T)$$
$$\text{asset2}: \quad S_0^2 = s_0^2, S_1^2, S_2^2, ..., S_n^2$$

$$\vdots \qquad \vdots$$

$$\text{asset}a: \quad S_0^a = s_0^a, S_1^a, S_2^a, ..., S_n^a$$

$$(3.61)$$

As before, we assume that the time 0 prices s_0^j are known constants.

This labeling strategy S_k^j keeps the notation as close as possible to our familiar single asset notation, adding only the superscript to pick out a particular risky asset in the market. Each of these random variables are defined on the same sample space Ω, and each can assume a finite number of values. It might help you to think of Ω as the set of all possible combined paths of all of the assets. We make no assumptions about the probability measure P on this sample space at this point. Each price process S^j has its own history of algebras $\mathcal{A}^j = \left(\mathcal{A}_k^j \right)_{k=0,1,2,...,n}$ that it generates, and we will assume the existence of an overall history of algebras:

$$\mathcal{A} = (\mathcal{A}_k)_{k=0,1,2,...,n}, \text{where } \mathcal{A}_0 \subset \mathcal{A}_1 \subset \cdots \subset \mathcal{A}_n \qquad (3.62)$$

to which each price process is **adapted**; that is:

$$S_k^j \text{ is } \mathcal{A}_k - \text{measurable for each asset } j = 1, 2, ..., a$$
$$\text{and each time index } k = 0, 1, ..., n. \qquad (3.63)$$

The intuitive interpretation of algebra $\mathcal{A}_k$ is the information that is known about all asset prices, and perhaps other extraneous information, up to and including clock time t_k.

There will also be a risk-free asset in the market S^0, whose initial value at time 0 will be scaled to be 1 monetary unit, and which grows by compound interest at constant rate r for each period of length Δt. Hence,

$$S_0^0 = 1, S_1^0 = (1+r), S_2^0 = (1+r)^2, ..., S_n^0 = (1+r)^n. \qquad (3.64)$$

Present value discounting of other asset values will be relative to this one; hence the kth value at time t_k of the discounted asset j process $\bar{S}^j$ will be:

$$\bar{S}_k^j = (1+r)^{-k} S_k^j. \tag{3.65}$$

The assumption of constant interest rate can also be eliminated with some work, but we will retain it to keep the presentation simpler.

To get at the issue of portfolios and replicating portfolios of these risky assets, we take our cue from Example 4 of Section 3.3. Investors are able to hold a number of shares θ_k^j in each asset j during the kth time interval $[t_{k-1}, t_k)$, and then they can rebalance their portfolios exactly at time t_k. That is, the value of a portfolio at time t_k, represented by the vector $\boldsymbol{\theta}_k = \left(\theta_k^0, \theta_k^1, \theta_k^2, ..., \theta_k^a\right)$, is:

$$\theta_k^0 (1+r)^k + \theta_k^1 \cdot S_k^1 + \theta_k^2 \cdot S_k^2 + \cdots + \theta_k^a \cdot S_k^a. \tag{3.66}$$

When we use the term **portfolio sequence** $\boldsymbol{\theta} = (\boldsymbol{\theta}_1, \boldsymbol{\theta}_2, ..., \boldsymbol{\theta}_n)$ in the remainder of the section, we refer to the portfolio choices that the investor makes at each time; $\boldsymbol{\theta}_1$ is the portfolio selected at time 0, $\boldsymbol{\theta}_2$ is the time t_1 choice, etc. At the end of the period, the investor has selected $\boldsymbol{\theta}_n$ at time t_{n-1} for the last interval $[t_{n-1}, t_n)$, and no more portfolio rebalancing is done at time t_n. We call such portfolio sequences **self-financing** if portfolio rebalancing never adds nor subtracts net value to their portfolios. Since the investor would rebalance the portfolio $\boldsymbol{\theta}_k$ at time t_k to a new one represented by a portfolio vector $\boldsymbol{\theta}_{k+1} = \left(\theta_{k+1}^0, \theta_{k+1}^1, \theta_{k+1}^2, ..., \theta_{k+1}^a\right)$, the self-financing condition at time t_k would be:

$$\begin{aligned} &\theta_k^0 (1+r)^k + \theta_k^1 \cdot S_k^1 + \theta_k^2 \cdot S_k^2 + \cdots + \theta_k^a \cdot S_k^a \\ &= \theta_{k+1}^0 (1+r)^k + \theta_{k+1}^1 \cdot S_k^1 + \theta_{k+1}^2 \cdot S_k^2 + \cdots + \theta_{k+1}^a \cdot S_k^a. \end{aligned} \tag{3.67}$$

Equations like these can be more concisely expressed using vector notation and the dot product. If we denote $\boldsymbol{S}_k = \left(S_k^0 = (1+r)^k, S_k^1, S_k^2, ..., S_k^a\right)$ as the vector of asset prices at time t_k, then the portfolio value at the end of time interval $[t_{k-1}, t_k)$ prior to portfolio rebalancing, that is formula (3.66), is:

$$\text{interval } k \text{ closing value} = \boldsymbol{\theta}_k \cdot \boldsymbol{S}_k, \tag{3.68}$$

and the self-financing condition (3.67) is:

$$\begin{aligned} \text{interval } k \text{ closing value} &= \text{interval } k+1 \text{ opening value} \\ \boldsymbol{\theta}_k \cdot \boldsymbol{S}_k &= \boldsymbol{\theta}_{k+1} \cdot \boldsymbol{S}_k. \end{aligned} \tag{3.69}$$

Other economic assumptions that we will make are that transactions can occur instantaneously without fees, and that there are no restrictions on short sales.

It will be convenient to introduce the **time k value function** of a portfolio $\theta = \left(\theta^0, \theta^1, ...\theta^a\right)$, which is just the total value of the portfolio at clock time t_k summed over all assets:

$$V_k(\boldsymbol{\theta}) = V_k\left(\theta^0, \theta^1, ...\theta^a\right) = \boldsymbol{\theta} \cdot \boldsymbol{S}_k = \sum_{j=0}^{a} \theta^j \cdot S_k^j. \qquad (3.70)$$

Then the self-financing condition becomes:

$$V_k\left(\boldsymbol{\theta}_k\right) = V_k\left(\boldsymbol{\theta}_{k+1}\right) \text{ for each } k = 1, ..., n-1. \qquad (3.71)$$

This says that for each instant k the total value at time t_k prior to portfolio rebalancing equals the value after rebalancing. It is easy to show (see Exercise 1) that for each k, V_k is linear as a function of $\boldsymbol{\theta}$; that is, for any two portfolio vectors $\boldsymbol{\theta}$ and $\boldsymbol{\delta}$ and constants c and d:

$$V_k(c \cdot \boldsymbol{\theta} + d \cdot \boldsymbol{\delta}) = c \cdot V_k(\boldsymbol{\theta}) + d \cdot V_k(\boldsymbol{\delta}). \qquad (3.72)$$

Lastly, we will be assuming that for each time index k, and each asset j, the new number of shares θ_{k+1}^j is $\mathcal{A}_k$-measurable. So at the time t_k of rebalancing, the investor knows the asset prices and any other extraneous information at that time, and on the basis of that, decides on a portfolio vector $\boldsymbol{\theta}_{k+1}$ to hold through time interval $[t_k, t_{k+1})$. This means that the **associated value process** for a portfolio sequence is:

$$V_0 = V_0\left(\boldsymbol{\theta}_1\right), V_1 = V_1\left(\boldsymbol{\theta}_2\right), V_2 = V_2\left(\boldsymbol{\theta}_3\right), ..., V_{n-1} = V_{n-1}\left(\boldsymbol{\theta}_n\right), V_n = V_n\left(\boldsymbol{\theta}_n\right). \qquad (3.73)$$

By the self-financing condition, for times t_k with k between 1 and $n-1$, we could also write $V_k = V_k\left(\boldsymbol{\theta}_{k+1}\right)$ as the value $V_k\left(\boldsymbol{\theta}_k\right)$ before rebalancing to the new portfolio $\boldsymbol{\theta}_{k+1}$.

Example 1. Consider a two-period binomial branch model with price tree as in Figure 3.16. This is the risky asset in Example 1 of Section 3.1, whose parameters are: $b = .07$, $a = -.02$, and $S_0 = \$200$. The risk-free asset was assumed in that example to have rate of return $r = .04$, and again, suppose that we have scaled the monetary unit in such a way that its initial value is $1. Let us track an investor, Jim, whose strategy is to put all of his wealth of $2000 into the risky asset initially, and then at time t_1 to let it ride if the asset makes a down move, in hopes that it will go up again, but if the risky asset makes an up move he will transfer half of his wealth to the risk-free asset. Trace the effect of the self-financing condition on Jim's transaction, and find what his portfolio wealth will be under each of the four possible scenarios.

Solution. At time 0, Jim can buy 10 shares of the risky asset with his $2000 investment, and he devotes nothing to the risk-free asset, so that his portfolio vector is $\boldsymbol{\theta}_1 = (0, 10)$. Since this is a constant random variable, and the initial algebra $\mathcal{A}_0$ of asset price information is:

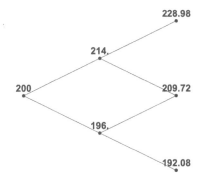

FIGURE 3.16 Two-period binomial branch process, $S_0 = \$200$, $b = .07$, $a = -.02$.

$$\mathcal{A}_0 = \{\emptyset, \Omega\} = \{\emptyset, \{(214, 228.98), (214, 209.72), (196, 209.72), (196, 192.08)\}\},$$

it follows that $\boldsymbol{\theta}_1$ is $\mathcal{A}_0$-measurable, since it is constant on the single atom of $\mathcal{A}_0$, namely Ω. The initial value is $V_0 = 0 \cdot \$1 + 10 \cdot \$200 = \$2000$.

Turning to time t_1, algebra $\mathcal{A}_1$ would contain information about whether the first transition was up or down; hence its atoms are:

$$\mathcal{A}_1 = \mathcal{A}(\{(214, 228.98), (214, 209.72)\}, \{(196, 209.72), (196, 192.08)\}).$$

Portfolio $\boldsymbol{\theta}_2$ has two different values depending on whether the first atom A_1 with time 1 price equal to 214, or the second atom A_2 with time 1 price equal to 196 occurred. In the former case, Jim sells off 5 shares to collect $5 \cdot \$214 = \$1070$, which he puts into the risk-free asset, whose value has grown to $\$1.04$. Hence he would own $\$1070/1.04 = 1028.85$ units of the risk-free asset in this case. If the risky asset goes down, that is, if A_2 occurs, he keeps his 10 shares of the risky asset. Thus, we have:

$$\boldsymbol{\theta}_2(\omega) = \begin{cases} (1028.85, 5) & \text{if } \omega \in A_1; \\ (0, 10) & \text{if } \omega \in A_2. \end{cases}$$

Hence $\boldsymbol{\theta}_2$ is $\mathcal{A}_1$-measurable. The value of his portfolio at time t_1 prior to transacting is:

$$\boldsymbol{\theta}_1 \cdot \boldsymbol{S}_1(\omega) = \boldsymbol{\theta}_1 \cdot (1.04, S_1(\omega)) = (0, 10) \cdot (1.04, S_1(\omega)) = \begin{cases} 2140 & \text{if } \omega \in A_1; \\ 1960 & \text{if } \omega \in A_2, \end{cases}$$

and the value after transacting is:

$$\boldsymbol{\theta}_2 \cdot \boldsymbol{S}_1(\omega) = \boldsymbol{\theta}_2 \cdot (1.04, S_1(\omega)) = \begin{cases} (1028.85, 5) \cdot (1.04, 214) = 2140 & \text{if } \omega \in A_1 \\ (0, 10) \cdot (1.04, 1960) = 1960 & \text{if } \omega \in A_2. \end{cases}$$

This verifies that the self-financing condition $\boldsymbol{\theta}_1 \cdot \boldsymbol{S}_1 = \boldsymbol{\theta}_2 \cdot \boldsymbol{S}_1$ holds, which is to be expected since Jim neither put more money into this portfolio nor withdrew any. Also, notice that the value $V_1 = \boldsymbol{\theta}_2 \cdot \boldsymbol{S}_1$ is $\mathcal{A}_1$-measurable, since it is constant on the atoms of $\mathcal{A}_1$.

To wrap up, $\mathcal{A}_2$ is the algebra of full path information, whose atoms are just the outcomes of Ω, which are:

$$\omega_1 = (214, 228.98), \omega_2 = (214, 209.72), \omega_3 = (196, 209.72), \omega_4 = (196, 192.08).$$

Jim's transactions have ceased. His final portfolio value is the random variable $V_2 = \boldsymbol{\theta}_2 \cdot \boldsymbol{S}_2$, which has four possible states, as below:

$$V_2(\omega) = \boldsymbol{\theta}_2 \cdot \boldsymbol{S}_2(\omega) = \begin{cases} (1028.85, 5) \cdot (1.04^2, 228.98) = 2257.70 & \text{if } \omega = \omega_1; \\ (1028.85, 5) \cdot (1.04^2, 209.72) = 2161.40 & \text{if } \omega = \omega_2; \\ (0, 10) \cdot (1.04^2, 209.72) = 2097.20 & \text{if } \omega = \omega_3; \\ (0, 10) \cdot (1.04^2, 192.08) = 1920.80 & \text{if } \omega = \omega_4. \end{cases}$$

Since V_2 is constant on the atoms of $\mathcal{A}_2$, V_2 is $\mathcal{A}_2$-measurable. The distribution of V_2 depends on how likely each path outcome ω_i is. In a typical binomial branch model, the probabilities would be $p^2, p(1-p), p(1-p)$, and $(1-p)^2$, respectively. ■

3.4.2   Gain, Arbitrage, and Attainability

In order to characterize arbitrage opportunities in this market, we need to introduce the concept of gain. An investor chooses a sequence of portfolios $\boldsymbol{\theta}_1, \boldsymbol{\theta}_2, ..., \boldsymbol{\theta}_n$ during the investment period $[0, T = t_n]$. Because of the self-financing condition, the overall investment gains in value only because of price changes between transaction times; the contribution to gain of asset j between times t_{i-1} and t_i is the number of shares held times the price change, namely $\theta_i^j \left(S_i^j - S_{i-1}^j \right)$. The **total net gain** from time 0 through time t_k is then computed as:

$$G_k = \sum_{i=1}^{k} V_i(\boldsymbol{\theta}_i) - V_{i-1}(\boldsymbol{\theta}_i) = \sum_{i=1}^{k} \sum_{j=0}^{a} \theta_i^j \left(S_i^j - S_{i-1}^j \right) = V_k(\boldsymbol{\theta}_k) - V_0(\boldsymbol{\theta}_1).$$

$$(3.74)$$

To see the last equation, if we write out the total net gain, we observe that:

$$\begin{aligned} G_k &= V_1(\boldsymbol{\theta}_1) - V_0(\boldsymbol{\theta}_1) + V_2(\boldsymbol{\theta}_2) - V_1(\boldsymbol{\theta}_2) + V_3(\boldsymbol{\theta}_3) - V_2(\boldsymbol{\theta}_3) \\ &\quad + \cdots + V_k(\boldsymbol{\theta}_k) - V_k(\boldsymbol{\theta}_{k-1}). \end{aligned}$$

By the self-financing condition, $V_1(\boldsymbol{\theta}_1) = V_1(\boldsymbol{\theta}_2)$, $V_2(\boldsymbol{\theta}_2) = V_2(\boldsymbol{\theta}_3)$, etc., so that all terms will subtract away except the V_0 and V_k terms shown.

We can also talk about the **discounted total net gain** $\bar{G}_k$ similarly, computing it as in formula (3.74), but with asset prices S_i^j and portfolio values V_i replaced by their discounted versions $\bar{S}_i^j = (1+r)^{-i}S_i^j$ and $\bar{V}_i = (1+r)^{-i}V_i$. Then also, $\bar{G}_k = \bar{V}_k(\boldsymbol{\theta}_k) - \bar{V}_0(\boldsymbol{\theta}_1)$.

As in the one-period case, the concept of arbitrage is vital. A careful definition in our new context follows now.

Definition 1. An **arbitrage strategy** is a self-financing portfolio sequence $\boldsymbol{\theta} = (\boldsymbol{\theta}_1, \boldsymbol{\theta}_2, ..., \boldsymbol{\theta}_n)$ such that there is no initial net investment $(V_0(\boldsymbol{\theta}_1) = 0)$, but there is some time t_k for which the total net gain G_k is always non-negative and has a positive probability of being strictly positive. ∎

It is not necessary for us to restrict arbitrage opportunities only to those situations in which there is a way of achieving such a gain G_n at the final time $t_n = T$, because if it is possible to do it at an earlier time t_k, then the investor can sell out all risky assets and invest the proceeds in the non-risky asset to secure the positive gain.

Example 2. Consider the simple case of a market with one risky asset satisfying the binomial branch model with $p = .5$, $b = .02$, $a = 0$, and $S_0 = \$50$, and a risk-free asset initially worth \$1 with rate of return $r = .02$ per period. An arbitrage strategy is easy to construct. Sell short a unit of the risky asset at time 0 to purchase 50 units of the risk-free asset. So $\boldsymbol{\theta}_1 = (50, -1)$ and the time 0 value is:

$$V_0 = 50(\$1) + (-1)(\$50) = 0.$$

Continue to hold this portfolio until time t_n, at which point it can be liquidated. The share of the risky asset can be worth no more than $\$50(1.02)^n$, in the case where all moves are up, which occurs with probability $.5^n$. The 50 units of the risk-free asset are certain to grow to $\$50(1.02)^n$. So the investor who plays this strategy cannot lose, and can gain the difference between the final value of the risk-free asset and the final value of the risky asset, which is strictly positive with probability $1 - .5^n$. In the notation of gain, $G_n \geq 0$ always, and $G_n > 0$ with probability greater than zero. Since $V_0 = 0$, this is arbitrage. ∎

We would like to have the most general view possible of a derivative in our model context. Thus, we introduce the following definition.

Definition 2. A **financial derivative** (also called a **contingent claim**) is a random variable X on the sample space Ω, which is measurable with respect to algebra $\mathcal{A}_n$. ∎

This means that a derivative may have a claim value that depends on one or all of the assets, and it may depend on the full path information for the assets, however its value is known by time $t_n = T$.

Definition 3. A derivative X is called **attainable** if there is a portfolio sequence $\theta = (\theta_1, \theta_2, ..., \theta_n)$ such that its value V_n at time $t_n = T$ equals the value of X for all outcomes. If this is the case, we say that the portfolio sequence **replicates** the derivative. A market is called **complete** if all derivatives are attainable, that is all derivatives have a replicating portfolio sequence. ■

Example 3. In the market of Example 1, define a derivative X to be the largest price that the risky asset assumes through times $t_0 = 0$, t_1, and t_2. This particular X depends upon the full path of the price process, not just its final value. Remember that the initial price was $200, and the full path outcomes were:

$$\omega_1 = (200, 214, 228.98), \omega_2 = (200, 214, 209.72),$$
$$\omega_3 = (200, 196, 209.72), \omega_4 = (200, 196, 192.08).$$

Then,

$$X(\omega) = \begin{cases} 228.98 & \text{if } \omega = \omega_1; \\ 214 & \text{if } \omega = \omega_2; \\ 209.72 & \text{if } \omega = \omega_3; \\ 200 & \text{if } \omega = \omega_4. \end{cases}$$

Can we construct a sequence of portfolios (θ_1, θ_2) that replicates this derivative?

The approach to take is to work backwards in the tree to find the intermediate values that the derivative takes on at each node, and the portfolio that would be required to achieve this. It is useful to expand the tree of Figure 3.16 so that all four distinct paths show, as in Figure 3.17. We also annotate nodes with the derivative values that we know in parentheses.

We need to select a portfolio $\theta_2 = (\theta_2^0, \theta_2^1)$ to satisfy the condition that the portfolio value at time 2 equals the claim value, in both the up and down cases. The risk-free asset has value 1.04^2 per unit at time t_2, and so the system appropriate to node 214 at level 1 is:

$$\begin{cases} \theta_2^0 \cdot 1.04^2 + \theta_2^1 \cdot 228.98 &= 228.98 \\ \theta_2^0 \cdot 1.04^2 + \theta_2^1 \cdot 209.72 &= 214 \end{cases} \implies \theta_2^0 = 47.0455, \theta_2^1 = .777778.$$

At time t_1 the risk-free asset is worth 1.04 per unit. Thus, the value of the portfolio at that node, which would be the derivative value there, is:

$$47.0455(1.04) + .7778(214) = 215.372.$$

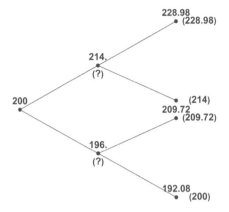

FIGURE 3.17 Unrecombined price tree for two-period binomial branch process.

Working similarly for node 196 at level 1:

$$\begin{cases} \theta_2^0 \cdot 1.04^2 + \theta_2^1 \cdot 209.72 = 209.72 \\ \theta_2^0 \cdot 1.04^2 + \theta_2^1 \cdot 192.08 = 200 \end{cases} \implies \theta_2^0 = 87.0562, \theta_2^1 = .55102.$$

The value of the derivative at that node is:

$$87.0562(1.04) + .55102(196) = 198.538.$$

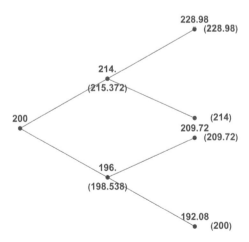

FIGURE 3.18 Price tree with time t_1 derivative values filled in.

The derivative values are updated at time t_1 in Figure 3.18. This allows us to work back to the portfolio $\boldsymbol{\theta}_1 = (\theta_1^0, \theta_1^1)$ to select at time 0. To replicate the derivative at t_1 we must have:

$$\begin{cases} \theta_1^0 \cdot 1.04 + \theta_1^1 \cdot 214 = 215.372 \\ \theta_1^0 \cdot 1.04 + \theta_1^1 \cdot 196 = 198.538 \end{cases} \implies \theta_1^0 = 14.6485, \theta_1^1 = .935222.$$

Then the value of the derivative at time 0 is:

$$14.6485 + .935222(200) = 201.693.$$

We appear to be finished, because we have a replicating portfolio and its initial value. But we should check that in each time t_1 case, the self-financing condition holds. This means:

$$\boldsymbol{\theta}_1 \cdot \boldsymbol{S}_1 = \boldsymbol{\theta}_2 \cdot \boldsymbol{S}_1 \iff$$
$$\begin{cases} (14.6485, .93522) \cdot (1.04, 214) = (47.0455, .7778) \cdot (1.04, 214) & \text{for } \{\omega_1, \omega_2\}; \\ (14.6485, .93522) \cdot (1.04, 196) = (87.0562, .55102) \cdot (1.04, 196) & \text{for } \{\omega_3, \omega_4\}. \end{cases}$$

Unsurprisingly, both left and right side values come out to 215.372 in the top case, and 198.538 in the bottom case; hence the self-financing condition checks. ■

3.4.3   Martingale Measures and the Fundamental Theorems

Next is the key definition upon which derivative valuation is based.

Definition 4. Let Ω be a discrete sample space with a history $\mathcal{A} = (\mathcal{A}_k)_{k=0,1,\dots,n}$ of algebras, and let there be price processes $S^j, j = 0, 1, \dots, a$ adapted to this history, as described above, where S^0 is the risk-free process. A probability measure Q on Ω is called a **martingale measure** if it attaches non-zero probability $Q(\omega) > 0$ to all outcomes, and if, for each asset j, the discounted price process $\left(\bar{S}_k^j\right)_{k=0,1,\dots,n} = \left((1+r)^{-k} S_k^j\right)_{k=0,1,\dots,n}$ is a martingale under Q. ■

Are there even such things as martingale measures? The next example constructs one in a fairly easy context.

Example 4. Construct a martingale measure on the sample space of all joint paths of two independent three-step binomial branch processes S^1 and S^2 as depicted in Figure 3.19. Suppose that the risk-free rate is $r = .02$. For S^1, the up and down rates are $b_1 = .05$ and $a_1 = .01$, and for S^2, the rates are $b_2 = .08$ and $a_2 = -.02$. The initial prices are $s_0^1 = \$40$, and $s_0^2 = \$80$.

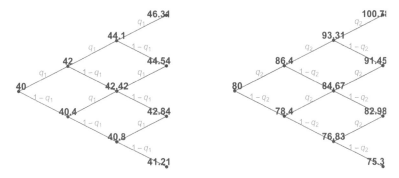

FIGURE 3.19 Three period dual asset binomial branch model.

Solution. The sample space should contain information about both processes, so we construct it as the set of all pairs of paths of length 3:

$$\Omega = \left\{ \left((40, s_1^1, s_2^1, s_3^1), (80, s_1^2, s_2^2, s_3^2) \right) \right\},$$

where there are eight legal sequences of s_1^1, s_2^1, s_3^1 values as shown in the tree on the left of the figure, and similarly eight legal sequences of s_1^2, s_2^2, s_3^2 values on the right. Thus, there are $8 \cdot 8 = 64$ possible outcomes in Ω. One example is:

$$\omega_1 = ((40, 42, 44.1, 46.305), (80, 78.4, 76.832, 75.295)),$$

in which the first asset goes up in each transition while the second goes down. A second example outcome is the following, in which the processes move in lockstep, with two up moves followed by a down move:

$$\omega_2 = ((40, 42, 44.1, 44.541), (80, 86.4, 93.312, 91.446)).$$

To fill in the formal structure, the random variables S_k^j are defined naturally on outcomes, picking out entry k of the jth quadruple of outcomes such as those listed above. For instance, for the first example outcome above, $S_2^1(\omega_1) = 44.1$, and for the second, $S_3^2(\omega_2) = 91.446$. The history of algebras $\mathcal{A}$ is the one generated by the two processes together, specifically:

$$\mathcal{A}_0 = \{\emptyset, \Omega\}, \mathcal{A}_1 = \mathcal{A}\left(S_1^1, S_1^2\right), \mathcal{A}_2 = \mathcal{A}\left(S_1^1, S_1^2, S_2^1, S_2^2\right),$$
$$\mathcal{A}_3 = \mathcal{A}\left(S_1^1, S_1^2, S_2^1, S_2^2, S_3^1, S_3^2\right).$$

We are used to finding a single risk-neutral probability of an up move to make a single discounted asset process into a martingale; here we just have two assets to deal with. The corresponding individual risk-neutral probabilities are:

$$\text{asset } S^1 : q_1 = \frac{r - a_1}{b_1 - a_1} = \frac{.02 - .01}{.05 - .01} = \frac{1}{4};$$

$$\text{asset } S^2 : q_2 = \frac{r - a_2}{b_2 - a_2} = \frac{.02 - (-.02)}{.08 - (-.02)} = \frac{2}{5}.$$

Since we require independence, probability should be assigned to outcomes by multiplication; for example, to outcome ω_1, consisting of three up moves for S^1 and three down moves for S^2, we would give probability:

$$(q_1)^3 (1 - q_2)^3 = \left(\frac{1}{4}\right)^3 \left(\frac{3}{5}\right)^3.$$

For our other example outcome, ω_2, with two ups and then a down for both processes, we would define its probability to be:

$$(q_1)^2 (1 - q_1) (q_2)^2 (1 - q_2) = \left(\frac{1}{4}\right)^2 \left(\frac{3}{4}\right) \left(\frac{2}{5}\right)^2 \left(\frac{3}{5}\right).$$

Our candidate martingale measure Q would work in this way to attach probabilities to all 64 outcomes. Specifically, $Q((a, b, c, d), (e, f, g, h))$ is a product $Q_1(a, b, c, d) \cdot Q_2(e, f, g, h)$ of marginal probability measures on the individual processes, determined by the risk-neutral probabilities q_1 and q_2. The marginal probabilities for each asset are exactly what they should be. For example, the probability that $S^1_2 = 44.1$ is:

$$\sum_{s^1_3} \sum_{s^2_1} \sum_{s^2_2} \sum_{s^2_3} Q \left((40, 42, 44.1, s^1_3), (80, s^2_1, s^2_2, s^2_3)\right)$$
$$= \sum_{s^1_3} Q_1 \left(40, 42, 44.1, s^1_3\right) \sum_{s^2_1} \sum_{s^2_2} \sum_{s^2_3} Q_2 \left(80, s^2_1, s^2_2, s^2_3\right)$$
$$= (q_1 q_1 q_1 + q_1 q_1 (1 - q_1)) \cdot 1$$
$$= (q_1)^2 (q_1 + 1 - q_1) = (q_1)^2.$$

Now we must check the martingale condition for each individual discounted asset. But this is easy. For asset 1, Q reduces to the marginal measure Q_1, which makes the discounted asset 1 process a martingale. Similarly for asset 2. ∎

Let us now plunge into the theory that we need. A preliminary result to our two major theorems follows below. It tells us that not only are the discounted asset processes themselves martingales under a martingale measure, but so are the value processes of self-financing portfolio sequences.

Theorem 1. A measure Q on sample space Ω is a martingale measure if and only if for all self-financing portfolio sequences $\boldsymbol{\theta} = (\boldsymbol{\theta}_1, \boldsymbol{\theta}_2, ..., \boldsymbol{\theta}_n)$, their associated discounted value functions $(\bar{V}_k)_{k=0,1,...,n}$ form $\mathcal{A}$-martingales.

Proof. First suppose that Q is a martingale measure. Let $\boldsymbol{\theta}$ be an arbitrary self-financing portfolio sequence. By equation (3.70), for each time index k, multiplying both sides by $(1 + r)^{-k}$ we have:

$$\bar{V}_k = \bar{V}_k \left(\boldsymbol{\theta}_{k+1}\right) = \boldsymbol{\theta}_{k+1} \cdot \overline{\boldsymbol{S}}_k = \sum_{j=0}^{a} \theta_{k+1}^j \cdot \bar{S}_k^j.$$

The $\mathcal{A}_k$-measurability of each asset price S_k^j and portfolio weight θ_{k+1}^j clearly implies measurability of $\bar{V}_k$ as well. By the linearity of conditional expectation and Theorem 7 of Section 3.3:

$$
\begin{aligned}
E_Q\left[\bar{V}_{k+1}|\mathcal{A}_k\right] &= E_Q\left[\sum_{j=0}^{a} \theta_{k+1}^j \cdot \bar{S}_{k+1}^j|\mathcal{A}_k\right] \\
&= \sum_{j=0}^{a} E_Q\left[\theta_{k+1}^j \cdot \bar{S}_{k+1}^j|\mathcal{A}_k\right] \\
&= \sum_{j=0}^{a} \theta_{k+1}^j \cdot E_Q\left[\bar{S}_{k+1}^j|\mathcal{A}_k\right].
\end{aligned}
$$

Because each discounted asset process is a martingale under Q, the conditional expectation inside the sum is simply $\bar{S}_k^j$. Also, the self-financing condition implies that $\sum_{j=0}^{a} \theta_{k+1}^j \cdot \bar{S}_k^j = \sum_{j=0}^{a} \theta_k^j \cdot \bar{S}_k^j$ for each asset j. Thus, we can continue as follows:

$$
\begin{aligned}
E_Q\left[\bar{V}_{k+1}|\mathcal{A}_k\right] &= \sum_{j=0}^{a} \theta_{k+1}^j \cdot E_Q\left[\bar{S}_{k+1}^j|\mathcal{A}_k\right] \\
&= \sum_{j=0}^{a} \theta_{k+1}^j \cdot \bar{S}_k^j \\
&= \sum_{j=0}^{a} \theta_k^j \cdot \bar{S}_k^j = \bar{V}_k.
\end{aligned}
$$

Therefore the discounted value process is a martingale.

For the converse, suppose that the discounted value functions $\left(\bar{V}_k\right)$ of all self-financing portfolio sequences $\boldsymbol{\theta}$ are martingales under a measure Q. By Exercise 9 of Section 3.3, conditional expectation satisfies the property that:

$$E[E[X|\mathcal{B}]] = E[X], \tag{3.75}$$

for all random variables X and algebras $\mathcal{B}$. In particular, for the final discounted value function $\bar{V}_n$ of a self-financing portfolio sequence, the martingale condition implies:

$$E_Q\left[\bar{V}_n|\mathcal{A}_0\right] = \bar{V}_0 = V_0 \implies E_Q\left[\bar{V}_n\right] = E_Q\left[E_Q\left[\bar{V}_n|\mathcal{A}_0\right]\right] = E_Q\left[V_0\right] = V_0. \tag{3.76}$$

Equation (3.76) further implies that the expected overall discounted gain of the portfolio sequence is zero for all self-financing portfolio sequences:

$$E_Q\left[\bar{G}_n\right] = E_Q\left[\bar{V}_n - \bar{V}_0\right] = E_Q\left[\bar{V}_n\right] - V_0 = 0. \tag{3.77}$$

Now to complete the proof that Q is a martingale measure, let us fix an arbitrary one of the assets, say asset j, and try to show the one-step martingale condition at an arbitrary fixed time t_k, namely:

$$E_Q\left[\bar{S}_{k+1}^j|\mathcal{A}_k\right] = \bar{S}_k^j.$$

To do this, let A be an arbitrary, fixed atom of $\mathcal{A}_k$. We construct a particular, very simple self-financing portfolio sequence $(\boldsymbol{\theta}_1, \boldsymbol{\theta}_2, ..., \boldsymbol{\theta}_n)$. For outcomes not in A, let all portfolios be $\mathbf{0}$. For outcomes $\omega \in A$, we do the following. Each portfolio vector $\boldsymbol{\theta}_1, \boldsymbol{\theta}_2, ..., \boldsymbol{\theta}_k$ that is held up to time t_k is zero. At time t_k, portfolio $\boldsymbol{\theta}_{k+1}$ is defined as follows: borrow cash (i.e. short sell the risk-free asset) to buy exactly one share of risky asset j, ignoring the other risky assets. The price of asset j at this time is $S_k^j(\omega)$, and the risk-free asset has appreciated to $(1+r)^k$, so we will need to borrow $S_k^j(\omega)/(1+r)^k = \bar{S}_k^j(\omega)$ units of the risk-free asset, leaving us with a portfolio:

$$\left(-\bar{S}_k^j(\omega), 0, 0, ..., 1, 0, ..., 0\right)$$

in which the 1 is in entry j. At time t_{k+1}, we sell the share of risky asset j, convert the proceeds to the risk-free asset, and leave the portfolio unchanged after that. The share of asset j will be worth $S_{k+1}^j(\omega)$ at time t_{k+1}, and the net value invested in the risk-free asset will have accrued to:

$$-\bar{S}_k^j(\omega)(1+r)^{k+1} + S_{k+1}^j(\omega).$$

Discounting this for $k+1$ periods gives a time t_{k+1} portfolio value of:

$$\bar{V}_{k+1}(\omega) = \frac{-\bar{S}_k^j(\omega)(1+r)^{k+1} + S_{k+1}^j(\omega)}{(1+r)^{k+1}} = -\bar{S}_k^j(\omega) + \bar{S}_{k+1}^j(\omega).$$

Since discounting occurs at the same rate as accumulation of value in the risk-free asset, this discounted value cannot change through the final time $T = t_n$. Therefore, for outcomes $\omega \in A$, :

$$\bar{V}_n(\omega) = -\bar{S}_k^j(\omega) + \bar{S}_{k+1}^j(\omega).$$

And note that for this strategy, $V_0 = 0$ for all outcomes, and $\bar{V}_n(\omega) = 0$ for outcomes ω that are not in A. The overall random variable $\bar{V}_n$ can therefore be written as $I_A \left(\bar{S}_{k+1}^j - \bar{S}_k^j\right)$. By our assumption, for this portfolio sequence,

$$E\left[\bar{V}_n\right] = V_0 = 0 \implies E\left[I_A \left(\bar{S}_{k+1}^j - \bar{S}_k^j\right)\right] = 0 \implies E\left[I_A \bar{S}_{k+1}^j\right] = E\left[I_A \bar{S}_k^j\right].$$

Because A was an arbitrary atom of $\mathcal{A}_k$, the last equation is sufficient for the desired martingale condition $E_Q\left[\bar{S}_{k+1}^j | \mathcal{A}_k\right] = \bar{S}_k^j$ to hold, which finishes the proof. ∎

We are now in a position to state the two Fundamental Theorems of Asset Pricing.

Theorem 2. (First Fundamental Theorem of Asset Pricing) In the market model of this section, the following two conditions are equivalent:

(a) There exist no arbitrage opportunities;
(b) There exists a martingale measure Q.

Proof. The implication (a) $\implies$ (b) is proved by an appeal to a theorem from linear algebra about linear subspaces of m-dimensional Euclidean space $\mathbb{R}^m$, which would take us farther afield than we would like to go here, so we will omit this part and direct the reader to Chapter 6 of Roman [17] for details.

For the direction (b) $\implies$ (a), let Q be the martingale measure. Suppose by contradiction that there is indeed a portfolio sequence $\boldsymbol{\theta} = (\boldsymbol{\theta}_1, \boldsymbol{\theta}_2, ..., \boldsymbol{\theta}_n)$ that gives rise to an arbitrage opportunity. Then $V_0(\boldsymbol{\theta}_1) = \bar{V}_0(\boldsymbol{\theta}_1) = 0$, but there is some time t_k for which the total discounted net gain $\bar{G}_k = \bar{V}_k(\boldsymbol{\theta}_k) - \bar{V}_0(\boldsymbol{\theta}_1) \geq 0$ for all outcomes, and it has a positive probability of being strictly positive. By the definition of martingale measure, $Q(\omega) > 0$ for all outcomes ω. But then,

$$E_Q\left[\bar{V}_k(\boldsymbol{\theta}_k)\right] = E_Q\left[\bar{V}_k(\boldsymbol{\theta}_k) - \bar{V}_0(\boldsymbol{\theta}_1)\right] = \sum_{\omega \in \Omega} \bar{G}_k(\omega)Q(\omega) > 0 = \bar{V}_0(\boldsymbol{\theta}_1).$$

This contradicts the martingale property of the discounted portfolio value process that was established in Theorem 1; hence there can be no such arbitrage opportunities in the market. ∎

Theorem 3. (Second Fundamental Theorem of Asset Pricing) Assume that there are no arbitrage trading strategies. Then the following two conditions are equivalent:

(a) There is a unique martingale measure;
(b) The model is complete (i.e. all contingent claims are attainable).

Proof. (a) $\implies$ (b). This is another case in which we need a major theorem of linear algebra about the existence of a vector perpendicular to all vectors in a proper subspace of Euclidean space. So we will not give the details but rather refer the reader to Roman, Chapter 6 [17].

To prove (b) $\implies$ (a), suppose that the market model satisfies completeness. By contradiction let us postulate the existence of two different martingale measures Q_1 and Q_2. (At least one does exist, by the assumption of no arbitrage and the First Fundamental Theorem.) We show that $Q_1(\omega) = Q_2(\omega)$ for all outcomes.

Since both of these measures are martingale measures, for any self-financing portfolio sequence, its associated discounted value process satisfies:

$$E_{Q_1}\left[\bar{V}_n\right] = \bar{V}_0 = E_{Q_2}\left[\bar{V}_n\right].$$

Multiplying both the first and last expressions by $(1 + r)^n$ we also have that $E_{Q_1}[V_n] = E_{Q_2}[V_n]$. Consider a particular but arbitrary outcome ω_0. The indicator random variable $X = I_{\{\omega_0\}}$, which pays \$1 if ω_0 occurs and zero otherwise, is a derivative that is attainable, since the market is complete. Let $\boldsymbol{\theta}$ be a portfolio sequence that culminates in the same final value V_n at time t_n as $I_{\{\omega_0\}}$ for all outcomes. Then, since the expected value of an indicator variable is the probability of the event being indicated, we can compute:

$$
\begin{aligned}
Q_1(\omega_0) &= E_{Q_1}\left[I_{\{\omega_0\}}\right] \\
&= E_{Q_1}[V_n] \\
&= E_{Q_2}[V_n] \\
&= E_{Q_2}\left[I_{\{\omega_0\}}\right] \\
&= Q_2(\omega_0).
\end{aligned}
$$

The two measures therefore agree on ω_0, and since this was an arbitrary outcome, they are the same probability measure. ∎

What does all this mean? In our market models, we are assuming that no arbitrage exists. In this case, the First Fundamental Theorem implies that there is a martingale measure Q, so that the initial value of any portfolio sequence, V_0, satisfies the equation $V_0 = \bar{V}_0(\boldsymbol{\theta}_1) = E_Q[\bar{V}_n(\boldsymbol{\theta}_n)]$. In other words, under Q, the initial value of the portfolio must equal its expected discounted final value. But what does this have to do with valuing derivatives? The Second Fundamental Theorem implies that, still assuming no arbitrage, if we can show that there is only one martingale measure, then all derivatives X are attainable; hence there is some self-financing portfolio sequence such that $X = V_n(\boldsymbol{\theta}_n)$ for all outcomes. The absence of arbitrage says that the initial value of the derivative must equal the initial value of the portfolio, since their final values agree. Plugging into the previous formula (and using V_0 to also denote the initial value of X), we have:

$$
V_0 = E_Q\left[\bar{V}_n(\boldsymbol{\theta}_n)\right] = E_Q[\bar{X}] = E_Q\left[(1 + r)^{-n}X\right]. \tag{3.78}
$$

Under these conditions (no arbitrage and a unique martingale measure), we can value a derivative at time 0 as the expected present value of its claim value at time t_n, which is the goal we sought.

Example 5. Returning to Example 4 concerning the two independent binomial branch assets, let us simplify by considering a problem in two time periods. The parameters of the two processes are the same: $s_0^1 = \$40$, $b_1 = .05$, $a_1 = .01$, $s_0^2 = \$80$, $b_2 = .08$, $a_2 = -.02$, and the risk-free rate is $r = .02$. We computed the risk-neutral probabilities as $q_1 = 1/4$ and $q_2 = 2/5$. The trees are shown in Figure 3.20. Assuming no arbitrage and also that independence of the two assets is required, find the initial value of a derivative X that returns $S_2^2 - S_2^1$ at time 2.

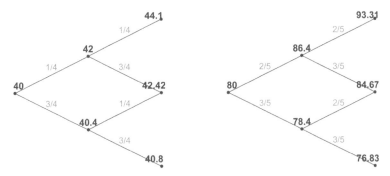

FIGURE 3.20 Two period, independent, dual asset binomial branch model.

Solution. As we saw in Example 4, there is only one choice of q_1 $(1/4)$ and one choice of q_2 $(2/5)$ that make the two processes martingales individually, and the independence assumption implies that there is only one way of combining these into joint probabilities on paths of the two processes to make a probability measure on the joint sample space $\Omega = \left\{\left(\left(40, s_1^1, s_2^1\right), \left(80, s_1^2, s_2^2\right)\right)\right\}$. So there is a unique martingale measure in this market which, by the discussion prior to this example, means that the initial value of this derivative can be computed as the expected present value:

$$E_Q\left[(1.02)^{-2}\left(S_2^2 - S_2^1\right)\right].$$

By linearity, this is rather easy to do, as follows:

$$
\begin{aligned}
E_Q\left[(1.02)^{-2}\left(S_2^2 - S_2^1\right)\right] &= (1.02)^{-2}\left(E_Q\left[S_2^2\right] - E_Q\left[S_2^1\right]\right) \\
&= (1.02)^{-2}\left(\tfrac{4}{25}(93.312) + 2\cdot\tfrac{6}{25}(84.672) + \tfrac{9}{25}(76.832)\right. \\
&\quad \left. -\tfrac{1}{16}(44.1) + 2\cdot\tfrac{3}{16}(42.42) + \tfrac{9}{16}(40.804)\right) \\
&= 40.
\end{aligned}
$$

It should not come as too big of a surprise that the value came out to be $40 = 80 - 40 = s_0^2 - s_0^1$, since the discounted processes are both martingales and we were computing the present value of the difference in expected prices at time 2. ∎

The approach to valuation suggested by the two Fundamental Theorems of Asset Pricing will prove very useful in the next section, where we consider some derivatives that are not as simple as European calls and puts.

Exercises 3.4

1. Show that for any two portfolio vectors $\boldsymbol{\theta}$ and $\boldsymbol{\delta}$ and constants c and d,

$$V_k(c\cdot\boldsymbol{\theta} + d\cdot\boldsymbol{\delta}) = c\cdot V_k(\boldsymbol{\theta}) + d\cdot V_k(\boldsymbol{\delta}).$$

2. As in Exercise 12 of Section 3.1, let there be one risky asset following a binomial branch model with down and up rates $a = .01$, $b = .10$, and initial price 350. There is also a risk-free asset whose value is 1 at time 0, and whose per period rate of return is $r = .03$. Compute the three-period price tree of the risky asset and, as in Example 1, identify the atoms of the algebras in the history generated by the price process. Also, compute the sequence of portfolios that an investor Marla uses if she plans to purchase 20 shares of the risky asset initially, and keep them until the price reaches 400, if ever, at which point she sells out and invests the proceeds in the risk-free asset. Verify that this portfolio sequence is self-financing.

3. An investor begins with $2000 which she splits evenly between a risky asset initially priced at $50 per share and a non-risky bank account. The account earns 1% per time period. She trades at times 2 and 4 as indicated in the table below, using her bank account to fund the transaction. If her trading strategy is self-financing, what is the value of her bank account at each time?

time	shares owned	risky price
0	20	$50
1	20	$48
2	25	$45
3	25	$47
4	30	$46
5	30	$48

4. In Example 1, show that for the risk-neutral probability q, $E_q[V_2] = (1+r)^2 V_0$.

5. Suppose that there is a binomial branch risky asset with initial price $25 and up and down rates $b = .08$, $a = .05$. If the risk-free rate is $r = .03$, find a self-financing portfolio sequence that results in arbitrage, showing clearly where the arbitrage enters in.

6. Extend Example 3 to a three-period model with the same parameters. Find a replicating portfolio sequence for the derivative which again returns the largest value along the path, and compute all intermediate values of the derivative.

7. Prove that if $\boldsymbol{\theta} = (\boldsymbol{\theta}_1, \boldsymbol{\theta}_2, ..., \boldsymbol{\theta}_n)$ and $\Gamma = (\boldsymbol{\gamma}_1, \boldsymbol{\gamma}_2, ..., \boldsymbol{\gamma}_n)$ are two self-financing portfolio sequences, then the component-wise sum $\Phi = \boldsymbol{\theta} + \Gamma$ is also self-financing.

8. Let a market consist of a risk-free asset scaled to an initial value of 1, with rate $r = .01$ per period, and the risky asset of Example 2 of Section 3.1. This asset was priced at 20 initially, and moved as a binomial branch process with parameters $b = .04$ and $a = -.04$. Find a replicating portfolio sequence for a

derivative X that returns the smallest price value of the risky asset along the path it traverses through time 3, and use this to compute the initial arbitrage-free value of X.

9. Consider a two-period, two-risky asset model in which risky asset 1 has initial price 50 and up and down rates $b_1 = .04$, $a_1 = -.01$, and asset 2 has initial price 30 and up and down rates $b_2 = .08$, $a_2 = -.04$. The risk-free asset, as usual, begins with value 1 and has rate of return $r = .02$ per period. List all outcomes in the sample space and all atoms of all algebras in the history generated jointly by the two price processes, and construct a martingale measure on the sample space of joint paths consistent with independence of the assets, computing the probability that it assigns to each outcome.

10. Consider a market with a risk-free asset with rate $r = .01$ per period, and one risky asset with initial value 100, whose price changes arithmetically in each period. Specifically, it either goes up by 4 or down by 2 monetary units. Find a martingale measure on this market, with respect to the history generated by the risky asset process. Find the initial value of a European call option with expiration time 3 and strike price 105.

11. A risky asset with initial price 50 moves as in the figure, following the binomial branch assumptions with up rate $b = .05$ and down rate $a = -.01$, except that at time 1 there is also the possibility that the price stays the same. The risk-free rate is $r = .02$. Find a martingale measure relative to the history generated by the risky asset process and decide whether it is unique.

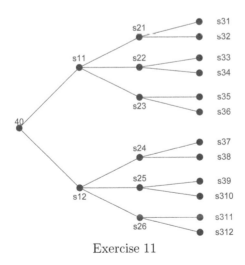

Exercise 11

12. Two risky assets move in two time periods according to the diagrams below. The two processes are defined on the same product sample space by means of the device of letting outcomes be pairs $\omega = (\omega_1, \omega_2)$, where ω_1 is the

path followed by the first asset and ω_2 is the path followed by the second. For example, the outcome $((U, U), (U, D))$ refers to the case where the first asset moves to 12 and 13.2 while the second moves to 22 and then 19.8. There will be 16 outcomes in this sample space. Assume for simplicity a risk-free asset that has zero rate of return. Does the market consisting of these assets have a martingale measure? If so, find it, and decide if it is unique. If the martingale measure exists and is not unique, find an additional assumption to make it so, and use the measure to value a derivative X that pays the difference of the two final values minus a threshold of 10, $S_2^2 - S_2^1 - 10$, or zero if that is negative.

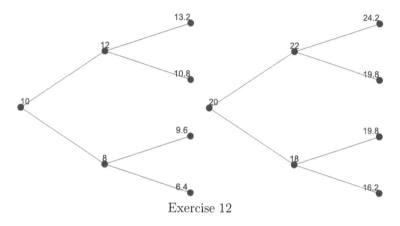

Exercise 12

13. A binomial branch process moves in 4 periods, with initial value 200, up rate $b = .04$, down rate $a = -.02$, and the risk-free rate is $r = .01$. A derivative X is defined as below; it is worth nothing unless the price ever reaches a cutoff value of 214 or above, and if it does, it is worth a fixed constant amount. Find the initial value of this derivative.

$$X = \begin{cases} 10 & \text{if } \max\{S_1, S_2, S_3, S_4\} \geq 214 \\ 0 & \text{otherwise.} \end{cases}$$

14. Suppose that a market has a risk-free asset with initial value 1 monetary unit and rate of return $r = .02$ per period, and three independent risky assets following binomial branch models. The up and down rates for these three are: $b_1 = .04, a_1 = 0$; $b_2 = .06, a_2 = -.02$; and $b_3 = .08, a_1 = -.04$. The initial prices are \$40, \$50, and \$60 respectively for the assets. A three-asset derivative X is defined as the average of the three final prices at time 3. Argue that there is a unique martingale measure relative to the history generated by the three risky assets together, and find the arbitrage free initial value of X.

15. In earlier examples and problems with multiple risky assets, we have made the simplifying assumption that the assets were independent. Here we ask the question: can a unique martingale measure be found in a 2-period, 2 asset

binomial branch model in which the second asset is required to have fixed conditional probabilities of transition given the first asset? For specificity, suppose that the risk-free rate is $r = .04$, the initial prices of both assets are 50, and the up and down rates for the two risky asset price processes are $b_1 = .08, a_1 = -.02; b_2 = .1, a_2 = -.04$. Suppose also that we require the conditional probability that asset 2 goes up given that asset 1 went up to be 2/3 on each transition. Find a martingale measure, and show that it is unique under the requirements of the last sentence.

3.5 Valuation of Non-Vanilla Options

Earlier in this chapter we developed effective computational techniques, namely chaining and martingale valuation, to value derivatives whose claim values depended only on the final values of the underlying assets on which the derivatives were based. For plain "vanilla" options such as European calls and puts, these techniques are sufficient. With the general results of Section 3.4 in hand, which permit us in particular to apply martingale valuation to path-dependent options and to multiple asset options, it is time to study a few of these kinds of derivative assets that are particularly common in markets.

Our first subsection shows the properties of **American options** in discrete time, and also briefly treats the related class of options called **Bermudan options** which can be executed at a few specified times prior to their termination time. Next, we will consider **barrier options**, which either come into effect or become void when the underlying asset passes a barrier. In the third subsection we show how to value **Asian options**, whose claim values depend upon the average price that the underlying asset took on throughout its motion. Finally, we will follow up on some examples and problems in Section 3.4 with a subsection on **two-asset options**.

Parts of this section draw heavily on the material in the prequel to this text, Hastings, Section 5.4 [11]. Again there are integer transition times 1, 2, 3, ... , and we will denote underlying asset price processes by $S_0 = s_0$ (known), $S_1, S_2, S_3, ...$. The risk-free asset that is present in the market will have a constant per period rate denoted by r.

3.5.1 American and Bermudan Options

Recall that, in contrast to European-style options, an American option may be executed at the discretion of its holder at any time prior to the expiration time n. This additional freedom appears to give American options extra value, and in the case of put options, it does. A surprising fact, though, is that American and European call options have the same initial value, as argued in the next theorem. (Both of the next two theorems are essentially edits of Section 5.4, Theorem 1 of Hastings [11].)

Theorem 1. Let X be a European call option expiring at time n with strike price E, hence $X = \max\{S_n - E, 0\}$. Let V_X denote its initial value. Let Y be an American call option with the same expiration time n and strike price E, which, if executed at time m, returns a value $Y = \max\{S_m - E, 0\}$. Denote its initial value by V_Y. Then under the assumption of no arbitrage:

$$V_X = V_Y. \tag{3.79}$$

Proof. To prove that $V_X \leq V_Y$, we argue by contradiction. If the contrary inequality $V_X > V_Y$ were true, at time 0 an investor could short one European call and buy one American call. Invest the difference $V_X - V_Y$ in the risk-free asset growing at rate r per period. The initial value of this strategy is $-V_X + V_Y + (V_X - V_Y) = 0$. Do not execute the American option prior to the final time n. At that time, the claim value of the option portion of the investor's portfolio has value:

$$Y - X = \max\{S_n - E, 0\} - \max\{S_n - E, 0\} = 0.$$

Meanwhile, the amount in the risk-free asset has grown to $(V_X - V_Y)(1+r)^n > 0$ with certainty. This is arbitrage, which we have assumed does not exist in this market; hence it must be that $V_X \leq V_Y$.

For the reverse inequality $V_X \geq V_Y$, we again work by contradiction, supposing that the American option has greater initial value, that is, $V_Y > V_X$. This inequality allows us to set up an arbitrage opportunity as follows. At time 0, short one American call, buy one European call, and invest the difference $V_Y - V_X$ in the risk-free asset. This portfolio has initial value $V_X - V_Y + (V_Y - V_X) = 0$. In being short, that is, issuing, the American call, our arbitraging investor does not have control over when the American call will be executed. If the American call is not executed prior to the expiration time n, then the final value of the portfolio of options is as above:

$$X - Y = \max\{S_n - E, 0\} - \max\{S_n - E, 0\} = 0.$$

In this case, the investment in the risk-free asset grows to $(V_Y - V_X)(1+r)^n > 0$ at time n, which is arbitrage. So the only challenge is if the American call is executed at a time $m < n$. To handle this case, the arbitrageur will short (that is, borrow) a share of the underlying asset at price S_m, deliver the share, or its monetary value, to the American call partner to fulfill that obligation, and invest the amount of E that was received from that person in the risk-free asset for $n - m$ periods. This sum will accrue to a value of $E(1+r)^{n-m} > E$ at time n. The arbitrageur can use this money to execute the European option, buying a share of the asset at a price E at time n (or market price S_n if that is cheaper) in order to resolve the short position. In this case, the investor has at time n not only the positive amount $(V_Y - V_X)(1+r)^n$ that has grown from time 0, but also the additional positive amount $E(1+r)^{n-m} - \min(S_n, E)$. This is an arbitrage strategy, which we forbid. Hence, it must be that $V_X \geq V_Y$. Since both inequalities are true, $V_X = V_Y$, which completes the proof of Theorem 1. ∎

Shortly we will work through a numerical example that confirms that the initial values for European and analogous American call options do come out the same. However, for put options the story is different.

Theorem 2. Let X be a European put option expiring at time n with strike price E; hence $X = \max\{E - S_n, 0\}$. Denote its initial value by V_X. Let Y be an American put option with the same expiration time n and strike price E, which, if executed at time $m < n$ returns a value $Y = \max\{E - S_m, 0\}$. Its initial value is denoted by V_Y. Then under the assumption of no arbitrage:

$$V_X \leq V_Y. \tag{3.80}$$

Proof. Suppose on the contrary that $V_X > V_Y$. Consider the portfolio that at time 0 is short one European put, long one American put, and holds $V_X - V_Y$ in the risk-free asset. This portfolio has initial value:

$$V_0 = -V_X + V_Y + (V_X - V_Y) = 0.$$

Maintain this portfolio until the expiration time n. The partner in the sale of the European put cannot act until this time. The final value of the portfolio is:

$$
\begin{aligned}
Y - X + (V_X - V_Y)(1+r)^n &= \max\{E - S_n, 0\} - \max\{E - S_n, 0\} \\
&\quad + (V_X - V_Y)(1+r)^n \\
&= 0 + (V_X - V_Y)(1+r)^n > 0.
\end{aligned}
$$

Without any initial net outlay on the investor's part, there is certainty of a positive ending value. This contradicts the no arbitrage assumption, so it must be that $V_X \leq V_Y$. ∎

Example 2 below shows that, as is typically the case, the European put value V_X can be strictly less than the American put value V_Y.

Remark. In the single-asset, no arbitrage, binomial branch world the work of the last section certainly applies, so that to value American options which are path dependent random variables Y, you can try to compute:

$$V_0 = E_Q\left[(1+r)^{-n} Y\right],$$

but it is not readily apparent how to do this. The way out of this conundrum is to think of the replicating portfolio for the option, which exists by the second Fundamental Theorem of Asset Pricing. Its discounted value process $\bar{V}_0, \bar{V}_1, ..., \bar{V}_n$ forms a martingale; in particular for each time l:

$$V_l = E_Q \left[(1+r)^{-1} V_{l+1} | \mathcal{A}_l \right]. \tag{3.81}$$

If the time l value of the derivative differs at all from the value of the portfolio, then arbitrage would enter in. This means that we can again use chaining to step backward from time n to value the derivative. The only difference for American options is that the investor chooses optimally which is greater: the immediate value of executing the derivative or the expected discounted value of waiting for one more time period. The procedure would be to initialize the claim values of the option at the final time n, and then to step back one period at a time to value the option at interior nodes of the option tree until reaching time 0. For an American call, the algorithm would be:

$$v_{nj} = \max \left(s_{n,j} - E, 0 \right); \tag{3.82}$$

$$v_{lj} = \max \left(\max \left(s_{l,j} - E, 0 \right), (1+r)^{-1} \left(q \cdot v_{l+1,j} + (1-q) \cdot v_{l+1,j+1} \right) \right),$$
$$l = n-1, n-2, ..., 1, 0. \tag{3.83}$$

As usual, q denotes the risk-neutral probability. The first quantity in the outer max is the value of executing the call immediately, and the second is the expected discounted optimal value of waiting for another period. Similarly, valuing an American put option would be done by:

$$v_{nj} = \max \left(E - s_{n,j}, 0 \right); \tag{3.84}$$

$$v_{lj} = \max \left(\max \left(E - s_{l,j}, 0 \right), (1+r)^{-1} \left(q \cdot v_{l+1,j} + (1-q) \cdot v_{l+1,j+1} \right) \right),$$
$$l = n-1, n-2, ..., 1, 0. \tag{3.85}$$

This process is illustrated in the next examples.

Example 1. Consider a market with a single risky asset following a binomial branch process with initial price $S_0 = 100$, and down and up rates $a = 0, b = .1$. Assume that the risk-free rate is $r = .04$. Find the initial value of an American call option expiring at time 3, with exercise price $E = 115$. Verify that it has the same initial value as a corresponding European call.

Solution. The risk-neutral probability that describes the martingale measure on the sample space of paths is:

$$q = \frac{r-a}{b-a} = \frac{.04 - 0}{.10 - 0} = .4.$$

The asset values, using the same notation for tree nodes as in Section 3.1, are:

level 0 : $S_0 = 100$;
level 1 : $s_{11} = 100(1.1) = 110$; $s_{12} = 100(1) = 100$;
level 2 : $s_{21} = 100(1.1)^2 = 121$; $s_{22} = 100(1.1)(1) = 110$;
            $s_{23} = 100(1)^2 = 100$;
level 3 : $s_{31} = 100(1.1)^3 = 133.1$; $s_{32} = 100(1.1)^2(1) = 121$;
            $s_{33} = 100(1.1)(1)^2 = 110$; $s_{34} = 100(1)^3 = 100$.

The asset tree, together with the underlying derivative tree, is shown in Figure 3.21.

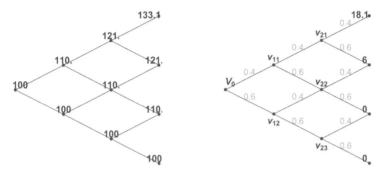

FIGURE 3.21 3-step binomial branch process and call option with martingale probabilities.

Because the strike price is $E = 115$, at level 3 only the top two nodes have non-zero claim values, namely $133.1 - 115 = 18.1$ and $121 - 115 = 6$, as shown in the figure. The European version of this call option would have value:

$$
\begin{aligned}
V_X &= (1+r)^{-3} E_q \left[\max \{ S_3 - E, 0 \} \right] \\
    &= (1.04)^{-3} \left((.4)^3 \cdot 18.1 + 3(.4)^2(.6) \cdot 6 \right) \\
    &= 2.566.
\end{aligned}
$$

For the American version of the option, from the remark above, we may chain backwards in the option tree, using the algorithm:

$$
v_{lj} = \max \left(\max \left(s_{l,j} - 115, 0 \right), (1.04)^{-1} \left(.4 \cdot v_{l+1,j} + .6 \cdot v_{l+1,j+1} \right) \right),
$$

that is, compare the value of immediate execution to the expected present value of waiting until the next time step, and choose the larger. The algorithm initializes with the level 3 values $v_{31} = 18.1$, $v_{32} = 6$, $v_{33} = 0$, and $v_{34} = 0$. At level 2 we can calculate:

$$
\begin{aligned}
v_{21} &= \max \left(\max(121 - 115, 0), (1.04)^{-1}(.4 \cdot (18.1) + .6 \cdot (6)) \right) \\
       &= \max(6, 10.4231) \\
       &= 10.4231;
\end{aligned}
$$

$$
\begin{aligned}
v_{22} &= \max\left(\max(110 - 115, 0), (1.04)^{-1}(.4 \cdot (6) + .6 \cdot (0))\right) \\
&= \max(0, 2.30769) \\
&= 2.30769;
\end{aligned}
$$

$$
\begin{aligned}
v_{23} &= \max\left(\max(100 - 115, 0), (1.04)^{-1}(.4 \cdot (0) + .6 \cdot (0))\right) \\
&= \max(0, 0) \\
&= 0.
\end{aligned}
$$

In all cases the action of waiting is more valuable than executing the option immediately. Moving back to time 1:

$$
\begin{aligned}
v_{11} &= \max\left(\max(110 - 115, 0), (1.04)^{-1}(.4 \cdot (10.4231) + .6 \cdot (2.30769))\right) \\
&= \max(0, 5.34024) \\
&= 5.34024;
\end{aligned}
$$

$$
\begin{aligned}
v_{12} &= \max\left(\max(100 - 115, 0), (1.04)^{-1}(.4 \cdot (2.30769) + .6 \cdot 0)\right) \\
&= \max(0, .887574) \\
&= .887574.
\end{aligned}
$$

Again it is not optimal for the investor to exercise the option early. Finally, for time 0:

$$
\begin{aligned}
V_0 &= \max\left(\max(100 - 115, 0), (1.04)^{-1}(.4 \cdot (5.34024) + .6 \cdot (.887574))\right) \\
&= \max(0, 2.566) \\
&= 2.566.
\end{aligned}
$$

The holder of the American option also will not execute at time 0. We have now confirmed that the initial value V_0 of the American option matches the European option value of 2.566. ■

Example 2. Now suppose that we have a market with one risky asset following a binomial branch model with initial price $S_0 = 40$, and parameters $b = .03$, $a = -.01$, and a risk-free asset whose rate is $r = .01$. Find the initial values of both a European and an American put option that expires at time 3, with strike price 41. Is it ever optimal for the option holder to execute the option early?

Solution. We calculate the risk-neutral probability as:

$$
q = \frac{r - a}{b - a} = \frac{.01 - (-.01)}{.03 - (-.01)} = \frac{.02}{.04} = .5.
$$

The stock price tree nodes are below, and are shown in Figure 3.22:

level 0 : $S_0 = 40$;
level 1 : $s_{11} = 40(1.03) = 41.2$; $s_{12} = 40(.99) = 39.6$;
level 2 : $s_{21} = 40(1.03)^2 = 42.436$; $s_{22} = 40(1.03)(.99) = 40.788$;
           $s_{23} = 40(.99)^2 = 39.204$;
level 3 : $s_{31} = 40(1.03)^3 = 43.7091$; $s_{32} = 40(1.03)^2(.99) = 42.0116$;
           $s_{33} = 40(1.03)(.99)^2 = 40.3801$; $s_{34} = 40(.99)^3 = 38.812$.

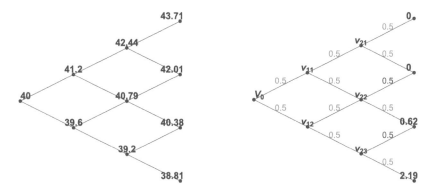

FIGURE 3.22 3-step binomial branch process and put option with martingale probabilities.

Since the strike price is 41, the put option has no value at the top two nodes v_{31} and v_{32} of the third level, and at the lower two nodes its values are:

$$v_{33} - \max(41 - 40.3801, 0) = .6199; \quad v_{34} = \max(41 - 38.812, 0) - 2.188.$$

This allows us to calculate the initial value of the European put:

$$\begin{aligned}
V_X &= (1+r)^{-3} E_q \left[\max\left\{E - S_3, 0\right\}\right] \\
&= (1.01)^{-3} \left(3(.5)^3 \cdot .6199 + (.5)^3 \cdot 2.188\right) \\
&= .4911.
\end{aligned}$$

Chaining back to time 2 we have:

$$\begin{aligned}
v_{21} &= \max\left(\max(41 - 42.436, 0), (1.01)^{-1}(.5 \cdot 0 + .5 \cdot 0)\right) \\
&= \max(0, 0) \\
&= 0; \\
v_{22} &= \max\left(\max(41 - 40.788, 0), (1.01)^{-1}(.5 \cdot 0 + .5 \cdot (.6199))\right) \\
&= \max(.212, .30688) \\
&= .30688; \\
v_{23} &= \max\left(\max(41 - 39.204, 0), (1.01)^{-1}(.5 \cdot (.6199) + .5 \cdot (2.188))\right) \\
&= \max(1.796, 1.390) \\
&= 1.796.
\end{aligned}$$

So we find that at node v_{23} it is better to exercise the option immediately to receive 1.796 than to wait and expect to receive a present value of 1.390. Next, we chain back to time 1.

$$
\begin{aligned}
v_{11} &= \max\left(\max(41-41.2,0),(1.01)^{-1}(.5\cdot 0+.5\cdot(.30688))\right) \\
&= \max(0,.15192) \\
&= .15192; \\
v_{12} &= \max\left(\max(41-39.6,0),(1.01)^{-1}(.5\cdot(.30688)+.5\cdot(1.796))\right) \\
&= \max(1.4,1.041) \\
&= 1.4.
\end{aligned}
$$

At the second node v_{12} at level 1, it is again the value of immediate exercise that prevails. For the initial value, we compute:

$$
\begin{aligned}
V_Y &= \max\left(\max(41-40,0),(1.01)^{-1}(.5\cdot(.15192)+.5\cdot(1.4))\right) \\
&= \max(1,.7683) \\
&= 1.
\end{aligned}
$$

At the initial node it is also optimal to exercise the option immediately, which makes the overall value of this American put equal to $V_Y = 1$, which is strictly greater than the value of the European analog of the put option. Oddly, under this configuration of coefficients, such a put option, amounting to an even exchange of \$1 between buyer and seller at time 0, would probably never be offered. ∎

A **Bermudan option** is a variation on an American option, in which the investor is allowed to exercise the option early, but only in certain periods determined by advance agreement between the option holder and the option issuer. Its value can be found similarly to that of a vanilla European-style option, by chaining in the ordinary European way in periods when the investor cannot execute the option, and in the American way (formulas (3.82)-(3.83) or (3.84)-(3.85)) when the investor must choose the better alternative: to execute the option or to wait. This is illustrated in Example 3.

Example 3. Suppose that an underlying asset has initial price 30 and satisfies the binomial branch model with parameters $b = .04$ and $a = -.02$. The risk-free rate is $r = .01$. Compute the initial value of a four-period Bermudan call option with strike price 32 which can be exercised only at times 2 and 4.

Solution. The risk-neutral probability is:

$$
q = \frac{r-a}{b-a} = \frac{.01-(-.02)}{.04-(-.02)} = .5.
$$

The asset price nodes are computed below and shown in Figure 3.23:

$S_0 = 30$;

$s_{11} = 30(1.04) = 31.2$; $\qquad\qquad$ $s_{12} = 30(.98) = 29.4$;

$s_{21} = 30(1.04)^2 = 32.448$; $\qquad$ $s_{22} = 30(1.04)(.98) = 30.576$;

$s_{23} = 30(.98)^2 = 28.812$;

$s_{31} = 30(1.04)^3 = 33.7459$; $\qquad$ $s_{32} = 30(1.04)^2(.98) = 31.799$;

$s_{33} = 30(1.04)(.98)^2 = 29.9645$; $\quad$ $s_{34} = 30(.98)^3 = 28.2358$;

$s_{41} = 30(1.04)^4 = 35.0958$; $\qquad$ $s_{42} = 30(1.04)^3(.98) = 33.071$;

$s_{43} = 30(1.04)^2(.98)^2 = 31.1631$; $\quad$ $s_{44} = 30(1.04)(.98)^3 = 29.3652$;

$s_{45} = 30(.98)^4 = 27.671$.

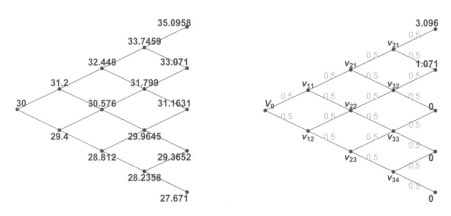

FIGURE 3.23 4-step asset tree and Bermudan call tree.

At time 4, only the top two nodes exceed the strike price, so that:

$$v_{41} = 35.0958 - 32 = 3.0958; \; v_{42} = 33.071 - 32 = 1.071;$$
$$v_{43} = 0 = v_{44} = v_{45}.$$

The investor cannot choose to execute the option at time 3, so the European chaining relationship $v_{3j} = (1+r)^{-1} (q \cdot v_{4,j} + (1-q) \cdot v_{4,j+1})$ is used to value the level 3 nodes from the level 4 option claim values. This gives:

$$
\begin{aligned}
v_{31} &= (1.01)^{-1} (.5 \cdot v_{41} + .5 \cdot v_{42}) \\
&= (1.01)^{-1}(.5 \cdot (3.09576) + .5 \cdot (1.071)) = 2.06275; \\
v_{32} &= (1.01)^{-1} (.5 \cdot v_{42} + .5 \cdot v_{43}) \\
&= (1.01)^{-1}(.5 \cdot (1.071) + .5 \cdot 0) = .530199; \\
v_{33} &= (1.01)^{-1} (.5 \cdot v_{43} + .5 \cdot v_{44}) = 0; \\
v_{34} &= (1.01)^{-1} (.5 \cdot v_{44} + .5 \cdot v_{45}) = 0.
\end{aligned}
$$

Time 2 is a time at which the Bermudan option can be executed, so in chaining, we must treat the option as an American option. Only the first node at level 2 has a value exceeding the strike price $E = 32$, however, so this is the only node at which a decision is necessary. As the computation below shows, the option is not executed at node 2,1, since $32.448 - 32 = .448$ is less than the chaining value of 1.28364.

$$
\begin{aligned}
v_{21} &= \max\left(\max(32.448 - 32, 0), (1.01)^{-1}(.5 \cdot (2.06275) + .5 \cdot (.530199))\right) \\
&= \max(.448, 1.28364) \\
&= 1.28364 \\
v_{22} &= \max\left(\max(30.576 - 32, 0), (1.01)^{-1}(.5 \cdot (.530199) + .5 \cdot 0)\right) = .262475; \\
v_{23} &= \max\left(\max(28.812 - 32, 0), (1.01)^{-1}(.5 \cdot 0 + .5 \cdot 0)\right) = 0.
\end{aligned}
$$

Since the Bermudan call option is not executable at time 1 and 0, we use European chaining to pull back to those levels:

$$
\begin{aligned}
v_{11} &= (1.01)^{-1} (.5 \cdot v_{21} + .5 \cdot v_{22}) \\
&= (1.01)^{-1}(.5 \cdot (1.28364) + .5 \cdot (.262475)) = .765403; \\
v_{12} &= (1.01)^{-1} (.5 \cdot v_{22} + .5 \cdot v_{23}) \\
&= (1.01)^{-1}(.5 \cdot (.262475) + .5 \cdot 0) = .129938; \\
V_0 &= (1.01)^{-1} (.5 \cdot v_{11} + .5 \cdot v_{12}) \\
&= (1.01)^{-1}(.5 \cdot (.765403) + .5 \cdot (.129938)) = .44324.
\end{aligned}
$$

Notice that it was never necessary to execute the option early. By no coincidence, the initial value of a plain European call option on this asset with the same strike price is:

$$
\begin{aligned}
E_q[(1.01)^{-4} \max(S_4 - 32, 0)] &= (1.01)^{-4}((.5)^4(3.0958) + 4 \cdot (.5)^3(.5)(1.071)) \\
&= .44324,
\end{aligned}
$$

which is the same as the value V_0 for the Bermudan option. Exercise 6 asks you to argue that for Bermudan call options with a single possible execution time prior to time n, the initial value is the same as the initial value of the European call option with the same parameters. ∎

In Section 3.1 we mentioned that the chaining procedure for European options was easily implemented on a spreadsheet. That is also true for American options. The only change necessary in the setup is to replace the formulas that calculate the option tree values by (3.83) or (3.85) according to whether an American call or put option is to be valued.

3.5.2   Barrier Options

Another category of options on a single asset is characterized by one of two conditions: either the option never comes into value unless its underlying asset crosses a **barrier** value, or the option permanently expires worthless if the asset crosses a barrier. Hence, barrier options come in two flavors: **knockin options** (down-and-in, or up-and-in), which achieve positive value only if at some point in their motion the asset price moves down below or up above the barrier; and **knockout options** (down-and-out, or up-and-out), which become permanently valueless if the asset price process moves down below the barrier or up above it. The possibility that the option could end up worthless ought to generally make it cheaper to buy than a corresponding European option, an attraction to the buyer, but it can be so constructed also to limit downside risk for the issuer, which implies that a market for these derivatives can exist (and does).

To see this more clearly, recall that a put option is very bad for the issuer of the option if the asset price sinks drastically, for then the option holder gains $E - S_n$, where E is the exercise price, and the option issuer correspondingly loses that amount. But if there is a barrier value B somewhere less than E and the option is defined as a down-and-out put option that becomes worthless if the asset price goes below B, then the issuer is protected from a disastrous crash of the underlying asset. For final prices S_n such that $B < S_n < E$, the option holder can execute the put, resulting in a loss of $E - S_n$ to the issuer, but that loss can be no more than $E - B$. If S_n falls below the barrier B, then the option becomes worthless. Thus, the holder of the barrier put option is gambling that the price of the asset will fall in the market, but not drastically, so that he is able to exercise the put at a profit.

As with American options, the claim value X of a barrier option will be path dependent, because it is affected by the asset price prior to the termination time, but it is a well-defined random variable on the space of paths. The same argument that we used in the last subsection, resting on the Fundamental Theorems, martingale valuation, and replicating portfolios, applies here. The initial value of a barrier option is the expected discounted value under the martingale measure of the claim value, and can be found by chaining. The only difference is that for any paths that move outside the barrier for a knockout option, X will have the value 0, and for a knockin option, for any paths that never move into the activation region determined by the barrier, X has value 0.

We show how to compute values of up-and-out barrier options first.

Example 4. Consider a market with a risk-free asset with rate of return $r = .01$ and one risky, binomial branch asset such that $S_0 = 60$, $b = .1$, and $a = -.05$. Compute the value of a three-period, up-and-out call option with strike price $E = 66$ and barrier $B = 70$. Compare this value to the value of an ordinary European call option on this asset with the same strike price.

Solution. The given rates of return imply that the risk-neutral probability is:

$$q = \frac{.01 - (-.05)}{.1 - (-.05)} = \frac{6}{15} = .4.$$

The risky asset nodes are:

$S_0 = 60$;

$s_{11} = 60(1.1) = 66$; $s_{12} = 60(.95) = 57$;

$s_{21} = 60(1.1)^2 = 72.6$; $s_{22} = 60(1.1)(.95) = 62.7$;

$s_{23} = 60(.95)^2 = 54.15$;

$s_{31} = 60(1.1)^3 = 79.86$; $s_{32} = 60(1.1)^2(.95) = 68.97$;

$s_{33} = 60(1.1)(.95)^2 = 59.565$; $s_{34} = 60(.95)^3 = 51.4425$.

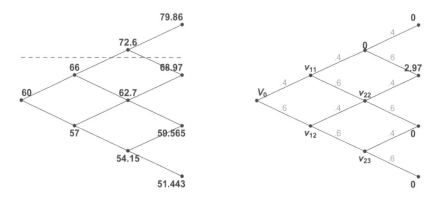

FIGURE 3.24 Asset tree with barrier at 70, and up-and-out barrier option tree.

The price tree for the underlying asset is in Figure 3.24, with a dashed line indicating the barrier at 70, so that if the asset ever goes to node $s_{21} = 72.6$ or $s_{31} = 79.86$, the barrier option is worthless. On the right of the figure we emphasize this by marking nodes v_{21} and v_{31} with zeros. Since the option is otherwise a call of European type, and node $s_{32} = 68.97$ is the only one at level 3 whose value exceeds the strike price of 66, it is the only level 3 node with a positive claim value of $68.97 - 66 = 2.97$.

There are a couple of ways to handle the valuation. The first is to appeal directly to martingale valuation. We have a small number of paths to deal with, because the two paths corresponding to 3 straight up transitions, and 2 ups followed by a down, give no claim value. As a matter of fact, the only two paths that do give a positive value are the up, down, up path and the down, up, up path. Therefore:

$$
\begin{aligned}
V_0 &= E_Q\left[(1.01)^{-3}X\right] \\
&= (1.01)^{-3}(.4 \cdot .6 \cdot .4 \cdot (2.97) + .6 \cdot .4 \cdot .4 \cdot (2.97)) \\
&= .553469.
\end{aligned}
$$

We can also try chaining, bearing in mind that we must force v_{21} to 0 instead of computing it using the chaining formula:

$$
\begin{aligned}
v_{21} &= 0; \\
v_{22} &= (1+r)^{-1} \left(q \cdot v_{32} + (1-q)v_{33} \right) \\
&= (1.01)^{-1}((.4) \cdot (2.97) + (.6) \cdot (0)) \\
&= 1.17624; \\
v_{23} &= (1+r)^{-1} \left(q \cdot v_{33} + (1-q)v_{34} \right) \\
&= (1.01)^{-1}((.4) \cdot (0) + (.6)(0)) \\
&= 0; \\
v_{11} &= (1+r)^{-1} \left(q \cdot v_{21} + (1-q)v_{22} \right) \\
&= (1.01)^{-1}((.4) \cdot (0) + (.6) \cdot (1.17624)) \\
&= .698755; \\
v_{12} &= (1+r)^{-1} \left(q \cdot v_{22} + (1-q)v_{23} \right) \\
&= (1.01)^{-1}((.4) \cdot (1.17624) + (.6)(0)) \\
&= .465837; \\
V_0 &= (1+r)^{-1} \left(q \cdot v_{11} + (1-q)v_{12} \right) \\
&= (1.01)^{-1}((.4) \cdot (.698755) + (.6) \cdot (.465837)) \\
&= .553469.
\end{aligned}
$$

The chaining value does match the value computed by finding the overall discounted expectation. As far as the value of the ordinary European option without the barrier, since node v_{31} would now give a claim value of $79.86 - 66 = 13.86$ the initial value would be:

$$
V_0 = E_Q \left[(1.01)^{-3} X \right] = (1.01)^{-3} \left(.4^3 \cdot (13.86) + 3 \cdot .4^2 \cdot .6 \cdot (2.97) \right) = 1.69116.
$$

This is substantially higher than the initial value of the up-and-out call, illustrating the potential initial savings for the investor who is interested in a call option, but at the cost of limiting the upside case of 13.86 if three straight ups happen. ∎

Now we illustrate the valuation of a down-and-in barrier option. Other option cases are in the exercises.

Example 5. In the same market as in Example 4, suppose that there is a down-and-in barrier put option on the asset, which activates only if the asset ever falls strictly below the initial price of 60. The strike price on the put is also 60. Find the initial value of the option and compare it to the initial value of a simple European put option with the strike price 60.

Solution. The asset price tree is just as before, shown in Figure 3.25 with a dashed line indicating the barrier; if the asset ever moves below the line it is active. Notice that since the barrier is 60, each of the paths corresponding to the sequence of moves (up, up, up), (up, up, down), and (up, down, up) never result in the option becoming active. Also, the sequence (down, up, up) will activate the option at time 1, but will not return any value for the put at time 3 because the asset price 68.97 exceeds the strike price. So only four paths: (up, down, down), (down, up, down), (down, down, up), and (down, down, down) return any value. The first three of these return $E - S_3 = 60 - 59.565 = .435$ at option tree node v_{33}, and the last returns $E - S_3 = 60 - 51.4425 = 8.5575$ at option node v_{34} as shown in the figure. The nodes v_{21}, v_{31}, and v_{32} cannot provide claim value for the barrier option and are zeroed out.

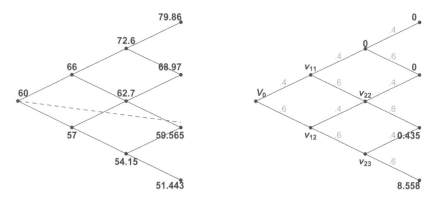

FIGURE 3.25 Asset tree with barrier at 60, and down-and-in barrier option tree.

Since the risk-neutral up probability is still $q = .4$, we can compute the martingale expectation as:

$$
\begin{aligned}
V_0 &= E_Q\left[(1.01)^{-3}X\right] \\
&= (1.01)^{-3}\left(3 \cdot (.6 \cdot .6 \cdot .4) \cdot (.435) + .6^3 \cdot (8.5575)\right) \\
&= 1.97645.
\end{aligned}
$$

Again we can confirm this by chaining, forcing the three nodes mentioned above to be zero:

$$
\begin{aligned}
v_{21} &= 0; \\
v_{22} &= (1+r)^{-1}\left(q \cdot v_{32} + (1-q)v_{33}\right) \\
&= (1.01)^{-1}((.4) \cdot (0) + (.6) \cdot (.435)) \\
&= .258416; \\
v_{23} &= (1+r)^{-1}\left(q \cdot v_{33} + (1-q)v_{34}\right) \\
&= (1.01)^{-1}((.4) \cdot (.435) + (.6)(8.5575)) \\
&= 5.25594; \\
v_{11} &= (1+r)^{-1}\left(q \cdot v_{21} + (1-q)v_{22}\right) \\
&= (1.01)^{-1}((.4) \cdot (0) + (.6) \cdot (.258416)) \\
&= .153514; \\
v_{12} &= (1+r)^{-1}\left(q \cdot v_{22} + (1-q)v_{23}\right) \\
&= (1.01)^{-1}((.4) \cdot (.258416) + (.6)(5.25594)) \\
&= 3.22468; \\
V_0 &= (1+r)^{-1}\left(q \cdot v_{11} + (1-q)v_{12}\right) \\
&= (1.01)^{-1}((.4) \cdot (.153514) + (.6) \cdot (3.22468)) \\
&= 1.97645.
\end{aligned}
$$

If we compute the initial value of the European put, we get, from the martingale formula:

$$
\begin{aligned}
V_0 &= E_Q\left[(1.01)^{-3}X\right] \\
&= (1.01)^{-3}\left(3\cdot(.6\cdot.6\cdot.4)\cdot(.435)+.6^3\cdot(8.5575)\right) \\
&= 1.97645.
\end{aligned}
$$

This is the same result as for the barrier option, because none of the paths that the barrier eliminated from consideration were paths that would have given positive claim value anyway. ∎

3.5.3 Asian Options

An **Asian option** is a derivative in which the claim value random variable X depends on (perhaps is) the average of the asset price values on the path that the asset takes. Since this is a perfectly good derivative, martingale valuation can find its initial value $V_0 = E_Q\left[(1+r)^{-n}X\right]$. The technique of looking at each possible path, illustrated in the examples of barrier option valuation, works well for small problems and can be implemented using appropriate technology for larger ones. It is important to note that not only does the final value of the asset count, but also the sequence of prices that the asset went through, so that it is best not to recombine paths in the price tree, but rather to leave them separate. Here is an illustration.

Example 6. Suppose that a risky asset moves according to a binomial branch process with initial value $S_0 = 70$, and parameters $a = -.05$, $b = .05$. The risk-free asset has rate of return $r = .02$. Find the initial value of a 3-period Asian derivative that returns the average value of the process minus the initial price. (So there is really no option here for the holder of this derivative. See the exercises for Asian-style derivatives with options.)

Solution. First, the risk-neutral probability is:

$$
q = \frac{.02-(-.05)}{.05-(-.05)} = .7.
$$

We can compute the asset prices as usual:

$$
\begin{aligned}
S_0 &= 70; \\
s_{11} &= 70(1.05) = 73.5; & s_{12} &= 70(.95) = 66.5; \\
s_{21} &= 70(1.05)^2 = 77.175; & s_{22} &= 70(1.05)(.95) = 69.825; \\
s_{23} &= 70(.95)^2 = 63.175; \\
s_{31} &= 70(1.05)^3 = 81.0338; & s_{32} &= 70(1.05)^2(.95) = 73.3163; \\
s_{33} &= 70(1.05)(.95)^2 = 66.3338; & s_{34} &= 70(.95)^3 = 60.0162.
\end{aligned}
$$

The value of this Asian derivative is:

$$V_0 = E_q \left[(1.01)^{-3} \left(\bar{S} - 70 \right) \right],$$

where $\bar{S} = \frac{1}{4} (70 + S_1 + S_2 + S_3)$. The table below displays all possible values of the average price $\bar{S}$ on all paths, the corresponding values of $\bar{S} - 70$, and their corresponding path probabilities. There are $2^3 = 8$ possible paths which have probability $q^k \cdot (1 - q)^{3-k}$ where $q = .7$ is the risk-neutral probability and k is the number of ups in the path. (Notice that if we had defined the derivative in the manner of a call option for instance, the claim values would be the larger of $\bar{S} - 70$ and 0, and the only change would be that the lowest five rows would have claim value 0.)

sequence	path	probability	average	claim value
uuu	$S_0, s_{11}, s_{21}, s_{31}$	$.7^3 = .343$	75.4272	5.4272
uud	$S_0, s_{11}, s_{21}, s_{32}$	$.7^2(.3) = .147$	73.4978	3.4978
udu	$S_0, s_{11}, s_{22}, s_{32}$	.147	71.6603	1.6603
duu	$S_0, s_{12}, s_{22}, s_{32}$	.147	69.9103	$-.0897$
dud	$S_0, s_{12}, s_{22}, s_{33}$	$.7(.3)^2 = .063$	68.1647	-1.8353
udd	$S_0, s_{11}, s_{22}, s_{33}$	.063	69.9147	-0.0853
ddu	$S_0, s_{12}, s_{23}, s_{33}$	.063	66.5022	-3.4978
ddd	$S_0, s_{12}, s_{23}, s_{34}$	$.3^3 = .027$	64.9228	-5.0772

Using the probabilities and claim values in the table, the initial value of this Asian option is:

$$V_0 = (1.02)^{-3}((5.4272)(.343) + \cdots + (-5.0772)(.027)) = 2.00539.$$

The chaining method would not be efficient here because to find the claim value of this derivative you would have to compute the path averages anyway. But you can check (see Exercise 12) that chaining does lead to the same value for V_0. ∎

3.5.4   Two-Asset Derivatives

In Section 3.4, Example 5 and Exercises 9, 12, and 14, we touched on valuation problems involving more than one risky asset. It is not necessary to do very much more here. But it is pertinent, in a section dealing with non-vanilla options, to give a nod to derivatives based on two assets, especially those that have claim values based on the larger of two final values. For simplicity, we will take only the independent case in our last example.

Example 7. Let us recycle the risky asset in Example 2, whose parameters were $S_0 = S_0^1 = 40$, $b = b_1 = .03$, $a = a_1 = -.01$, and again suppose that the risk-free rate is $r = .01$. Append to this market another risky asset S^2, independent of the first, with parameters $S_0^2 = 40, b_2 = .05, a_2 = -.03$. Find the initial value of a derivative that expires at time 3 and returns the larger of S_3^1 and S_3^2.

Solution. The risk-neutral probability for asset 1 was calculated in Example 2 to be $q = q_1 = .5$. For the second asset, we have:

$$q_2 = \frac{r - a_2}{b_2 - a_2} = \frac{.01 - (-.03)}{.05 - (-.03)} = \frac{.04}{.08} = .5.$$

The stock prices for asset 1 were found in Example 2, and are shown again below on the left tree in Figure 3.26.

$$S_0 = S_0^1 = 40;$$
$$s_{11} = 41.2; \quad s_{12} = 39.6;$$
$$s_{21} = 42.436; \quad s_{22} = 40.788; \quad s_{23} = 39.204;$$
$$s_{31} = 43.7091; \quad s_{32} = 42.0116; \quad s_{33} = 40.3801; \quad s_{34} = 38.812.$$

For our new asset 2, we compute the node values as follows:

$$S_0^2 = 40;$$
$$s_{11} = 40(1.05) = 42; \quad s_{12} = 40(.97) = 38.8;$$
$$s_{21} = 40(1.05)^2 = 44.1; \quad s_{22} = 40(1.05)(.97) = 40.74;$$
$$s_{23} = 40(.97)^2 = 37.636;$$
$$s_{31} = 40(1.05)^3 = 46.305; \quad s_{32} = 40(1.05)^2(.97) = 42.777;$$
$$s_{33} = 40(1.05)(.97)^2 = 39.5178; \quad s_{34} = 40(.97)^3 = 36.5069.$$

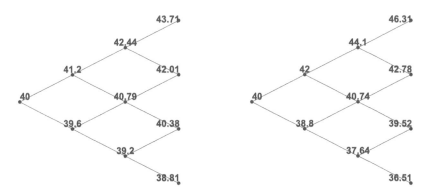

FIGURE 3.26 Two independent risky asset processes.

The fact that the assets are independent means that the probabilities attached to the time 3 values are the products of the individual probabilities for the two assets, and both happened to have risk-neutral probability $1/2$. We list all of the 16 possible pairs of values (S_3^1, S_3^2) in the table below, with the derivative claim value $X = \max\{S_3^1, S_3^2\}$ and the probabilities for each pair.

(S_3^1, S_3^2)	$X = \max\{S_3^1, S_3^2\}$	probability
$(43.7091, 46.305)$	46.305	$(.5)^3(.5)^3$
$(43.7091, 42.777)$	43.7091	$3(.5)^3(.5)^3$
$(43.7091, 39.5178)$	43.7091	$3(.5)^3(.5)^3$
$(43.7091, 36.5069)$	43.7091	$(.5)^3(.5)^3$
$(42.0116, 46.305)$	46.305	$3(.5)^3(.5)^3$
$(42.0116, 42.777)$	42.777	$9(.5)^3(.5)^3$
$(42.0116, 39.5178)$	42.0116	$9(.5)^3(.5)^3$
$(42.0116, 36.5069)$	42.0116	$3(.5)^3(.5)^3$
$(40.3801, 46.305)$	46.305	$3(.5)^3(.5)^3$
$(40.3801, 42.777)$	42.777	$9(.5)^3(.5)^3$
$(40.3801, 39.5178)$	40.3801	$9(.5)^3(.5)^3$
$(40.3801, 36.5069)$	40.3801	$3(.5)^3(.5)^3$
$(38.812, 46.305)$	46.305	$(.5)^3(.5)^3$
$(38.812, 42.777)$	42.777	$3(.5)^3(.5)^3$
$(38.812, 39.5178)$	39.5178	$3(.5)^3(.5)^3$
$(38.812, 36.5069)$	38.812	$(.5)^3(.5)^3$

Combining probabilities for the various distinct values of X, we get that the initial value of the derivative is:

$$
\begin{aligned}
V_0 &= E_Q\left[(1.01)^{-3}X\right] \\
&= (1.01)^{-3}(46.305 \cdot 8 \cdot (.5)^6 + 43.7091 \cdot 7 \cdot (.5)^6 \\
&\quad + 42.777 \cdot 21 \cdot (.5)^6 + 40.3801 \cdot 12 \cdot (.5)^6 + 42.0116 \cdot 12 \cdot (.5)^6 \\
&\quad + 39.5178 \cdot 3 \cdot (.5)^6 + 38.812 \cdot 1 \cdot (.5)^6) \\
&= 41.262. \ \blacksquare
\end{aligned}
$$

Exercises 3.5

1. Consider a risky asset with initial price $S_0 = 50$ and parameters $b = .08, a = -.04$. Assume that the risk-free rate is $r = .02$. Find the value at all nodes of an American call option on this asset expiring at time 3 with exercise price 50, and make sure that it is never optimal for the holder to exercise it early.

2. Suppose that a market consists of a risky asset following a binomial branch process with initial price $S_0 = 40$ and parameters $b = .03$ and $a = -.02$, and a risk-free process with rate of return $r = .01$. Find the initial value of an American put option with strike price 40 that expires in two periods, making note of the nodes at which it is optimal to exercise the option early.

3. It is possible for an American put option to have the same value as its European counterpart. Compute an American put option value at all nodes

of the option tree so as to verify that the option is never exercised early in the following situation: exercise time $n = 2$, risk-free rate $r = .001$, up and down rates $b = .08$, $a = -.005$, initial asset price $S_0 = 40$, strike price $E = 42$.

4. What would go wrong in the proof of Theorem 2 if you tried to argue, similarly to Theorem 1, that the value of the American put cannot be strictly greater than the value of the European put?

5. A risky asset with initial price 100 moves arithmetically as shown in the figure, going up by 5 or down by 5 in each period. There is also a risk-free asset with rate $r = .03$ in the market. Find the initial value of a 3-period American put option with strike price 97.

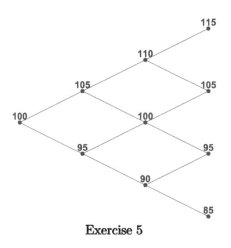

Exercise 5

6. Argue that for Bermudan call options with a single possible execution time prior to time n, the initial value is the same as the initial value of the European call option with the same parameters.

7. Consider the market of Example 2. Find the initial value of a 3-period Bermudan put option with the same strike price $E = 41$ as in the example, which is able to be executed at time 1 or at time 3.

8. A barrier call option expiring at time 3 has the following features. The initial price of the asset the option is based on is 50, and the strike price is 52. If the price of the underlying asset ever goes below 48, the option becomes worthless. The binomial branch parameters of the asset are $b = .05$, $a = -.05$, and the risk-free rate of interest per period is $r = .01$. Find the initial value of this barrier call option. Check that you get the same result with chaining as with martingale valuation.

9. Construct a scenario in which an up-and-out knockout barrier option might serve the interest of both the holder and the issuer.

10. An underlying asset in binomial branch motion has initial price 40, up rate $b = .05$, and down rate $a = -.04$. A risk-free asset yields constant rate of return $r = .02$. Use martingale valuation (not chaining) to compute the initial value of a 3-period up-and-in put option with barrier 41 and strike price 39.

11. In Example 2, there was an underlying asset with initial price $S_0 = \$40$, and up and down parameters $b = .03$, and $a = -.01$, and the risk-free asset had rate $r = .01$. The initial values of a 3-period European and American put option with strike price $41 on the asset were computed as .4911 for the European put and 1 for the American put. Suppose now that a barrier put option of up-and-in type is defined on this underlying asset with barrier $B = \$41$. Use chaining to compute the initial value of the barrier option, making note of whether it is cheaper than the ordinary European option. Explain the result.

12. Using the computations that were made in Example 6, verify that chaining leads to the same value for V_0. Use the exploded tree rather than the recombined tree, so that level 3 has 8 nodes, level 2 has 4 nodes, and level 1 has 2. (Be careful to order the time 3 nodes properly, with the uu paths followed by the ud paths, then the du paths, and lastly the dd paths)

13. Find the initial value of a 3-period Asian put option that pays the larger of $80 - \bar{S}$ and 0. The underlying asset on which the option is based has initial value 70, and the up and down rates are $b = .06, a = .01$, with a risk-free rate of .03.

14. Suppose that there is a binomial branch asset with initial price 100 and parameters $a = -.02$, $b = .04$. Suppose also that the risk-free asset has rate $r = .01$. By itemizing all paths, find the initial value of an Asian call option with claim value equal to $\max \{ (S_0 + S_1 + S_2)/3 - 100, 0 \}$.

15. Consider again the risky asset of Example 3, with initial price 30 and parameters $b_1 = .04$ and $a_1 = -.02$. Again the rate of return of the risk-free asset is .01. Suppose there is another risky asset, independent of the first, with initial price 28, up rate $b_2 = .05$, and down rate $a_2 = -.05$. Find the initial value of a derivative that returns the smaller of the two asset prices at time 2.

16. For the market of Exercise 15, find the initial value of a derivative that returns the average of the two asset values at time 3.

3.6   Derivatives Pricing by Simulation

For plain derivatives such as European call and put options, computation of initial values is relatively routine, either via the martingale formula or by the technique of chaining. But when we move to non-vanilla, path-dependent options such as those in the last section, especially Asian-style and barrier options, as the number of time intervals gets larger, exact computation can become difficult to carry out.

By the Fundamental Theorems of Asset Pricing, initial values of a large assortment of derivatives can be computed as present values of expected claim values at expiration under the martingale measure Q. Any expectation can be estimated with high precision by a sample mean of a random sample. Therefore, an alternative to exact valuation is to simulate a large number of paths of the underlying asset, compute the claim values and their present values for each of these paths, and average the present values.

Also, by the Central Limit Theorem, regardless of the exact probability distribution of the final claim values, the sample mean of both final and initial simulated claim values is approximately normally distributed. This enables us to give approximate 95% confidence intervals for our estimates, which are roughly two sample standard deviations below and above the sample mean value of the derivative. The reader can check any standard introductory statistics book for details; we will just informally use a two-standard deviation interval about the mean as a yardstick for the precision of our estimates.

We will mostly show pseudo-code algorithms for the simulations that we need, and report the results. In the electronic version of this text are full *Mathematica* programs that implement the algorithms. These are used to generate the simulation results that we look at. The pseudo-code is easy to transport to programs in any high-level computer language, including several of the most popular at the time of this writing: R, Python, Java, C++, and Excel's macro language.

3.6.1   Setup and Algorithm

As before, we are assuming that there is one underlying risky asset moving as a binomial branch process, on which a derivative is to be based. The asset has known initial price s_0, up rate b, and down rate a. The derivative expires at time n, and the risk-free rate of return per period is r. The claim value of the derivative is a well-defined function V_n of the path of the underlying asset. One particular path will be denoted, as usual, by:

$$S_0 = s_0, S_1, S_2, S_3, ..., S_n. \tag{3.86}$$

Depending on what kind of derivative we have, the final claim value will be a specific function:

$$V_n = V_n\left(s_0, s_1, s_2, ..., s_n\right) \tag{3.87}$$

of the particular path $s_0, s_1, s_2, ..., s_n$ that is traversed. Its present value is $V_0 = (1+r)^{-n} V_n$.

We will replicate the experiment of simulating a path and computing V_0 a total of m times, where m is large. Denote the final claim values that we simulate as:

$$V_{n,1}, V_{n,2}, ..., V_{n,m},$$

and similarly for the initial claim values $V_{0,i} = (1+r)^{-n} V_{n,i}$. We might choose to retain all of the values of $V_{0,i}$ in a list if we are interested in the empirical distribution of the final claim value, and simply compute their sample mean and standard deviation at the end. Or, we might retain and update only the sum and sum of squares of V_0 values as we go, to avoid saving the previous information. Of course, after we are done, the sample mean is the sum divided by the number of replications:

$$\bar{V}_0 = (V_{0,1} + V_{0,2} + \cdots + V_{0,m})/m,$$

and the sample variance can be computed either using the defining formula or by the well-known computational formula as:

$$S_{V_0}^2 = \frac{1}{m-1} \sum_{i=1}^{m} (V_{0,i} - \bar{V}_0)^2 = \frac{1}{m-1} \left(\sum_{i=1}^{m} V_{0,i}^2 - \frac{1}{m} \cdot \left(\sum_{i=1}^{m} V_{0,i} \right)^2 \right).$$
$$(3.88)$$

The sample standard deviation S_{V_0} is the square root of the sample variance, and the approximate 95% confidence interval for the true value of V_0 is:

$$\bar{V}_0 \pm 2 \cdot \frac{S_{V_0}}{\sqrt{m}}. \qquad (3.89)$$

To produce a simulation algorithm for derivative valuation is now a relatively simple thing. One needs a "wrapper program" that mostly just directs traffic as follows:

Main Simulation Valuation Algorithm

Input: r, a, b, s_0, n, m

1. Compute the risk-neutral probability $q = \frac{r-a}{b-a}$;

2. Initialize an empty list of V_0 values;

3. Do the following steps m times:

(a) Simulate a path of the underlying asset using q;
(b) Compute the value of V_n for that path;
(c) Compute $V_0 = (1+r)^{-n}V_n$;
(d) Append the new V_0 to the list of V_0 values;

Output:
(a) Final list of V_0 values;
(b) Average value $\bar{V}_0$ of the m simulated V_0s;
(c) Sample variance $S_{V_0}^2$ of the m simulated V_ns ;
(d) Precision estimate $2 \cdot \frac{S_{V_0}}{\sqrt{m}}$.

Step 3(a) needs a simple procedure to simulate paths of the price process, which the main program will call on. In pseudo-code this would be as below.

Path Simulation Algorithm

Input: s_0, a, b, n, q

1. Initialize list of states as $\{s_0\}$;

2. For step $i = 1$ to step n:
   (a) Simulate an up move with probability q, or a down move with probability $1 - q$;
   (b) If the move is up, adjoin the state $s_i = (1 + b)s_{i-1}$ to the list; otherwise adjoin state $s_i = (1 + a)s_{i-1}$ to the list;

3. Return the list $\{s_0, s_1, s_2, ..., s_n\}$.

Figure 3.27 shows one particular simulated path of a binomial branch process with $p = .5$, $a = -.05/20$, $b = .05/20$, initial price $s_0 = 50$, and 20 steps. The idea behind the choice of these parameters is that a and b are now seen as rates per unit of continuous clock time, and time runs from 0 to 1, so that in each of the twenty periods, the up and down rates are divided by 20 to get the per period rate. (Section 3.7 will elaborate on this idea.) In the electronic version of the text, the *Mathematica* code is given that carries out this simulation.

These two tools may be used generally for the approximate simulation valuation of any derivative. The devil is in the details of step 3(b) of the main algorithm: given a path, determine the final value of the derivative. If the derivative is a simple function of the final price value, as with European options, this is easy. We will see that Asian-style options are also no challenge, but we will have to think harder about American, Bermudan, and barrier options.

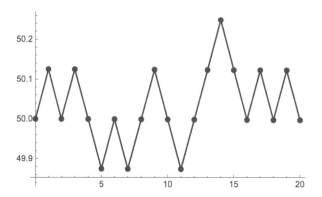

FIGURE 3.27 A 20-step simulated binomial branch process.

3.6.2   Examples

Let us look more closely at how claim values of some derivatives might be implemented as functions of the path of the price process, and check the performance of simulation in some cases where we already know the true initial value of the derivative. Then, we will look at an example in many time periods, which would be difficult to do exactly.

Example 1. In Example 1 of Section 3.1 there was an underlying asset following the binomial branch model with parameters $b = .07$, $a = -.02$, and $S_0 = 200$, and a risk-free asset with rate $r = .04$. An ordinary two-period European put option was available, with exercise price $E = 198$. We computed that the initial price of the option was $V_0 = .608154$. Approximate this via simulation. How many replications does it take so that we can be reasonably confident that the estimate is within .10 of the true value?

Solution. Once again, the risk-neutral probability is:

$$q = \frac{r - a}{b - a} = \frac{.04 - (-.02)}{.07 - (-.02)} = \frac{.06}{.09} = \frac{2}{3}.$$

The function that gives the final claim value is:

$$V_n = \max(198 - S_n, 0),$$

and plugging this into the simulation algorithm allows us to make the computations.

One run of a simulation program with 100 replications produced a point estimate $\bar{V}_0 = .602$, which is very close to the true value of about .608, but the precision estimate came out to $2 \cdot \frac{S_{\bar{V}_0}}{\sqrt{m}} = .344$. This means that we can only say that we are highly confident that the true price is in the interval $[.602 - .344, .602 + .344] = [.258, .946]$. This is a rather wide range, so we should try increasing m to get a more precise estimate. Using the approach of doubling the sample size repeatedly, the results in the table below were obtained.

# reps	$\bar{V}_0$	precision	interval
100	.602	.344	$[.258, .946]$
200	.684	.257	$[.427, .941]$
400	.602	.171	$[.431, .773]$
800	.636	.124	$[.512, .760]$
1600	.612	.086	$[.526, .698]$

All of the intervals contain the true value of .608, but the precision only fell beneath .1 on the last experiment, with $m = 1600$ trials. Instead of fishing in this way, one smaller experimental run could be made to estimate the standard deviation of V_0, and then we can do some algebra to solve for the required m. The first run, with $m = 100$, gave an estimate $S_{V_0} \approx 1.72$ (the bigger experiments produced similar results). Accepting this standard deviation estimate, we would like:

$$\text{precision} \leq .1 \implies 2 \cdot \frac{S_{V_0}}{\sqrt{m}} \leq .1$$
$$\implies m \geq \left(\frac{2S_{V_0}}{.1}\right)^2 \approx \left(\frac{2 \cdot 1.72}{.1}\right)^2 = 1183.36.$$

This is only an approximation to the required number of replications m, but we should be reasonably confident if we run 1200 replications. This was actually done, and the point estimate of V_0 was .657, with a precision of about .103 in the simulation. ∎

Next we try a barrier option.

Example 2. In Example 4 of Section 3.5 we had a market with one risky, binomial branch asset such that $s_0 = 60$, $b = .1$, and $a = -.05$. The risk-free rate of return was $r = .01$. A three-period, up-and-out call option with strike price $E = 66$ and barrier $B = 70$ was valued at $V_0 = .553469$. The asset tree and option tree are repeated here as Figure 3.28. Estimate V_0 by simulation of 50, 100, 200, and 1000 replications and compare the interval estimates to the true value.

Solution. As in the previous example, the risk-neutral probability is:

$$q = \frac{.01 - (-.05)}{.1 - (-.05)} = \frac{6}{15} = .4.$$

The path simulation program and main algorithm are reusable for any derivative; what we must do is find the function that gives the final claim value of this particular derivative. As an up-and-out barrier call option, the claim value will be zero if any price rises to 70 or greater, which we can check with the maximum function applied to the list of prices:

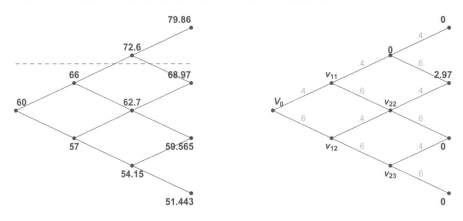

FIGURE 3.28 Asset tree with barrier at 70, and up-and-out barrier option tree.

If $\max\left(S_0, S_1, S_2, S_3\right) \geq 70$, then $V_n = 0$, else $V_n = \max\left(S_3 - 66, 0\right)$.

The results of one particular simulation experiment are listed in the table below.

#reps	$\bar{V}_0$	precision	interval
50	.577	.329	$[.248, .906]$
100	.490	.218	$[.272, .708]$
200	.504	.155	$[.349, .659]$
1000	.525	.070	$[.455, .595]$

The true V_0 value of about .553 is contained in all of these intervals, which is comforting, but the precision is only tolerable in the last case with $m = 1000$ replications. Our computational experience so far is indicating that because of the variability in the simulated values of V_0, a rather large value of m should be our starting point. ∎

At the beginning of the section a claim was made that the large sample distribution of the approximating values $\bar{V}_0$ of the initial claim value should be normal, by the Central Limit Theorem. Let us look at this in the context of the barrier option of Example 2.

The idea would be to use the main algorithm some large number of times M, picking out the simulated $\bar{V}_0$ each time and forming a list of data with them. Then we would plot a histogram of this list of data, with bar heights equal to relative frequencies divided by category lengths, and compare it to a normal density function.

For the up-and-out barrier call option above, a total of $M = 300$ simulation experiments, each of which consisted of $m = 500$ replications, were run to form a data list of 300 $\bar{V}_0$ values. The sample mean and standard deviation of this data list turned out to be .556909 and .0556957, respectively. Figure 3.29 superimposes a properly scaled histogram and a graph of the normal density with this mean and this standard deviation. The fit is satisfactory, but we see clearly that there is much variability in the simulated $\bar{V}_0$, leading to uncertainty about the actual price, especially if you look at the standard deviation relative to the average price.

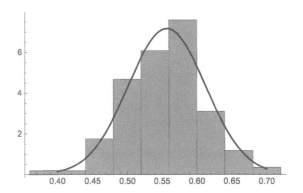

FIGURE 3.29 Normality of estimates $\bar{V}_0$ for up-and-out barrier option.

We have not seen an example with an American option yet, for which there is good reason. It is difficult to know, from looking at the sequence of states only, what the holder of the option receives. This is because the option holder makes an optimal decision, at each time and for the current state, whether to execute the option or to wait. We do not know what those decisions are exactly, without going through the backward recursive process of valuing the option at future nodes, since the user is to choose the larger, at each time l, of the immediate exercise value $\max(E - S_l, 0)$ and the expected discounted future value of the option:

$$(1 + r)^{-1} \left(q \cdot V_{l+1}^{\text{up}} + (1 - q)V_{l+1}^{\text{down}} \right).$$

The literature contains approaches to this problem, among which is Longstaff and Schwartz [13], in which path simulations are used to estimate, backwards from time $n - 1$, the average value of not executing the option. But the computational difficulty of doing this is similar to, perhaps even greater than, the backstepping process that we illustrated in the previous section, which is easily implemented on a spreadsheet. So simulation is not the best approach in this case.

But simulation is an ideal approach to the difficult problem of valuing Asian-style derivatives, as we will see in the following example.

Example 3. Recall Example 6 of Section 3.5, in which the parameters of a risky asset were $s_0 = 70, b = .05$, and $a = -.05$. We assumed a risk-free rate of $r = .02$, from which the risk-neutral probability came out to $q = .7$. In that example we computed, by itemizing paths, that the initial value of a 3-period Asian derivative that gives the holder the average value of the price process minus the initial price was $V_0 = 2.00539$. Use simulation with 2000 replications to estimate this V_0. Further, suppose instead that the derivative expires in 10 periods. The pathwise calculation would require us to consider 2^{10} paths, which is more than 1000 of them. Apply simulation to estimate the initial value of this longer Asian derivative.

Solution. This time the function that gives the claim value of the derivative is easy:

$$V_n = \bar{S} - 70.$$

Inserting this into step 3(b) of the main algorithm, and using 2000 replications, we obtain for one particular simulation:

$$\bar{V}_0 = 1.960, \text{precision} = .128, \text{interval} = [1.832, 2.088].$$

So we have done quite well, coming within about .04 of the true value with our point estimator, with an estimated precision of less than .13.

To answer the second question, it is enough to just rerun the simulation with $n = 10$ instead. For the longer-duration derivative we get:

$$\bar{V}_0 = 6.141, \text{precision} = .233, \text{interval} = [5.908, 6.374]. \ \blacksquare$$

By now you should be reasonably confident that, as long as the number of replications is large and it is not too difficult to express the claim value directly as a function of the path of the price process, simulation is a useful tool for estimating option values. Now, as a final example in which it is very difficult to compute the option value directly, let's try to value a knock-in put option with many periods.

Example 4. Suppose that a 10-period down-and-in European barrier call option with barrier $B = 39$ and strike price $E = 41$ is defined on a risky asset with initial price $s_0 = 40$, and up and down rates $b = .01$ and $a = -.02$ per period. The risk-free rate is $r = .002$. Estimate the initial value of this derivative using simulation, to within about .005.

Solution. The problem conditions do make sense, because the smallest and largest possible values of the underlying asset after 10 periods are:

$$s_0(1 + a)^{10} = 40(.98)^{10} = 31.8658;$$
$$s_0(1 + b)^{10} = 40(1.01)^{10} = 43.0803.$$

So it is possible for the down-and-in option to sink below the barrier price at 39, and it is possible for it to rise above the strike price of 41 by the end. What the barrier serves to do is to cut out paths that mostly consist of up moves from the beginning; hence it makes the call strike price harder to achieve than it would be without the barrier. This design should make the option relatively cheap.

The risk-neutral probability to use for simulating paths is:

$$q = \frac{r - a}{b - a} = \frac{.002 - (-.02)}{.01 - (-.02)} = .73333.$$

The usual path simulation tool and main program work here, and so we just have to define the claim value in terms of the path. Since the derivative is basically a European call, if it ever becomes active it returns the larger of $S_n - E$ and 0. But in order to become active, somewhere the price must sink below 39, which is the same as saying that the minimum price is less than 39; otherwise the option has no value. So the appropriate claim value function would be:

if $\min (s_0, S_1, S_2, ..., S_{10}) < 39$ then $V_{10} = \max (S_{10} - E, 0)$, else $V_{10} = 0$.

After some experimentation, an average value of V_0 in 1000 replications turned out to be $\bar{V}_0 = .00939$ with a precision of .0047, which meets the requirement of the problem. Incidentally, 4000 replications gave revised estimates of .00998 and .0024 for the value of V_0 and the precision, so we should be quite confident that the option has a tiny initial value somewhere between about .008 and .012. ∎

Exercises 3.6

1. In an algorithm in which you are only interested in returning the mean of the V_0 values, and you do not want to retain a list of all of these simulated values, show that the mean can be updated from the $(k - 1)^{\text{st}}$ path to the kth by the formula:

$$\text{new } \bar{V}_0 \text{ after } k^{\text{th}} \text{ path} = \frac{(k - 1) \left(\text{old } \bar{V}_0 \text{ after } (k - 1)^{\text{st}} \text{ path}\right) + V_0 \text{ for new path}}{k}$$

2. What would you expect to happen to the width of an approximate confidence interval for V_0 if the number of replications is tripled? Is your answer an exact answer, or is it subject to some amount of random variability?

3. (Technology required) In Exercise 3.1-7, we had a binomial branch asset with initial price 50, and up and down rates $b = .05$, $a = -.02$. The risk-free rate was assumed to be $r = .02$. A European call option with exercise time 2 and strike price $54 was defined on this asset. If you have not yet done this exercise, find the initial value of the option exactly, and then run simulations with 500, 1000, and 2000 replications, checking the accuracy of your estimated

values.

4. (Technology required) In the setting of the previous exercise, perform the experiment of simulating 200 replications to produce an estimate $\bar{V}_0$ a total of 200 times. Plot a histogram of the distribution of $\bar{V}_0$ and comment on whether the distribution seems to be approximately normal.

5. (Technology required) In Example 5 of Section 3.5, we assumed a market with a risk-free rate of return $r = .01$ and a risky asset such that $s_0 = 60$, $b = .1$, and $a = -.05$. There, we computed that the value of a three-period, down-and-in barrier put option on this asset with strike price 60 and lower barrier 60 (which becomes active if the asset ever falls strictly below the initial price of 60) is $V_0 = 1.97645$. Using an initial sample of 500, determine the number of replications that you will need to estimate V_0 to within .08, and then run a simulation with at least that many replications.

6. (Technology required) Suppose that a risky asset has parameters $s_0 = 100$, $b = .06$, and $a = -.02$. The risk-free rate is $r = .02$. An 8-period Asian-style put option is available that pays the larger of 96 minus the average price on the path and 0. Use a 200-replication trial to estimate the standard deviation of V_0, use that value to solve for m large enough to estimate V_0 to within .05, and finally run such a simulation.

7. (Technology required) Does the distribution of the simulated V_0 values themselves approach normality? Why or why not? Test this in the context of the down-and-in barrier put option of Exercise 5 by simulating 500 V_0 values and plotting a histogram.

8. (Technology required) Consider the risky asset of Exercise 13 of Section 3.5, which had parameters $b = .06, a = .01$, and initial price 70. The rate of return on the risk-free asset was .03. In that exercise we were able to use exact techniques to find the initial value of a 3-period Asian put option that pays the larger of $80 - \bar{S}$ and 0. Now, use simulation to estimate the value of a 5-period Asian put option with the same payout. Use enough replications to achieve a precision of about .05.

9. (Technology required) A risky asset is undergoing binomial branch motion with up and down parameters $b = .05, a = -.03$, and initial price 50. The risk-free rate in the market is $r = .01$. Create and run a complete program that can simulate the underlying asset over 4 periods and return one simulated value of a down-and-out knockout call option with lower barrier 48 and strike price 52. (In your program, you should leave parameters general, and then call the program with the parameter values given in this problem.)

10. (Technology required) Suppose that a risky asset moves according to a binomial branch process with initial value $s_0 = 20$, up rate $b = .03$, and down rate $a = -.01$. Assume that the risk-free asset has rate of return $r = .01$. A derivative pays the largest positive difference between the asset price and 21 over the first five periods (or zero if there are no positive differences). Use simulation to value this derivative to within .02.

11. (Technology required) In Exercises 15 and 16 of Section 3.5 there were two independent risky assets: one with initial price 30 and parameters $b_1 = .04$ and $a_1 = -.02$, and the other with initial price 28, $b_2 = .05$ and $a_2 = -.05$. The risk-free rate was $r = .01$. The derivative (a) in Exercise 15 returned the smaller of the two asset prices at time 2, and derivative (b) in Exercise 16 returned the average of the two asset values at time 3. Use simulation to approximate the initial values of these derivatives.

12. (Technology required) Consider a derivative which is a portfolio of a call and a put option on the same underlying asset. We will assume that the asset is in binomial branch motion with parameters $b = .05, a = -.02$. Its initial price is 80, and the risk-free rate is .02. The call option has strike price 82, the put option has strike price 78, and both options expire at time 6. Write a program to simulate 2000 replications of this portfolio and estimate its initial value.

3.7 From Discrete to Continuous Time (A Preview)

When we set up our market model in Section 3.4, we broke up the time interval $[0, T]$ into n equally sized subintervals, each with length $\Delta t = \frac{T}{n}$, whose partition points were labeled:

$$t_0 = 0, t_1, t_2, ..., t_{n-1}, t_n = T,$$

where $t_k = k\Delta t = k\frac{T}{n}$. Figure 3.15 from that section displayed the time axis, which is reproduced here as Figure 3.30. Our goal over the next few chapters is to extend our discrete-time market model to continuous time, and to discover the natural generalization of the binomial branch process for risky asset price modeling. With that generalization in hand, we will reconsider the derivative valuation problem in the continuous setting.

FIGURE 3.30 Time axis for asset motion.

We will approach the problem via a limiting process; specifically, fixing initial time 0 and final time T, we will let the number of subintervals n approach infinity. This brief section is meant to set the stage for that process.

First we need to modify our treatment of the motion of the risk-free asset. So far, we have assumed a constant rate of return r for the asset over each subinterval of length Δt. Then the value of one monetary unit would grow to $(1+r)^n$ by time T. We cannot simply keep r constant and let $n \longrightarrow \infty$; otherwise this value would increase to infinity, which is not reasonable. To resolve this difficulty, we appropriate the letter r for a different use: the rate of return per unit of clock time. In all of the earlier formulas, the risk-free rate r per discrete period would be replaced by $r\Delta t$. In this modified framework, the value of one unit of the risk-free asset at time T would now be:

$$(1 + r\Delta t)^n = \left(1 + \frac{rT}{n}\right)^n. \tag{3.90}$$

This expression does reach a well-known limit:

$$\lim_{n \to \infty} (1 + r\Delta t)^n = \lim_{n \to \infty} \left(1 + \frac{rT}{n}\right)^n = e^{rT}. \tag{3.91}$$

(Do Exercise 1 to convince yourself of this limit.) We recognize the expression e^{rT} as the continuous compound interest growth factor, with continuous rate of return r. In particular, any present values that we want to calculate, discounted from a continuous clock time t to 0, would entail multiplying the future value by the factor e^{-rt}.

The next question is: what will happen to the binomial branch process as the number of subintervals becomes very large, but the length of each subinterval approaches 0? At a typical level n in the price tree, we have seen that there are $n + 1$ nodes, so we are gaining precision in the modeling of possible prices as we use shorter subintervals. Figure 3.31 shows a 10-step process on time interval $[0, 1]$, so that $\Delta t = .1$. The initial price is the known constant s_0, and using our usual notation $s_{l,j}$ for the jth node in level l, we see that there are 11 possible prices at level 10, which is clock time $T = 1$, namely:

$$s_{10,1} = (1 + b)^{10}(1 + a)^0 s_0,$$
$$s_{10,2} = (1 + b)^9(1 + a)s_0,$$
$$\vdots$$
$$s_{10,11} = (1 + b)^0(1 + a)^{10} s_0.$$

But again an adjustment will need to be made in our model. If there are n steps, the maximum value of the asset process after these n steps (that is, at time T) is $s_0(1 + b)^n$ which will approach infinity as n does. In general, the

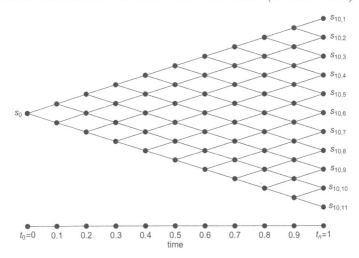

FIGURE 3.31 A 10-step binomial branch process on $[0,1]$.

j^{th} node at level n is:

$$s_{n,j} = s_0(1+b)^{n-j+1}(1+a)^{j-1} = s_0 \left(\frac{1+a}{1+b}\right)^{j-1}(1+b)^n, \qquad (3.92)$$

which will also approach ∞ as n does, for each fixed $j \leq n$. (See Exercise 2 for the case $j = n + 1$.) So we must modify our use of the parameters a and b so that, like the per period risk-free rate, they approach 0 as the time increment Δt approaches 0.

We will return to this issue shortly, but first let us consider the binomial branch model in a different way, and compare it to the way that real price data behave. The price ratios $R_l = S_l/S_{l-1}$, $l = 1, 2, ..., n$ are independent and identically distributed with two possible values, $1 + b$ and $1 + a$. But now we can multiplicatively telescope:

$$S_T = S_n = s_0 \cdot \frac{S_1}{s_0} \cdot \frac{S_2}{S_1} \cdot \frac{S_3}{S_2} \cdots \frac{S_n}{S_{n-1}} = s_0 \cdot R_1 \cdot R_2 \cdot R_3 \cdots R_n. \qquad (3.93)$$

Hence the random price S_T at time T is not a sum, but rather a product of independent identically distributed random variables. The Central Limit Theorem should not be expected to apply directly to S_T, so we should not expect S_T to be approximately normally distributed for large n.

But if we apply the natural logarithm to both sides of equation (3.93), then:

$$\log(S_T) = \log(s_0 \cdot R_1 \cdot R_2 \cdot R_3 \cdots R_n) = \log(s_0) + \sum_{i=1}^{n} \log(R_i). \qquad (3.94)$$

Therefore, the log of S_T is a constant plus the sum of n independent, identically distributed random variables $\log(R_l)$, each with two possible values. Call those values $c = \log(1+b)$ and $d = \log(1+a)$ for the moment. As clock time grows, more such random variables are added on; for instance in the time interval $[0,1]$ we have:

$$\log(S_{t_1}) = \log(s_0) + \log(R_1);$$
$$\log(S_{t_2}) = \log(s_0) + \log(R_1) + \log(R_2);$$
$$\log(S_{t_3}) = \log(s_0) + \log(R_1) + \log(R_2) + \log(R_3);$$
$$\vdots$$

One important thing to note is that the successive differences of logged prices have something in common. We have:

$$\log(S_{t_1}) - \log(s_0) = \log(R_1);$$
$$\log(S_{t_2}) - \log(S_{t_1}) = \log(R_2);$$
$$\log(S_{t_3}) - \log(S_{t_2}) = \log(R_3);$$
$$\vdots$$

so that these successive differences are independent and identically distributed. (See Exercise 3 for a generalization.)

Each of the logged ratios has mean:

$$E[\log(R_i)] = p \cdot c + (1-p) \cdot d,$$

and variance:

$$
\begin{aligned}
\mathrm{Var}(\log(R_i)) &= E\left[(\log(R_i))^2\right] - (E[\log(R_i)])^2 \\
&= p \cdot c^2 + (1-p) \cdot d^2 - (p \cdot c + (1-p) \cdot d)^2 \\
&= (p - p^2)c^2 - 2p(1-p)cd + ((1-p) - (1-p)^2)d^2 \\
&= p(1-p)c^2 - 2p(1-p)cd + (1-p)(1-(1-p))d^2 \\
&= p(1-p)(c^2 - 2cd + d^2) \\
&= p(1-p)(c-d)^2.
\end{aligned}
$$

For the nth logged price, that is for $\log(S_{t_n}) = \log(S_T)$, the mean and variance would be:

$$E[\log(S_T)] = \log(s_0) + n(p \cdot c + (1-p) \cdot d);$$
$$\mathrm{Var}(\log(S_T)) = np(1-p)(c-d)^2. \tag{3.95}$$

Now, how should we define the possible values c and d of each $\log(R_i)$ so that the mean and variance of $\log(S_T)$ reach a well-defined limit as the number of subintervals n approaches infinity, and so that the price model reflects how real assets actually behave?

The graph in Figure 3.32 is very suggestive. It displays the logs of price values of General Electric stock, observed weekly at Friday closing for the calendar years 2010 through 2016, as a function of week. The dark line that is superimposed is a best-fitting line to the data obtained by ordinary least squares estimation. There is weekly variability, but we find evidence here that real logged prices may have means that increase linearly with time. It is not clear exactly what the variances do from this picture, but since price values in the far future are less certain than price values in the near future, we might suspect that the variance of logged prices grows with time, and perhaps is also proportional to time.

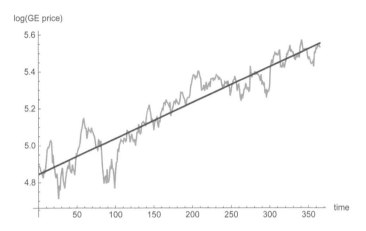

FIGURE 3.32 Log prices vary linearly with time, with a random disturbance.

To correct the problem we encountered with unbounded price values, the most obvious strategy to try is what we did with the risk-free asset, that is, to let the possible up and down rates of return for a period of length Δt be $b\Delta t$ and $a\Delta t$. Then the ratios R_i have possible values $1 + b\Delta t$ and $1 + a\Delta t$, and the logged ratios $\log(R_i)$ have possible values $c = \log(1 + b\Delta t)$ and $d = \log(1 + a\Delta t)$. Recall the first order Taylor approximation $\log(1 + x) \approx x$, which becomes better and better as x approaches zero. This gives us the right behavior for the mean of $\log(S_T)$, at least on an approximate basis, because:

$$
\begin{aligned}
E\left[\log\left(S_T\right)\right] &= \log\left(s_0\right) + n(p \cdot c + (1 - p) \cdot d) \\
&= \log\left(s_0\right) + n(p \cdot \log(1 + b\Delta t) + (1 - p) \cdot \log(1 + a\Delta t)) \\
&\approx \log\left(s_0\right) + n(p \cdot b\Delta t + (1 - p) \cdot a\Delta t) \\
&= \log\left(s_0\right) + n\Delta t(pb + (1 - p)a) \\
&= \log\left(s_0\right) + (pb + (1 - p)a) \cdot T.
\end{aligned}
$$

$$(3.96)$$

Thus, the mean log price is linear as a function of time T, which is consistent with the behavior of real prices. But the variance will not work; in fact, in Exercise 4 you will show the equation:

$$\text{Var}\left(\log\left(S_T\right)\right) \quad = \quad p(1-p)(b-a)^2(\Delta t)\cdot T. \tag{3.97}$$

Unfortunately, this will approach 0 as $n \to \infty$ and $\Delta t \to 0$, which is an unsatisfactory model.

In Section 5.1 we will show how to model the price ratios slightly differently in order to sidestep this problem. To be specific, we will investigate the implications of reparameterizing the parameters b and a in the original discrete model by:

$$b = \mu\Delta t + \sigma\sqrt{\Delta t}; \ \ a = \mu\Delta t - \sigma\sqrt{\Delta t}, \tag{3.98}$$

where μ and σ are constants. Under this new form of the binomial branch model, the probability distribution of the price variable S_T can be summarized. Using k to represent the number of up moves among the n, the possible states of S_T are:

$$s_0\left(1+\mu\Delta t + \sigma\sqrt{\Delta t}\right)^k\left(1+\mu\Delta t - \sigma\sqrt{\Delta t}\right)^{n-k}, k = 0, 1, 2, ..., n. \tag{3.99}$$

The probabilities for those states are binomial probabilities with parameter p:

$$\binom{n}{k}p^k(1-p)^{n-k}, k = 0, 1, 2, ..., n. \tag{3.100}$$

Also, using the linear approximation $\log(1+x) \approx x$, the states of $\log\left(S_T\right)$ are approximately:

$$\begin{aligned}
&\log\left(s_0\right) + k\log\left(1+\mu\Delta t + \sigma\sqrt{\Delta t}\right) + (n-k)\log\left(1+\mu\Delta t - \sigma\sqrt{\Delta t}\right)\\
&\approx \log\left(s_0\right) + k\left(\mu\Delta t + \sigma\sqrt{\Delta t}\right) + (n-k)\left(\mu\Delta t - \sigma\sqrt{\Delta t}\right)\\
&= \log\left(s_0\right) + n\mu\Delta t + 2k\sigma\sqrt{\Delta t}\\
&= \log\left(s_0\right) + \mu T + 2\sigma\sqrt{\Delta t}\cdot k, k = 0, 1, 2, ..., n.
\end{aligned} \tag{3.101}$$

These states are linear functions of the variable k in the binomial mass function and carry the same binomial probabilities in formula (3.100). Thus, the distribution of $\log\left(S_T\right)$ will have the same shape as the binomial, which is known to approximate the normal density function for large n.

A case is plotted in Figure 3.33, in which the interval $[0, 1]$ is divided into $n = 20$ subintervals of length $\Delta t = .05$ each. Parameters $s_0 = 50$, $\mu = .05$, and $\sigma = .1$ were used to calculate the approximate states in formula (3.101). Thus, as we move to the continuous analog of the binomial branch process, we expect the logged price at time T to be normally distributed with a mean that is a linear function of time and variance that is proportional to time. We shall meet such a process in Chapter 4.

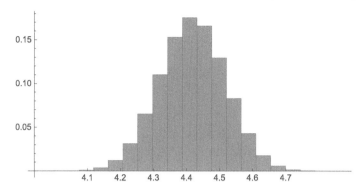

FIGURE 3.33 Distribution of $\log(S_1)$, $n = 20$ subintervals, $s_0 = 50$ $\mu = .05, \sigma = .1$.

Exercises 3.7

1. Verify the limit in formula (3.91) by first taking the log of both sides and then applying L'Hopital's Rule:

$$\lim_{x \to \infty} \frac{f(x)}{g(x)} = \lim_{x \to \infty} \frac{f'(x)}{g'(x)} \text{ if both } f \text{ and } g \text{ approach } 0 \text{ or both approach } \infty$$

2. In reference to formula (3.92), what limits are possible for $j = n + 1$?

3. Use formula (3.94) to argue that all differences of logged prices $D_{m,m+n} = \log(S_{t_{m+n}}) - \log(S_{t_m})$ are independent of, and identically distributed with, differences $D_{l,l+n} - \log(S_{t_{l+n}}) - \log(S_{t_l})$ as long as the time intervals $(t_m, t_{m+n}]$ and $(t_l, t_{l+n}]$ are disjoint.

4. Derivation (3.96) shows that if the price ratios R_i have possible values $1 + b\Delta t$ and $1 + a\Delta t$, then the mean log price is linear as a function of time T, which is consistent with the behavior of real prices. However, verify formula (3.97), which implies that the variance of the logged price approaches 0 as the spacing Δt approaches 0. (Use the linear approximation $\log(1+x) \approx x$.)

5. In a new binomial branch model with time dependent coefficients $b\Delta t$ and $a\Delta t$ replacing b and a respectively, what is the value of the risk-neutral probability? (Remember that we use $r\Delta t$ for the per period risk-free rate.) Show that if the parameter translations in formula (3.98) are used, then:

$$q = \frac{1}{2} + \frac{(r - \mu)\sqrt{\Delta t}}{2\sigma}.$$

6. (Technology required) Suppose that a risky asset moves as a binomial branch process with initial state $s_0 = 40$ in such a way that its μ parameter is .03 and its σ parameter is .02. (See formula (3.98).) The risk-free rate of return is .01 per unit of clock time. If there are 20 time steps in the interval $[0, 1]$ and the up probability is $p = .5$, find the probability that the logged process takes on a value less than 3.7 at time 1.

7. (Technology required) Consider the risky asset process in Exercise 6, and note the formula for the risk-neutral q in Exercise 5. A European call option expiring at time 1 has a strike price of 43.5. What is the probability, under the risk-neutral q, that the option is in the money at the end (i.e. it has a claim value greater than 0 at time 1)?

4

Continuous Probability Models

To understand the limiting behavior of the binomial branch process, as the number of periods becomes large while the clock time per period becomes small, we must know some things about continuous-time, continuous state space models in probability and stochastic processes. So the rest of this text will concentrate on continuous financial mathematics, for which we would benefit from a brief review of key ideas in continuous probability. Some of this has already been alluded to in Chapter 1, but in this chapter we will go into more detail on the subjects of continuous random variables and distributions, expectation, independence and dependence, distributions related to the normal distribution, and stochastic processes related to Brownian motion.

4.1 Continuous Distributions and Expectation

4.1.1 Densities and Cumulative Distribution Functions

Not all random phenomena take on finitely many possible values, or even countably many. In simulation we talk about selecting a random real number from [0,1]. The lifetime of a living thing or a mechanical device is some non-negative real number. The amount of snowfall in a year is real-valued, etc. Here we have in mind an idealization in which we can observe random numerical outcomes to arbitrary precision. In finance, we look at asset price and rate of return models, which practically speaking are only recorded to a finite number of decimal places. But the number of possible values may be very large, making computations very cumbersome, so we may derive some benefit by supposing that the values take place in a continuum.

The ***random variable*** concept, in which a random variable is a function from a sample space of outcomes Ω of a random phenomenon into a set called its state space, is still useful. The state space can be taken to be the real line, or a subset of the real line, or a subset of n-dimensional Euclidean space as in our first definition.

Definition 1. Let E be a subset of $\mathbb{R}^n$. A random variable $X : \Omega \longrightarrow E$ is said to have ***probability distribution*** P_X if, for any subset B of E:

$$P_X(B) = P[X \in B]. \ \blacksquare \qquad\qquad (4.1)$$

A probability distribution P_X works as a probability measure on the state space E. It must satisfy the usual properties that $P_X(E) = 1$, $P_X(B) \geq 0$ for all $B \subset E$, and P_X must be additive over disjoint sets. Actually, as we will see in a later section, we are not necessarily free to apply P_X to just any subset of Euclidean space, so we will modify the "for any subset" phrase accordingly later.

A continuous probability distribution on $\mathbb{R}$ can be determined in a way that should be familiar to us:

Definition 2. Suppose that a random phenomenon can result in possible outcomes $E \subset \mathbb{R}$. A ***probability density function*** (p.d.f.) on E is a function $f(x)$ such that $f(x) \geq 0$ for all x and $\int_E f(x)dx = 1$. A continuous random variable X taking values in E is said to have this density function if for all subsets B of E:

$$P_X(B) = P[X \in B] = \int_B f(x)dx. \ \blacksquare \qquad\qquad (4.2)$$

Thus, the probability that X takes a value in a subset of its state space, such as the interval $[c, d]$ in Figure 4.1, is the area beneath the p.d.f. in that interval. It follows that the probability that X assumes a constant value x identically is zero for all x. This also means that in the density model of continuous probability, it does not matter if endpoints of intervals are included; for example, $P[X \in [c, d]] = P[X \in [c, d)]$, since the probability that $X = d$ is zero .

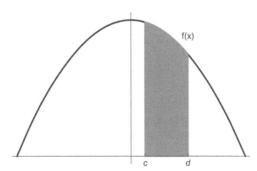

FIGURE 4.1 $P[X \in [c, d]] = $ area under p.d.f. between c and d.

An alternative characterization of a probability distribution of a random variable taking values in $\mathbb{R}$ is the c.d.f., defined below.

Definition 3. A *cumulative distribution function* (c.d.f.) F on $\mathbb{R}$ is a non-decreasing, non-negative function such that:

$$\lim_{x \to -\infty} F(x) = 0 \text{ and } \lim_{x \to +\infty} F(x) = 1.$$

A continuous random variable X is said to have c.d.f. F if for all $x \in \mathbb{R}$:

$$P_X\left((-\infty, x]\right) = P[X \le x] = F(x). \blacksquare \tag{4.3}$$

Geometrically, the value $F(x)$ of the c.d.f. at a point x represents the area under the density function to the left of x. By additivity of probability, the c.d.f. determines the action of the probability distribution on intervals by the formula:

$$P[c < X \le d] = P[X \le d] - P[X \le c]. \tag{4.4}$$

Both the p.d.f. and c.d.f. have multivariate versions that we will consider in Section 4.2. Graphs of densities and c.d.f.s look like those in Figure 4.2(a) and (b). In this case the distribution happens to be a member of the important family of *lognormal distributions* that we will study in a later subsection.

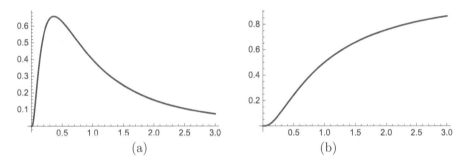

FIGURE 4.2 (a) A continuous p.d.f; (b) a continuous c.d.f.

The c.d.f. is the more general object than the density, since it covers both discrete and mixed discrete-continuous distributions as well as continuous ones. When a distribution does have a density, the following relationships hold.

$$F(x) = P[X \le x] = \int_{-\infty}^{x} f(t)\, dt; F'(x) = f(x). \tag{4.5}$$

The second equation is an immediate consequence of the first, together with the Fundamental Theorem of Calculus.

Although most of the probability distributions that we will use in this book are related to the normal distribution, two other distributions are worthy of note. The *uniform distribution* on interval $[a, b] \subset \mathbb{R}$ (symbolized $U[a, b]$) is the distribution whose density is constant on that interval and zero otherwise:

$$f(x) = \begin{cases} \frac{1}{b-a} & \text{if } x \in [a, b]; \\ 0 & \text{otherwise.} \end{cases} \tag{4.6}$$

Notice that $f(x) \geq 0$ for all x, and the constant is chosen to ensure that $\int_a^b f(x)\,dx = 1$. The **exponential distribution** with parameter $\lambda > 0$ (symbolized $\exp(\lambda)$) has density function:

$$f(x) = \begin{cases} \lambda e^{-\lambda x} & \text{if } x \geq 0; \\ 0 & \text{otherwise.} \end{cases} \tag{4.7}$$

This function also has the properties that $f(x) \geq 0$ for all x and $\int_0^\infty f(x)\,dx = 1$, which are necessary for it to qualify as a p.d.f. Pictures of two particular cases are in Figure 4.3.

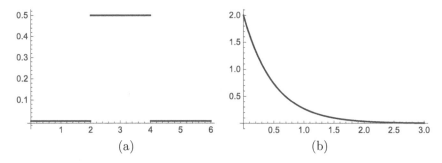

FIGURE 4.3 (a) $U[2,4]$ density; (b) $\exp(2)$ density.

Example 1. Suppose that the time T of the increase in the interest rate of a credit card, measured in years from the beginning of a particular year, is an exponentially distributed random variable with parameter 2, and that the amount of the increase is a uniformly distributed random variable A on the interval $[0, .5]$ (in units of percent). Then the probability that the rate will not increase this year is:

$$P[T > 1] = \int_1^\infty 2e^{-2t}\,dt = -e^{-2t}\big|_1^\infty = e^{-2} \approx .135.$$

The probability that, when an increase does happen, it will be by no more than .2 percent, is:

$$P[A \leq .2] = \int_0^{.2} \frac{1}{.5}\,da = 2 \cdot (.2) = .4.$$

We could have computed this as the length of the subinterval $[0, .2]$ specified by the event, divided by the length of the interval $[0, .5]$ that makes up the state space of random variable A. A general formula for the c.d.f. of T is:

$$F(x) = P[T \leq x] = \int_0^x 2e^{-2t} dt = -e^{-2t}|_0^x = 1 - e^{-2x}, x \geq 0.$$

For $x < 0$, the c.d.f. would be zero. Similarly, the c.d.f. of A would be:

$$G(x) = P[A \leq x] = \int_0^x \frac{1}{.5} da = \int_0^x 2 \, da = 2x, 0 \leq x < .5.$$

Also, $G(x) = 0$ for $x < 0$, and $G(x) = 1$ for $x \geq .5$. ∎

4.1.2 Expectation

The main ideas and results regarding expectation are similar to the discrete case. Suppose that a continuous random variable X has p.d.f. $f(x)$. Then, as noted above, the probability that X identically equals a constant value c is zero, since:

$$P[X = c] = \int_c^c f(x) \, dx = 0. \tag{4.8}$$

This is, of course, a key difference between discrete and continuous probability distributions. Density is not the same as probability. But a left-hand approximation for the probability that X is in a short interval to the right of c is

$$P[X \in [c, c + \Delta x]] = \int_c^{c+\Delta x} f(x) \, dx \approx \int_c^{c+\Delta x} f(c) \, dx = f(c) \cdot \Delta x. \tag{4.9}$$

This yields the interpretation that the density value $f(c)$ gives us the rate at which probability is accumulating at point c as we sweep left to right in the state space. This observation is also consistent with the calculus result that $F'(x) = f(x)$.

Turning to expectation, if we discretize a continuous random variable X by splitting its state space into intervals $[x_i, x_{i+1}]$ of small, equal width Δx, and create a discrete random variable $\tilde{X}$ whose states are the left endpoints x_i of these intervals and whose probability weights are the probabilities $f(x_i) \cdot \Delta x$ as estimated above, then the expectation of the discretized version of X is:

$$E\left[\tilde{X}\right] = \sum_i x_i \cdot f(x_i) \cdot \Delta x. \tag{4.10}$$

As the subinterval length Δx shrinks to 0, we recognize a Riemann sum for the function $x \cdot f(x)$, which converges to the integral in the following definition.

Definition 4. The *expected value* (or *expectation*, or **mean**) of a continuous random variable X with density function f is the following integral, if it converges:

$$\mu = E[X] = \int_{-\infty}^{+\infty} x \cdot f(x)dx. \quad \blacksquare \tag{4.11}$$

In the integral in formula (4.11), the limits of integration adjust according to the domain of values on which the density f is non-zero. Specifically, if $f > 0$ only on an interval $[a, b]$, then the limits $-\infty$ and $+\infty$ are replaced by a and b, respectively, in formula (4.11).

Example 2. Returning to Example 1, the expected value of the time of change in interest rate is:

$$E[T] = \int_0^\infty t \cdot f(t)\, dt = \int_0^\infty t \cdot 2e^{-2t}dt.$$

Apply integration by parts to this integral with $u = t$, $dv = 2e^{-2t}dt$, and hence $du = dt$, $v = -e^{-2t}$ to get:

$$
\begin{aligned}
E[T] &= \int_0^\infty t \cdot 2e^{-2t}dt \\
&= t \cdot \left(-e^{-2t}\right)|_0^\infty - \int_0^\infty \left(-e^{-2t}\right)\, dt \\
&= 0 + \int_0^\infty e^{-2t}\, dt \\
&= \left(-\frac{e^{-2t}}{2}\right)|_0^\infty \\
&= \frac{1}{2}
\end{aligned}
\tag{4.12}
$$

The time units are years, so a change in rate is expected in 6 months. The expected amount of the increase, in units of percent, is:

$$E[A] = \int_0^{.5} a \cdot \frac{1}{.5}\, da = 2 \cdot \int_0^{.5} a \cdot da = 2 \cdot \frac{a^2}{2}|_0^{.5} = (.5)^2 = .25. \quad \blacksquare \tag{4.13}$$

It is easy to generalize the computations in (4.12) and (4.13) (see Exercises 5 and 6) to derive that if a random variable X has the $\exp(\lambda)$ distribution, then its mean is:

$$E[X] = \frac{1}{\lambda}, \tag{4.14}$$

and if X has the $U([a, b])$ distribution, then:

$$E[X] = \frac{a+b}{2}. \tag{4.15}$$

We will be able to show in Section 4.2 that in the continuous case as in the discrete case, linearity holds:

$$E[aX + bY] = aE[X] + bE[Y]. \tag{4.16}$$

This enables us to talk about variance and its computational formula.

Definition 5. The **variance** of a continuous random variable X with density function f is the following integral, if it converges:

$$\sigma^2 = \text{Var}(X) = E\left[(X - \mu)^2\right] = \int_{-\infty}^{+\infty} (x - \mu)^2 \cdot f(x)dx. \quad \blacksquare \qquad (4.17)$$

As in the discrete case, variance measures the spread of the distribution about its mean μ. The symbol σ is, as usual, the square root of the variance and is called the **standard deviation** of the distribution.

In general, the expected value of a function g of a continuous random variable X is:

$$E[g(X)] = \int_{-\infty}^{+\infty} g(x) \cdot f(x)\, dx, \qquad (4.18)$$

where f is the probability density function of X. Thus, the expectation that defines the variance $E\left[(X - \mu)^2\right]$ has meaning. We can use linearity to expand and get the well-known computational formula:

$$\begin{aligned}
\text{Var}(X) &= E\left[(X - \mu)^2\right] \\
&= E\left[X^2 - 2\mu X + \mu^2\right] \\
&= E\left[X^2\right] - 2\mu \cdot E[X] + \mu^2 \\
&= E\left[X^2\right] - \mu^2.
\end{aligned} \qquad (4.19)$$

Example 3. For the time of interest rate change T in Example 1, we had computed that $\mu = E[T] = 1/\lambda = 1/2$. Using the computational formula (4.19), we have that the variance is the expected square of T minus $(1/2)^2$. The expected square can be found using two applications of integration by parts. In the first, $u = t^2$, $dv = 2e^{-2t}dt$, and hence $du = 2tdt$, $v = -e^{-2t}$, and we get:

$$\begin{aligned}
E\left[T^2\right] &= \int_0^\infty t^2 \cdot 2e^{-2t}dt \\
&= t^2 \cdot \left(-e^{-2t}\right)\big|_0^\infty - \int_0^\infty \left(-2te^{-2t}\right) dt \\
&= 0 + \int_0^\infty t \cdot 2e^{-2t}dt.
\end{aligned} \qquad (4.20)$$

But the last integral on the right is just the integral that represents $E[T]$, so we don't have to carry out the second stage of integration by parts. The answer will be $1/\lambda = 1/2$. Thus,

$$\text{Var}(T) = E\left[T^2\right] - \mu^2 = \frac{1}{2} - \left(\frac{1}{2}\right)^2 = \frac{1}{4}. \qquad (4.21)$$

The standard deviation is therefore $\sigma = \sqrt{1/4} = 1/2$ year. In general, it can be shown similarly that the variance of the $\exp(\lambda)$ distribution is $1/\lambda^2$. $\blacksquare$

4.1.3   Normal and Lognormal Distributions

By far the most important continuous distribution for financial mathematics is the normal distribution.

Definition 6. A continuous random variable X taking values throughout the real line $\mathbb{R}$ is said to have the ***normal density function*** with parameters μ and σ^2 if, for all subsets B of $\mathbb{R}$:

$$P[X \in B] = \int_B \frac{1}{\sqrt{2\pi\sigma^2}} e^{-\frac{(x-\mu)^2}{2\sigma^2}} \, dx. \qquad (4.22)$$

(We use the shorthand $X \sim N\left(\mu, \sigma^2\right)$ if this is the case.) ∎

It can be shown that the parameters are well-labeled; μ is the mean of the distribution and σ^2 is the variance. A sketch of the p.d.f. in the integrand in the special case $\mu = 0$, $\sigma^2 = 1$, called the ***standard normal density***, appears in Figure 4.4. Note the symmetry about the vertical line $x = \mu = 0$, and also note that beyond three standard deviations $3\sigma = 3 \cdot 1 = 3$ from the mean, there is very little probability weight. These properties carry through for normal distributions with other means and variances as well.

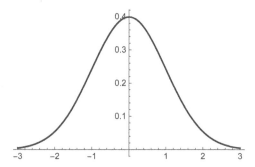

FIGURE 4.4 $N(0,1)$ density function.

Example 4. One of the reasons that the normal distribution is so important in financial mathematics is that it is used as a model for rate of return behavior of risky assets. It is mathematically convenient to work with, but does it fit reality in any way?

The answer is: yes and no. In Figure 4.5(a) we display a histogram of weekly rates of return for ExxonMobil stock on the New York Stock Exchange covering the years 2012-2016. The sample mean and standard deviation of those rates are $\bar{R} = .00010$ and $S_R = .01112$. The data histogram shows a shape that is consistent with the bell-shape of a normal density. But in part (b) of the figure, we superimpose a normal density with parameters equal to this sample mean and standard deviation, and we find that the density does not fit the data very well. Data is clustered around the mean more than expected, and there are more extreme observations than expected. There are a few reasons why this could be. Perhaps in real markets, price change activity is very quiet most of the time, resulting in numerous rates of change near zero, more, perhaps, than a normal model would predict.

In this case, there may be another explanation, though. In Figure 4.6 we repeat the histogram of weekly rates, superimposing a normal density with the same mean but with standard deviation $S_R - .0025$ (arrived at through some trial and error). This time the fit is slightly better through the region $[-.02, .02]$ where the large majority of the data lie. What seems to have happened with ExxonMobil during this time period is an unusual number of extreme changes, both up and down, in which rates of return were less than $-.02$ or greater than $.02$. It may be that a probability distribution with fatter tails, such as the non-central t distribution, would provide a better fit. But the empirical observation that "most of the time" the rates of return are normal suggests a more complicated model, which is to suppose that the generic rate of return variable has a normal density, but with occasional discrete shocks. Much of the classical theory, however, is built upon assumptions of normality, and for most of the ensuing development in this text, we will continue along these lines. ■

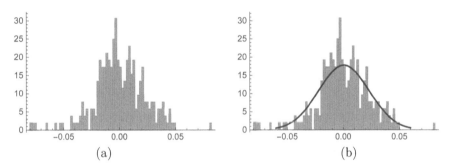

FIGURE 4.5 (a) Rates of return for ExxonMobil, 2012-2016; (b) with normal density.

Recall from probability theory the standardization theorem, stated below.

Theorem 1. If a random variable X has the $N\left(\mu, \sigma^2\right)$ distribution, then the random variable

$$Z = \frac{X - \mu}{\sigma} \tag{4.23}$$

has the standard normal distribution. ■

You are led through a proof of this fact in Exercise 10.

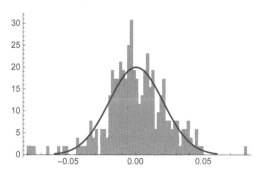

FIGURE 4.6 ExxonMobil rates with normal density, reduced variance.

Example 5. Suppose that the rate of return R that an investor receives on her $1000 investment is normally distributed with mean 5% and standard deviation 2%. Find the probability that the investment will be worth between $980 and $1050 at the end of one period.

Solution. The relationship between initial value, final value, and rate of return is:

$$\text{final value} = \text{initial value}(1 + \text{rate of return}).$$

Therefore, the final value is between $980 and $1050 if and only if:

$$980 \le 1000(1 + R) \le 1050 \quad \Longleftrightarrow \quad .98 \le 1 + R \le 1.05$$
$$\Longleftrightarrow \quad -.02 \le R \le .05.$$

To find $P[-.02 \le R \le .05]$, we can numerically integrate the $N\left(.05, .02^2\right)$ density between the endpoints $-.02$ and $.05$. The result turns out to be about .4998. Or, we could standardize the inequality and use the standardization theorem.

$$
\begin{aligned}
P[-.02 \le R \le .05] &= P\left[\tfrac{-.02-.05}{.02} \le \tfrac{R-.05}{.02} \le \tfrac{.05-.05}{.02}\right] \\
&= P[-3.5 \le Z \le 0] \\
&= P[Z \le 0] - P[Z \le -3.5] \\
&= .5 - P[Z \le -3.5].
\end{aligned}
$$

A standard normal table or appropriate technology will confirm that the probability is .4998 to the fourth decimal place. ∎

The price of a risky asset is often assumed to have a distribution called the **lognormal distribution**. The name comes from the fact that if Y is such a random variable, the assumption is that $X = \log(Y)$ is normal with some mean μ and variance σ^2. In other words, a lognormal random variable Y is represented as $Y = e^X$ where X is normal. In the definition below we write the form of the density.

Definition 7. A random variable Y has the *lognormal distribution* with parameters μ and σ^2 (for which we write $Y \sim \mathcal{LN}(\mu, \sigma^2)$) if its logarithm has the $N(\mu, \sigma^2)$ density. The p.d.f. of Y is:

$$f_Y(y; \mu, \sigma^2) = \frac{1}{y} \frac{1}{\sqrt{2\pi\sigma^2}} e^{-\frac{(\log(y)-\mu)^2}{2\sigma^2}}, \, y > 0. \ \blacksquare \qquad (4.24)$$

The form of the density comes from the following reasoning. Recall that the derivative of a c.d.f. is the density with which it is associated. Consider the c.d.f. F_Y of a log-normal random variable Y, whose logarithm $X = \log(Y)$ is normally distributed. Inverting the relationship, $Y = e^X$. Then:

$$
\begin{aligned}
F_Y(y) &= P[Y \le y] \\
&= P[e^X \le y] \\
&= P[X \le \log(y)] \\
&= F_X(\log(y)),
\end{aligned}
$$

where F_X is the normal c.d.f. of random variable X. Differentiating with respect to y on both sides, we have, by the chain rule:

$$f_Y(y) = F_Y'(y) = F_X'(\log(y)) \cdot \frac{1}{y} = f_X(\log(y)) \cdot \frac{1}{y}.$$

In the last formula, f_X is the normal density of X, and thus substitution into that formula produces the form for the density of Y that is in (4.24).

The lognormal density function is right-skewed and focuses all of its weight on positive values, as shown in Figure 4.7 in the case where $\mu = 0$ and $\sigma^2 = 1$. Increases in the value of the μ parameter shift the distribution to the right and flatten it.

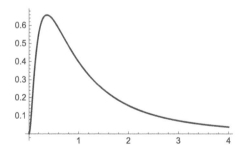

FIGURE 4.7 $\mathcal{LN}(0,1)$ density function.

The importance of the lognormal distribution to finance is this. We work similarly to the development of Section 3.7. Let a risky asset have price values

S_i at successive time periods. If $R_1, R_2, R_3, \ldots$ are the rates of return, so that $S_i = (1 + R_i) S_{i-1}, i = 1, 2, 3, \ldots$ then the asset price at time n is:

$$S_n = s_0 \cdot \frac{S_1}{s_0} \cdot \frac{S_2}{S_1} \cdot \frac{S_3}{S_2} \cdots \frac{S_n}{S_{n-1}} = s_0 \left(1 + R_1\right) \left(1 + R_2\right) \left(1 + R_3\right) \cdots \left(1 + R_n\right).$$

(4.25)

This implies that the log of the price at time n is:

$$
\begin{aligned}
\log\left(S_n\right) &= \log\left(s_0 \left(1 + R_1\right) \left(1 + R_2\right) \left(1 + R_3\right) \cdots \left(1 + R_n\right)\right) \\
&= \log\left(s_0\right) + \sum_{l=1}^{n} \log\left(1 + R_l\right).
\end{aligned}
$$

(4.26)

If n is large and the R_i are independent and identically distributed, the Central Limit Theorem implies that $\log\left(S_n\right)$ is approximately normal. Hence S_n itself is lognormally distributed.

Our final example takes advantage of such a model.

Example 6. Suppose that the final price of a share of stock, measured as a multiple of its initial price, is lognormally distributed with parameters $\mu = .45$ and $\sigma = .3$. An investor has an option to buy a share of the stock at 1.2 times its initial price, which is profitable only if the final price exceeds this (in which case the investor can buy the stock at the lower option price and immediately resell at the higher market price for a profit). What is the likelihood of profit?

Solution. Let $X = S_n/s_0$ be the final price divided by the initial price; then we are assuming that $\log(X)$ is normal with $\mu = .45$, $\sigma = .3$. The likelihood of profit is:

$$P[X > 1.2] = P[\log(X) > \log(1.2)].$$

Since $\log(X)$ is normal with the given mean and standard deviation, we can standardize both sides of the last inequality to get:

$$
\begin{aligned}
P[X > 1.2] &= P\left[\frac{\log(X) - .45}{.3} > \frac{\log(1.2) - .45}{.3}\right] \\
&= P[Z > -.8923] \\
&= 1 - P[Z \le -.8923].
\end{aligned}
$$

A standard normal table, or an appropriate technology such as Excel, R, or *Mathematica*, gives the latter probability as about .1861, so that the probability of profit is about $1 - .1861 = .8139$. ∎

Exercises 4.1

1. Suppose that a probability density function has the form $f(x) = cx^2, x \in [0, 1]$. What value must the constant c have? If a random variable X has this distribution, compute $P[X > .5]$.

2. Find the mean and variance of the distribution in Exercise 1.

3. Suppose that a probability density function has the form $f(x) = cxe^{-2x}$, for $x \in [0, \infty)$. What value must the constant c have? If a random variable X has this distribution, compute $P[X \leq 1]$.

4. Find the mean of the distribution in Exercise 3.

5. Verify formula (4.14) for the mean of the $\exp(\lambda)$ distribution.

6. Verify formula (4.15) for the mean of the $U([a, b])$ distribution.

7. In derivation (4.19), was the general theorem on linearity of expectation really necessary to prove the result, or is there a more direct approach?

8. Compute the variance of the continuous uniform distribution on the interval $[a, b]$.

9. The V-shaped function in the figure constitutes a valid density function. Find the maximum value b. If a random variable X has this density, find the probability that $X > .5$. Finally, find the mean and variance of this distribution.

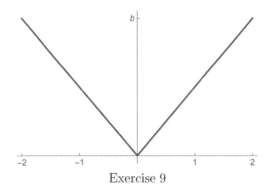

Exercise 9

10. Prove Theorem 1. (Hint: Write the defining expression for the c.d.f. of Z, and substitute the formula for Z. Isolate X in the inequality. Then write the result in terms of the c.d.f. of X and differentiate with respect to z.)

11. (Technology or tables required) If random variable X has the normal distribution with parameters $\mu = 1.2$ and $\sigma^2 = 4$, find:
   (a) $P[X > -1]$;
   (b) a number x such that $P[X > x] = .3$;
   (c) $P\left[e^{2X} \leq 7\right]$.

12. (Technology required) In Example 5, what is the smallest mean rate of return that still makes a probability of at least 60% that the investment will be worth at least $1020? Assume the the standard deviation is still 2%.

13. (Technology required) In Example 6, what is the expected value of the profit on the option?

4.2   Joint Distributions

In this section we review the most important concepts regarding continuous joint distributions of random variables. More detail can be found in Hastings [10], from which we draw heavily. Particularly, we are interested in independence and dependence issues, leading up to an understanding of multivariate normal distributions. Prerequisite material here is multiple integration of functions of several variables. Basically, multiple integration is just repeated single integration over the individual variables. But even if that idea is not very familiar to you, we show computations in some detail, which should let you review in context and then operate with some confidence.

4.2.1   Basic Ideas

You should already have strong familiarity with joint distributions of discrete random variables from past work in the discrete case. We can combine those concepts with the concept of density function in the continuous case to produce the following basic definition.

Definition 1. Let $X_1, X_2, ..., X_n$ be continuous random variables defined on a common sample space Ω, and let E be the space of all possible values of the n-dimensional random vector $\boldsymbol{X} = (X_1, X_2, ..., X_n)$. These random variables are said to have ***joint probability density function*** $f(x_1, x_2, ..., x_n)$ if for any subset B of E,

$$P\left[(X_1, X_2, ..., X_n) \in B\right] = \int \int_B \cdots \int f(x_1, x_2, ..., x_n)\, dx_n dx_{n-1} \cdots dx_1.$$
$$(4.27)$$

Here, f must be a non-negative function that integrates to 1 over all of E. ∎

The same caveat about which subsets B of the joint state space E are possible to use applies here as in the single-dimensional case. We will have more to say about that in Section 4.3. In this section, we will only deal with simple subsets of $\mathbb{R}^n$ such as Cartesian products of intervals. The concepts are well-illustrated in the two-dimensional case, and most of the time we will be in that problem domain.

The related notion of **joint cumulative distribution function** is defined in a natural way as the cumulative probability to the left of all random variables. For example, in the two-dimensional case we would have:

$$F(x_1, x_2) = P[X_1 \le x_1, X_2 \le x_2] = \int_{-\infty}^{x_1} \int_{-\infty}^{x_2} f(t_1, t_2)\, dt_2 dt_1. \qquad (4.28)$$

Example 1. Suppose that three random variables X_1, X_2, and X_3 have the joint density below. Find $P[X_1 > 2, X_2 < 1, X_3 \in [2,3]]$.

$$f(x_1, x_2, x_3) = \frac{1}{12} \cdot e^{-x_2}, x_1 \in [0,3], x_2 \in (0, +\infty), x_3 \in [1,5].$$

Solution. In light of the specified regions for the three random variables, to compute the joint probability, x_1 would be integrated from 2 to 3, x_2 from 0 to 1, and x_3 from 2 to 3:

$$
\begin{aligned}
P[X_1 > 2, X_2 < 1, X_3 \in [2,3]] &= \int_2^3 \int_0^1 \int_2^3 f(x_1, x_2, x_3)\, dx_3 dx_2 dx_1 \\
&= \tfrac{1}{12} \cdot \int_2^3 \int_0^1 e^{-x_2} \int_2^3 1\, dx_3 dx_2 dx_1 \\
&= \tfrac{1}{12} \cdot \int_2^3 \int_0^1 e^{-x_2} \cdot 1\, dx_2 dx_1 \\
&= \tfrac{1}{12} \cdot \int_2^3 (-e^{-x_2})\,|_0^1 dx_1 \\
&= \tfrac{1}{12} \cdot \int_2^3 (1 - e^{-1})\, dx_1 \\
&= \tfrac{1}{12} \cdot (1 - e^{-1}) \cdot 1 \\
&= \tfrac{1}{12} \cdot (1 - e^{-1}) \approx .0527. \ \blacksquare
\end{aligned}
$$

Example 2. Suppose that the joint density of two random variables X_1 and X_2 is:

$$f(x_1, x_2) = c(x_1 + 2x_2), x_1, x_2 \in [0,1],$$

where c is a constant. Then:
(a) Find the value of c that makes f a legitimate density;
(b) Find the joint c.d.f. of X_1 and X_2; and
(c) Compute $P[X_1 \le X_2]$.

Solution. (a) Since f is non-negative (as long as c is) for all values in the state space $E = [0,1] \times [0,1]$, we must choose c so that the double integral of f over this square is exactly 1. To do so, we compute:

$$
\begin{aligned}
\int_0^1 \int_0^1 c(x_1 + 2x_2)\, dx_2 dx_1 &= c \cdot \int_0^1 (x_1 x_2 + x_2^2)\,|_0^1 dx_1 \\
&= c \cdot \int_0^1 (x_1 + 1)\, dx_1 \\
&= c \cdot \left(\frac{x_1^2}{2} + x_1\right)|_0^1 \\
&= c \cdot \tfrac{3}{2}.
\end{aligned}
$$

Therefore, c must be $\frac{2}{3}$.

(b) A similar computation yields:

$$
\begin{aligned}
F\left(x_{1}, x_{2}\right) &= P\left[X_{1} \leq x_{1}, X_{2} \leq x_{2}\right] \\
&= \int_{0}^{x_{1}} \int_{0}^{x_{2}} \frac{2}{3}\left(t_{1}+2 t_{2}\right) d t_{2} d t_{1} \\
&= \frac{2}{3} \cdot \int_{0}^{x_{1}}\left(t_{1} t_{2}+t_{2}^{2}\right)\left.\right|_{0}^{x_{2}} d t_{1} \\
&= \frac{2}{3} \cdot \int_{0}^{x_{1}}\left(t_{1} x_{2}+x_{2}^{2}\right) d t_{1} \\
&= \frac{2}{3} \cdot\left(\frac{t_{1}^{2} x_{2}}{2}+t_{1} x_{2}^{2}\right)\left.\right|_{0}^{x_{1}} \\
&= \frac{2}{3} \cdot\left(\frac{x_{1}^{2} x_{2}}{2}+x_{1} x_{2}^{2}\right) .
\end{aligned}
$$

(c) To compute $P\left[X_{1} \leq X_{2}\right]$, we integrate the joint density as x_{2} ranges from x_{1} to 1, then as x_{1} ranges from 0 to 1:

$$
\begin{aligned}
P\left[X_{1} \leq X_{2}\right] &= \int_{0}^{1} \int_{x_{1}}^{1} \frac{2}{3}\left(x_{1}+2 x_{2}\right) d x_{2} d x_{1} \\
&= \frac{2}{3} \cdot \int_{0}^{1}\left(x_{1} x_{2}+x_{2}^{2}\right)\left.\right|_{x_{1}}^{1} d x_{1} \\
&= \frac{2}{3} \cdot \int_{0}^{1}\left(\left(x_{1}+1\right)-\left(x_{1}^{2}+x_{1}^{2}\right)\right) d x_{1} \\
&= \frac{2}{3} \cdot \int_{0}^{1}\left(-2 x_{1}^{2}+x_{1}+1\right) d x_{1} \\
&= \frac{2}{3} \cdot\left(-\frac{2 x_{1}^{3}}{3}+\frac{x_{1}^{2}}{2}+x_{1}\right)\left.\right|_{0}^{1} \\
&= \frac{2}{3} \cdot\left(-\frac{2}{3}+\frac{1}{2}+1\right)=\frac{5}{9} . \ \blacksquare
\end{aligned}
$$

A 3-D plot of the joint density function in Example 2 as a function of the two variables x_{1} and x_{2} is in Figure 4.8. The probability of a set B in the $x_{1}-x_{2}$ plane is the volume of the solid bounded above by the surface, which is a plane here, and below by B. The answer of 5/9 to part (c) of the example is not unreasonable when the set B is the triangle of points where $0 \leq x_{1} \leq x_{2} \leq 1$, which is the northwest half of the unit square.

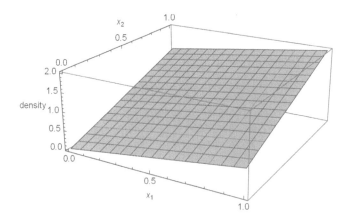

FIGURE 4.8 Joint p.d.f. $f\left(x_{1}, x_{2}\right)=c\left(x_{1}+2 x_{2}\right)$.

Expectation in the multivariate case follows along the lines of single variable continuous expectation.

Definition 2. Let $X_1, X_2, ..., X_n$ be continuous random variables defined on a common sample space Ω with joint probability density function $f(x_1, x_2, ..., x_n)$ on state space E. The **expected value** of a function g of the random variables is defined by:

$$E\left[g\left(X_1, X_2, ..., X_n\right)\right] = \int \int_E \cdots \int g\left(x_1, x_2, ..., x_n\right) \cdot f\left(x_1, x_2, ..., x_n\right) dx_n dx_{n-1} \cdots dx_1, \tag{4.29}$$

if the multiple integral converges. ∎

We can now easily verify the linearity of expectation in the continuous case. For two random variables, we would have:

$$\begin{aligned} E\left[cX_1 + dX_2\right] &= \int_E \int \left(cx_1 + dx_2\right) \cdot f\left(x_1, x_2\right) dx_2 dx_1 \\ &= \int_E \int \left(cx_1\right) \cdot f\left(x_1, x_2\right) dx_2 dx_1 \\ &\quad + \int_E \int \left(dx_2\right) \cdot f\left(x_1, x_2\right) dx_2 dx_1 \\ &= c \cdot \int_E \int x_1 \cdot f\left(x_1, x_2\right) dx_2 dx_1 \\ &\quad + d \cdot \int_E \int x_2 \cdot f\left(x_1, x_2\right) dx_2 dx_1 \\ &= cE\left[X_1\right] + dE\left[X_2\right]. \end{aligned} \tag{4.30}$$

The derivation for three, four, or more random variables is similar, or see Exercise 4 for an easy alternative.

Example 3. In Example 1, compute $E\left[X_1 \cdot X_2 \cdot X_3\right]$.

Solution. Plugging into the definition, we see that the integral can be split into the product of three single integrals as follows:

$$\begin{aligned} E\left[X_1 \cdot X_2 \cdot X_3\right] &= \int_0^3 \int_0^\infty \int_1^5 x_1 x_2 x_3 \cdot \frac{1}{12} \cdot e^{-x_2} dx_3 dx_2 dx_1 \\ &= \frac{1}{12} \cdot \int_0^3 x_1 dx_1 \cdot \int_0^\infty x_2 \cdot e^{-x_2} dx_2 \cdot \int_1^5 x_3 dx_3. \end{aligned}$$

The two outer integrals will be easy. For the middle integral with respect to x_2, we note that it is just the mean of an exponential random variable with parameter $\lambda = 1$; hence its value is $1/\lambda = 1$. Finishing the computation, we find that:

$$\begin{aligned} E\left[X_1 \cdot X_2 \cdot X_3\right] &= \frac{1}{12} \cdot \int_0^3 x_1 dx_1 \cdot 1 \cdot \int_1^5 x_3 \, dx_3 \\ &= \frac{1}{12} \cdot \left(\frac{x_1^2}{2} \Big|_0^3\right) \cdot \left(\frac{x_3^2}{2} \Big|_1^5\right) \\ &= \frac{1}{12} \cdot \frac{9}{2} \cdot \left(\frac{25}{2} - \frac{1}{2}\right) = \frac{9}{2}. \; \blacksquare \end{aligned}$$

4.2.2   Marginal and Conditional Distributions

Sometimes we want to draw conclusions about a single random variable even when there are others in the model. The density of a single random variable is easy to derive, by integrating out all others from the joint density.

Definition 3. Let $X_1, X_2, ..., X_n$ be random variables defined on a common sample space Ω and let $E_1, E_2, ..., E_n$ be the respective state spaces of possible values of the random variables. Suppose that the joint probability density function is $f(x_1, x_2, ..., x_n)$. The *marginal probability density function of X_i* is:

$$f_i(x_i) = \int_{E_1} \int_{E_2} \cdots \int_{E_{i-1}} \int_{E_{i+1}} \cdots \int_{E_n} f(x_1, x_2, ..., x_n)$$
$$dx_n \cdots dx_{i+1} dx_{i-1} \cdots dx_2 dx_1. \quad \blacksquare \tag{4.31}$$

The idea is easiest to see in the two-variable case. If X_1 and X_2 are random variables with joint density $f(x_1, x_2)$, and E_2 is the complete state space of X_2, then:

$$\begin{aligned} P[X_1 \in B] &= P[X_1 \in B, X_2 \in E_2] \\ &= \int_B \left(\int_{E_2} f(x_1, x_2)\, dx_2 \right) dx_1 \\ &= \int_B f_1(x_1)\, dx_1. \end{aligned} \tag{4.32}$$

Similarly, for X_2, $P[X_2 \in B] = \int_B f_2(x_2)\, dx_2$, so the marginal densities completely determine all probabilities of events involving individual random variables.

Joint marginals of subgroups of the full set of random variables are defined and computed similarly. To compute a joint marginal density, integrate out the complete joint density function over all the state spaces of random variables that are not in the desired subgroup. For instance, for three random variables, the joint marginal density of the first two would be:

$$f_{12}(x_1, x_2) = \int_{E_3} f(x_1, x_2, x_3)\, dx_3. \tag{4.33}$$

Here is a quick illustration of the computation and use of marginal densities.

Example 4. In Example 2, we had two random variables X_1 and X_2 with joint density function:

$$f(x_1, x_2) = \frac{2}{3}(x_1 + 2x_2), x_1, x_2 \in [0, 1].$$

Find the marginal p.d.f.s of each random variable. Use them to compute the probabilities $P[X_1 < .25]$ and $P[X_2 > .5]$.

Solution. The marginal density of X_1 is:

$$
\begin{aligned}
f_1(x_1) &= \int_{E_2} f(x_1, x_2)\, dx_2 \\
&= \int_0^1 \tfrac{2}{3}(x_1 + 2x_2)\, dx_2 \\
&= \tfrac{2}{3}(x_1 x_2 + x_2^2)\,\big|_0^1 \\
&= \tfrac{2}{3}(x_1 + 1).
\end{aligned}
$$

The set of possible values of X_1 is still the interval $[0, 1]$. It is easy to check that this f_1 is non-negative and integrates to 1 over $[0, 1]$; therefore it is indeed a valid density. Using it, we can compute:

$$
P[X_1 < .25] = \int_0^{.25} \tfrac{2}{3}(x_1 + 1)\, dx_1 = \tfrac{2}{3}\left(\tfrac{x_1^2}{2} + x_1\right)\Big|_0^{1/4} = \tfrac{2}{3}\left(\tfrac{1}{32} + \tfrac{1}{4}\right) = \tfrac{3}{16}.
$$

For X_2, we integrate out the joint density over x_1 values from 0 to 1, to obtain:

$$
\begin{aligned}
f_2(x_2) &= \int_{E_1} f(x_1, x_2)\, dx_1 \\
&= \int_0^1 \tfrac{2}{3}(x_1 + 2x_2)\, dx_1 \\
&= \tfrac{2}{3}\left(\tfrac{x_1^2}{2} + 2x_2 x_1\right)\Big|_0^1 \\
&= \tfrac{2}{3}\left(2x_2 + \tfrac{1}{2}\right).
\end{aligned}
$$

This function does integrate to 1 over the full state space $[0, 1]$ of X_2. Then the desired probability is:

$$
P[X_2 > .5] = \int_{.5}^1 \tfrac{2}{3}\left(2x_2 + \tfrac{1}{2}\right) dx_2 = \tfrac{2}{3}\left(x_2^2 + \tfrac{1}{2}x_2\right)\Big|_{1/2}^1 = \tfrac{2}{3}\left(\tfrac{3}{2} - \tfrac{1}{2}\right) = \tfrac{2}{3}. \ \blacksquare
$$

Remember from basic discrete conditional probability that for two discrete random variables X_1 and X_2:

$$
P[X_2 = x_2 | X_1 = x_1] = \frac{P[X_1 = x_1, X_2 = x_2]}{P[X_1 = x_1]}.
$$

In words, the conditional probability mass function of X_2 given X_1 is the joint mass function divided by the marginal p.m.f. of X_1. Replacing mass functions by density functions in the continuous case, we generate the following definition.

Definition 4. Let X_1, X_2 be two random variables defined on a common sample space Ω, with joint probability density function $f(x_1, x_2)$. Let $f_1(x_1)$ be the marginal probability density function of X_1. Define the **conditional p.d.f. of X_2 given $X_1 = x_1$** to be:

$$f_{2|1}(x_2|x_1) = \frac{f(x_1, x_2)}{f_1(x_1)}. \tag{4.34}$$

Similarly, if $f_2(x_2)$ is the marginal p.d.f. of X_2, define the **conditional p.d.f. of X_1 given $X_2 = x_2$** as:

$$f_{1|2}(x_1|x_2) = \frac{f(x_1, x_2)}{f_2(x_2)}. \ \blacksquare \tag{4.35}$$

Then to compute probabilities of events involving one continuous random variable given a particular value of another, the conditional p.d.f. is integrated over appropriate limits. For more than two random variables the idea is similar. A conditional (joint) density of a subgroup of random variables given its complementary subgroup is the overall joint density divided by the marginal (joint) density of the subgroup of variables being conditioned on. For example, in the case of three random variables, the conditional density of X_3 given both $X_1 = x_1$ and $X_2 = x_2$ is:

$$f_{3|1,2}(x_3|x_1, x_2) = \frac{f(x_1, x_2, x_3)}{f_{12}(x_1, x_2)}. \tag{4.36}$$

For instance, in Example 4 we computed the marginal density of X_1. Using this, the conditional density of X_2 given a particular value x_1 is:

$$f_{2|1}(x_2|x_1) = \frac{f(x_1, x_2)}{f_1(x_1)} = \frac{\frac{2}{3}(x_1 + 2x_2)}{\frac{2}{3}(x_1 + 1)} = \frac{(x_1 + 2x_2)}{(x_1 + 1)}, x_2 \in [0, 1].$$

It is common for a conditional p.d.f. to depend upon the observed value of the random variable being conditioned on, as it does here, and in some situations even the state space may change depending on this observed value. But in the next example, a special condition is illustrated.

Example 5. Example 1 introduced three random variables X_1, X_2, and X_3 with joint density:

$$f(x_1, x_2, x_3) = \frac{1}{12} \cdot e^{-x_2}, x_1 \in [0, 3], x_2 \in (0, +\infty), x_3 \in [1, 5].$$

Let us find the conditional density of X_3 given both $X_1 = x_1$ and $X_2 = x_2$, and use it to compute the conditional probability:

$$P[X_3 \geq 3 | X_1 = 1, X_2 = 2].$$

Solution. By formula (4.36) we must first find the joint marginal f_{12} for each particular pair of values x_1 and x_2 of the first two random variables. To do this we integrate out x_3:

$$f_{12}(x_1, x_2) = \int_1^5 \frac{1}{12} \cdot e^{-x_2} \, dx_3 = \frac{1}{12} \cdot e^{-x_2} \cdot \int_1^5 1 \, dx_3 = \frac{4}{12} \cdot e^{-x_2}.$$

Then:

$$f_{3|1,2}(x_3|x_1, x_2) = \frac{f(x_1, x_2, x_3)}{f_{12}(x_1, x_2)} = \frac{\frac{1}{12} \cdot e^{-x_2}}{\frac{4}{12} \cdot e^{-x_2}} = \frac{1}{4}, x_3 \in [1, 5].$$

The conditional density does not depend on either x_1 or x_2; in fact, we can recognize it as the uniform density on the interval $[1, 5]$. It is now easy to find the probability that was requested:

$$P[X_3 \geq 3 | X_1 = 1, X_2 = 2] = \frac{\text{length of } [3, 5]}{\text{length of } [1, 5]} = \frac{1}{2}.$$

As a final note that will transition us to the next subsection, we can compute the marginal density of X_3 by integrating out the other two variables:

$$
\begin{aligned}
f_3(x_3) &= \int_0^3 \int_0^\infty f(x_1, x_2, x_3) \, dx_2 dx_1 \\
&= \int_0^3 \int_0^\infty \frac{1}{12} \cdot e^{-x_2} \, dx_2 dx_1 \\
&= \frac{1}{12} \cdot \int_0^3 (-e^{-x_2} |_0^\infty) \, dx_1 \\
&= \frac{1}{12} \cdot \int_0^3 1 \, dx_1 \\
&= \frac{3}{12} = \frac{1}{4}, x_3 \in [1, 5].
\end{aligned}
$$

Therefore the conditional density of X_3 given the other variables matches the marginal, that is, the unconditional density. In symbols, $f_{3|1,2}(x_3|x_1, x_2) = f_3(x_3)$ for all x_1, x_2, x_3. In Exercise 8 you will check that the same can be said for the conditionals of X_1 and X_2 given the complementary pair of random variables. This is the property of **independence**, which is explored in the next subsection. ∎

Remark. We will have more to say about conditional expectation in Section 4.3, but for now, we just define it to be the expectation using the conditional density as the integrating factor, specifically:

$$E[g(X_2)|X_1 = x_1] = \int_{E_2} g(x_2) f_{2|1}(x_2|x_1) \, dx_2, \tag{4.37}$$

where E_2 is the state space of X_2. (Note that this state space may change depending on the particular x_1 value.) For instance, the conditional density of X_2 given X_1 in Example 4 was found to be:

$$f_{2|1}(x_2|x_1) = \frac{(x_1 + 2x_2)}{(x_1 + 1)}, x_2 \in [0, 1].$$

Therefore the conditional mean of X_2 given $X_1 = x_1$ is:

$$\begin{aligned}
E\left[X_2 | X_1 = x_1\right] &= \int_{E_2} x_2 \cdot f_{2|1}\left(x_2 | x_1\right) dx_2 \\
&= \int_0^1 x_2 \cdot \frac{(x_1 + 2x_2)}{(x_1 + 1)} dx_2 \\
&= \frac{1}{x_1 + 1} \int_0^1 (x_1 x_2 + 2x_2^2) dx_2 \\
&= \frac{1}{x_1 + 1} \left(x_1 \cdot \frac{x_2^2}{2} + \frac{2x_2^3}{3} \right)\Big|_0^1 \\
&= \frac{1}{x_1 + 1} \left(\frac{1}{2} x_1 + \frac{2}{3} \right).
\end{aligned}$$

4.2.3 Independence

Here is the definition of independence of continuous random variables.

Definition 5. Continuous random variables $X_1, X_2, ..., X_n$ are called **mutually independent** if their joint p.d.f. factors into the product of their marginal p.d.f.s, i.e.

$$f\left(x_1, x_2, ..., x_n\right) = f_1\left(x_1\right) f_2\left(x_2\right) \cdots f_n\left(x_n\right). \quad \blacksquare \qquad (4.38)$$

Although we will not show it here, it is true that such continuous random variables are independent if and only if the joint c.d.f. factors into the product of the marginal c.d.f.s:

$$\begin{aligned}
F\left(x_1, x_2, ..., x_n\right) &= P\left[X_1 \le x_1, X_2 \le x_2, ..., X_n \le x_n\right] \\
&= P\left[X_1 \le x_1\right] \cdot P\left[X_2 \le x_2\right] \cdots P\left[X_n \le x_n\right] \qquad (4.39) \\
&= F_1\left(x_1\right) F_2\left(x_2\right) \cdots F_n\left(x_n\right).
\end{aligned}$$

More generally, independence occurs if for any collection of subsets B_i of the respective state spaces E_i of the random variables, the following factorization holds:

$$\begin{aligned}
P\left[X_1 \in B_1, X_2 \in B_2, ..., X_n \in B_n\right] &= P\left[X_1 \in B_1\right] \cdot P\left[X_2 \in B_2\right] \\
&\qquad \cdots P\left[X_n \in B_n\right].
\end{aligned} \qquad (4.40)$$

In Example 5, we made an observation about independence and the relationship between conditional and marginal distributions. The next example is an illustration of a general principle.

Example 6. Suppose that random variables X_1, X_2, X_3, and X_4 are independent. Show that the conditional joint distribution of X_3 and X_4 given both X_1 and X_2 is equal to the marginal joint distribution of X_3 and X_4.

Solution. The independence property means that the joint density of the four random variables factors into the product of their marginal densities, that is:

$$f\left(x_1, x_2, x_3, x_4\right) = f_1\left(x_1\right) f_2\left(x_2\right) f_3\left(x_3\right) f_4\left(x_4\right).$$

The marginal joint density of X_3 and X_4 is:

$$
\begin{aligned}
f_{34}(x_3, x_4) &= \int_{E_1} \int_{E_2} f(x_1, x_2, x_3, x_4)\, dx_2 dx_1 \\
&= \int_{E_1} \int_{E_2} f_1(x_1) f_2(x_2) f_3(x_3) f_4(x_4)\, dx_2 dx_1 \\
&= f_3(x_3) f_4(x_4) \cdot \int_{E_1} f_1(x_1)\, dx_1 \cdot \int_{E_2} f_2(x_2)\, dx_2 \\
&= f_3(x_3) f_4(x_4) \cdot 1 \cdot 1 = f_3(x_3) f_4(x_4).
\end{aligned}
$$

A very similar computation shows that the joint marginal density $f_{12}(x_1, x_2)$ of X_1 and X_2 is $f_1(x_1) f_2(x_2)$. Then:

$$
\begin{aligned}
f_{34|12}(x_3, x_4 | x_1, x_2) &= \frac{f(x_1, x_2, x_3, x_4)}{f_{12}(x_1, x_2)} \\
&= \frac{f_1(x_1) f_2(x_2) f_3(x_3) f_4(x_4)}{f_1(x_1) f_2(x_2)} \\
&= f_3(x_3) f_4(x_4),
\end{aligned}
$$

and therefore the conditional joint density of variables 3 and 4 given variables 1 and 2 is the same as the marginal joint density $f_{34}(x_3, x_4)$. ∎

No matter how many random variables are involved, and which of them are being conditioned on which others, computations like the one in Example 6 show that mutual independence implies that the conditional distributions equal the corresponding marginals. In other words, knowledge of the values of a subcollection of a group of independent random variables does not change the probability distributions of other random variables in the group when independence holds.

A classical result from probability theory, which we do not prove here, is that a sum of independent normally distributed random variables is also normally distributed. Since the mean of a sum is the sum of the means, and since, under independence, the variance of a sum is the sum of the variances (see for instance Theorem 4 of Section 1.3), we also can find the mean and variance of a sum of independent normal random variables, as in the next example.

Example 7. Suppose that a discrete-time, continuous state asset price process $(X_t)_{t=0,1,2,3,\ldots}$ moves in time in such a way that the increments of the process $X_{t+1} - X_t$ are independent and identically distributed normal random variables with mean 1 and variance .5. The initial state X_0 is known to be 20. Find the distribution, mean, and variance of X_3. Use the results to compute $P[X_3 > 21]$.

Solution. Telescoping, we can write:

$$
X_3 = X_0 + (X_1 - X_0) + (X_2 - X_1) + (X_3 - X_2).
$$

Therefore X_3 is the constant $X_0 = 20$ plus the sum of three independent $\mathcal{N}(1, .5)$ random variables. This means that X_3 has the normal distribution with mean $20 + 3(1) = 23$ and variance $.5 + .5 + .5 = 1.5$. For the last question, we can standardize X_3 and consult tables or technology for the appropriate standard normal probability as follows:

$$
\begin{aligned}
P\left[X_3 > 21\right] &= 1 - P\left[X_3 \leq 21\right] \\
&= 1 - P\left[Z \leq \frac{21-23}{\sqrt{1.5}}\right] \\
&= 1 - P[Z \leq -1.633] = .9588. \ \blacksquare
\end{aligned}
$$

4.2.4 Covariance and Correlation

There is little new to say about covariance and correlation. The definitions and main properties are identical to the discrete case, but computations of means and variances are done via integration, rather than summation. For completeness, here are the main properties.

Definition 6. Let X_1 and X_2 be random variables. The **covariance** between X_1 and X_2 is:

$$
\sigma_{12} = \text{Cov}\left(X_1, X_2\right) = E\left[(X_1 - \mu_1) \cdot (X_2 - \mu_2)\right], \tag{4.41}
$$

where μ_1 and μ_2 are the means of X_1 and X_2, respectively. The **correlation** between X_1 and X_2 is:

$$
\rho_{12} = \text{Corr}\left(X_1, X_2\right) = E\left[\frac{(X_1 - \mu_1)}{\sigma_1} \cdot \frac{(X_2 - \mu_2)}{\sigma_2}\right] = \frac{\text{Cov}\left(X_1, X_2\right)}{\sigma_1 \cdot \sigma_2}, \tag{4.42}
$$

where σ_1 and σ_2 are the standard deviations of X_1 and X_2, respectively. $\blacksquare$

We have the computational formula for covariance:

$$
\text{Cov}\left(X_1, X_2\right) = E\left[X_1 X_2\right] - \mu_1 \mu_2, \tag{4.43}
$$

which follows directly from linearity of expectation. Covariance is clearly commutative in its two arguments, i.e. $\text{Cov}\left(X_1, X_2\right) = \text{Cov}\left(X_2, X_1\right)$ and similarly for correlation. Correlation is bounded between -1 and 1:

$$
-1 \leq \rho \leq 1, \text{equivalently } |\rho| \leq 1. \tag{4.44}
$$

Rearranging the defining equation for the correlation gives:

$$
\text{Cov}\left(X_1, X_2\right) = \rho_{12} \sigma_1 \sigma_2. \tag{4.45}
$$

If X_1 and X_2 are independent random variables, then $\text{Cov}\left(X_1, X_2\right) = 0$ and $\rho = 0$. In the independent case we also have:

$$
\text{Var}\left(c_1 X_1 + c_2 X_2\right) = c_1^2 \sigma_1^2 + c_2^2 \sigma_2^2, \tag{4.46}
$$

but if the two random variables are dependent, then the variance becomes:

$$
\text{Var}\left(c_1 X_1 + c_2 X_2\right) = c_1^2 \sigma_1^2 + c_2^2 \sigma_2^2 + 2 c_1 c_2 \text{Cov}\left(X_1, X_2\right). \tag{4.47}
$$

Constant coefficients factor out of the covariance, in either component:

$$\text{Cov}\left(c_1 X_1, X_2\right) = c_1 \text{Cov}\left(X_1, X_2\right); \text{Cov}\left(X_1, c_2 X_2\right) = c_2 \text{Cov}\left(X_1, X_2\right). \quad (4.48)$$

More generally, covariance is **bilinear**:

$$
\begin{aligned}
\text{Cov}\left(c_1 X_1 + c_2 X_2, c_3 X_3 + c_4 X_4\right) = {}& c_1 c_3 \text{Cov}\left(X_1, X_3\right) + c_1 c_4 \text{Cov}\left(X_1, X_4\right) \\
& + c_2 c_3 \text{Cov}\left(X_2, X_3\right) \\
& + c_2 c_4 \text{Cov}\left(X_2, X_4\right).
\end{aligned}
$$
$$(4.49)$$

Example 8. Consider two random variables X_1 and X_2 with joint p.d.f. below. Find the value of the constant c that makes f a valid density, and compute the covariance and correlation between X_1 and X_2.

$$
f\left(x_1, x_2\right) = \begin{cases} c x_1 x_2 & \text{if } 0 \leq x_2 \leq x_1 \leq 2; \\ 0 & \text{otherwise.} \end{cases}
$$

Solution. Since f is in the form of a product of two individual functions of x_1 and x_2, it is tempting to assume that X_1 and X_2 are independent. But the state space of X_2 depends on the value of x_1, specifically, $0 \leq x_2 \leq x_1$, and the state space of X_1 depends on x_2, that is, $x_2 \leq x_1 \leq 2$. The joint state space is shown in Figure 4.9. Observe that f vanishes off of a non-rectangular region, and so it cannot fully factor into a product of marginals. For instance, the joint density at $(x_1, x_2) = (1, 1.5)$ is 0, but X_1 does have positive density at 1 and X_2 has positive density at 1.5.

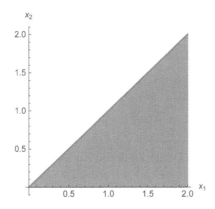

FIGURE 4.9 Joint state space of f.

The computations are routine, but lengthy. To solve for c, we set up the equation:

$$
\begin{aligned}
1 &= \int \int_E f(x_1, x_2)\, dx_2 dx_1 \\
&= c \cdot \int_0^2 \int_0^{x_1} x_1 x_2 dx_2 dx_1 \\
&= c \cdot \int_0^2 x_1 \left(\frac{x_2^2}{2}\right)_0^{x_1} dx_1 \\
&= c \cdot \int_0^2 \left(\frac{x_1^3}{2}\right) dx_1 \\
&= c \cdot \left(\frac{x_1^4}{8}\right)_0^2 = c \cdot 2.
\end{aligned}
$$

Therefore the coefficient c must be $1/2$.

We need the marginals of both variables in order to calculate the individual means and variances. These are done as follows.

$$
\begin{aligned}
f_1(x_1) &= \frac{1}{2} \cdot \int_0^{x_1} x_1 x_2 dx_2 \\
&= \frac{1}{2} x_1 \cdot \left(\frac{x_2^2}{2}\right)_0^{x_1} \\
&= \frac{1}{4} x_1^3, \; x_1 \in [0, 2];
\end{aligned}
$$

$$
\begin{aligned}
f_2(x_2) &= \frac{1}{2} \cdot \int_{x_2}^{2} x_1 x_2 dx_1 \\
&= \frac{1}{2} x_2 \cdot \left(\frac{x_1^2}{2}\right)_{x_2}^{2} \\
&= \frac{1}{2} x_2 \cdot \left(2 - \frac{x_2^2}{2}\right) \\
&= x_2 - \frac{1}{4} x_2^3, \; x_2 \in [0, 2];
\end{aligned}
$$

Then the means are:

$$
\begin{aligned}
E[X_1] &= \int_0^2 x_1 \cdot \frac{1}{4} x_1^3 dx_1 \\
&= \left(\frac{1}{20} x_1^5\right)_0^2 = \frac{32}{20} = \frac{8}{5};
\end{aligned}
$$

$$
\begin{aligned}
E[X_2] &= \int_0^2 x_2 \cdot \left(x_2 - \frac{1}{4} x_2^3\right) dx_2 \\
&= \int_0^2 \left(x_2^2 - \frac{1}{4} x_2^4\right) dx_2 \\
&= \left(\frac{1}{3} x_2^3 - \frac{1}{20} x_2^5\right)_0^2 = \frac{8}{3} - \frac{32}{20} = \frac{16}{15}.
\end{aligned}
$$

We will also need the expected squares, in order to compute the variances and standard deviations. Omitting details, it is easy to derive that:

$$
E[X_1^2] = \int_0^2 x_1^2 \cdot \frac{1}{4} x_1^3 dx_1 = \frac{8}{3}; E[X_2^2] = \int_0^2 x_2^2 \cdot \left(x_2 - \frac{1}{4} x_2^3\right) dx_2 = \frac{4}{3}.
$$

Then the variances are:

$$
\begin{aligned}
\sigma_1^2 &= \text{Var}(X_1) = E[X_1^2] - (E[X_1])^2 = \frac{8}{3} - \left(\frac{8}{5}\right)^2 = \frac{8}{75}; \\
\sigma_2^2 &= \text{Var}(X_2) = E[X_2^2] - (E[X_2])^2 = \frac{4}{3} - \left(\frac{16}{15}\right)^2 = \frac{44}{225}.
\end{aligned}
$$

For the covariance we must compute the expected product:

$$
\begin{aligned}
E\left[X_1 X_2\right] &= \int \int_E x_1 x_2 f\left(x_1, x_2\right) dx_2 dx_1 \\
&= \frac{1}{2} \cdot \int_0^2 \int_0^{x_1} x_1 x_2 \left(x_1 x_2\right) dx_2 dx_1 \\
&= \frac{1}{2} \cdot \int_0^2 \int_0^{x_1} x_1^2 x_2^2 dx_2 dx_1 \\
&= \frac{1}{2} \cdot \int_0^2 x_1^2 \left(\frac{x_2^3}{3}\right)_0^{x_1} dx_1 \\
&= \frac{1}{2} \cdot \int_0^2 \left(\frac{x_1^5}{3}\right) dx_1 \\
&= \frac{1}{2} \cdot \left(\frac{x_1^6}{18}\right)_0^2 = \frac{64}{36} = \frac{16}{9}.
\end{aligned}
$$

This implies that:

$$
\mathrm{Cov}\left(X_1, X_2\right) = E\left[X_1 X_2\right] - E\left[X_1\right] E\left[X_2\right] = \frac{16}{9} - \left(\frac{8}{5}\right)\left(\frac{16}{15}\right) = \frac{16}{225},
$$

and:

$$
\rho = \mathrm{Corr}\left(X_1, X_2\right) = \frac{\mathrm{Cov}\left(X_1, X_2\right)}{\sigma_1 \sigma_2} = \frac{16/225}{\sqrt{\frac{8}{75}}\sqrt{\frac{44}{225}}} = .4924. \ \blacksquare
$$

When two, three, or more random variables are involved in a problem, it can be useful to summarize the pairwise covariance or correlation information in a matrix. We are led to the following definition.

Definition 7. Let $X_1, X_2, ..., X_n$ be random variables. The **covariance matrix** is the $n \times n$ matrix whose $i-j$ component is the covariance $\sigma_{ij} = \mathrm{Cov}\left(X_i, X_j\right)$. Hence the diagonal entry in position i is $\sigma_{ii} = \mathrm{Cov}\left(X_i, X_i\right) = \sigma_i^2$, and the covariance matrix has the form:

$$
\Sigma = \begin{pmatrix}
\sigma_1^2 & \sigma_{12} & \sigma_{13} & \cdots & \sigma_{1n} \\
\sigma_{12} & \sigma_2^2 & \sigma_{23} & \cdots & \sigma_{2n} \\
\sigma_{13} & \sigma_{23} & \sigma_3^2 & \cdots & \sigma_{3n} \\
\vdots & \vdots & \vdots & \ddots & \vdots \\
\sigma_{1n} & \sigma_{2n} & \sigma_{3n} & \cdots & \sigma_n^2
\end{pmatrix}. \tag{4.50}
$$

The **correlation matrix** of the random variables is the matrix whose $i-j$ component is the correlation $\rho_{ij} = \mathrm{Corr}\left(X_i, X_j\right)$. Since the correlation of each variable with itself is 1, the correlation matrix has the form:

$$
\rho = \begin{pmatrix}
1 & \rho_{12} & \rho_{13} & \cdots & \rho_{1n} \\
\rho_{12} & 1 & \rho_{23} & \cdots & \rho_{2n} \\
\rho_{13} & \rho_{23} & 1 & \cdots & \rho_{3n} \\
\vdots & \vdots & \vdots & \ddots & \vdots \\
\rho_{1n} & \rho_{2n} & \rho_{3n} & \cdots & 1
\end{pmatrix}. \ \blacksquare \tag{4.51}
$$

Covariance and correlation matrices will always be symmetric, because of the commutativity of $\mathrm{Cov}\left(X_1, X_2\right)$ and $\mathrm{Corr}\left(X_1, X_2\right)$ in the two arguments.

An important special case is illustrated in the next example.

Example 9. Find the covariance matrix for the three random variables in Example 1, whose joint density is:

$$f(x_1, x_2, x_3) = \frac{1}{12} \cdot e^{-x_2}, x_1 \in [0,3], x_2 \in (0, +\infty), x_3 \in [1,5].$$

Solution. In Example 5 the marginal density of X_3 was computed, which showed that X_3 had the uniform distribution on $[1,5]$. We leave it to the reader to show that also X_1 is uniformly distributed on $[0,3]$, and that X_2 has the exponential distribution with parameter $\lambda = 1$. But then the product of the marginals is:

$$f_1(x_1) f_2(x_2) f_3(x_3) = \frac{1}{3} \cdot e^{-x_2} \cdot \frac{1}{4} = \frac{1}{12} \cdot e^{-x_2} = f(x_1, x_2, x_3),$$

on the state space described by $x_1 \in [0,3], x_2 \in (0, +\infty), x_3 \in [1,5]$. Hence the random variables are mutually independent. Their covariances with each other are 0. It is a routine computation to show that the variance of the uniform distribution on an interval $[a,b]$ is $(b-a)^2/12$, and the $\exp(\lambda)$ variance is $1/\lambda^2$. Therefore,

$$\sigma_1^2 = \frac{(3-0)^2}{12} = \frac{3}{4}; \ \sigma_2^2 = \frac{1}{1^2} = 1; \ \sigma_3^2 = \frac{(5-1)^2}{12} = \frac{4}{3}.$$

The covariance matrix of X_1, X_2, and X_3 is:

$$\Sigma = \begin{pmatrix} 3/4 & 0 & 0 \\ 0 & 1 & 0 \\ 0 & 0 & 4/3 \end{pmatrix}.$$

Thus, the correlation matrix is the 3×3 identity matrix, that is, there are 1's on the diagonal and 0's elsewhere. ■

It is clear from Example 9 that when random variables are mutually independent, their covariance matrix will be of diagonal structure, and the correlation matrix will be an identity matrix of the appropriate size.

4.2.5 Bivariate Normal Distribution

One of the most important joint distributions for financial applications is the two-variable generalization of the normal distribution, which we introduce in this subsection. We begin with an example data set.

Example 10. The figures below are based on daily rate of return data for the first six months of 2016 on two fast food restaurant giants: McDonald's and Yum! Brands, Inc., which includes chains such as Pizza Hut and Taco Bell.

The scaled histograms in Figure 4.10 show a shape that is characteristic of the single-variable normal distribution (except, as noted in Section 4.1, there are a few more extreme rates of return than may be predicted by normality, particularly on the left side). But the plot of one against the other in Figure 4.11 reveals something new: the data points distribute themselves roughly in an elliptical cloud, whose axes of symmetry are tilted with respect to the coordinate axes. Rates of return on these two companies are clearly not independent.

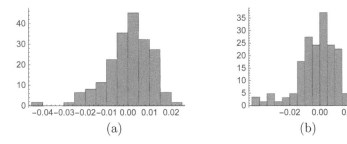

(a)   (b)

FIGURE 4.10 (a) McDonald's rates of return; (b) Yum! rates of return, 2016.

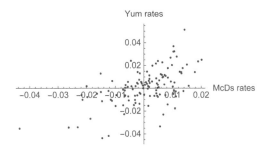

FIGURE 4.11 Scatter plot of Yum! rates vs. McDonald's rates, 2016.

Some elementary data analysis on sample means, standard deviations, and correlation, labeling McDonald's as asset 1 and Yum! as asset 2, yields:

$$\bar{X}_1 = .000240; \ \bar{X}_2 = .001234;$$
$$S_1 = .010303; S_2 = .015431;$$
$$R_{12} = .584455.$$

A reasonable probabilistic model for the generic pair of rates of return (X_1, X_2) might be one with a joint density that yields normal marginal densities for each of the individual companies, with means and standard deviations estimated from the corresponding sample statistics. But the sample correlation R_{12} indicates clearly that we should not be content to assume independence.

What kind of model do we need to account for non-zero correlations? ■

We are now ready to define the generalization of the normal distribution to two variables. It will depend on the two means, the two variances, and the correlation between the two random variables.

Definition 8. A pair of random variables X_1, X_2 is said to have the **bivariate normal density,** with mean vector $\boldsymbol{\mu} = (\mu_1, \mu_2)$ and covariance matrix

$$\Sigma = \begin{pmatrix} \sigma_1^2 & \rho\sigma_1\sigma_2 \\ \rho\sigma_1\sigma_2 & \sigma_2^2 \end{pmatrix}$$

if its joint density is:

$$
\begin{aligned}
f(x_1, x_2) &= \frac{1}{2\pi\sigma_1\sigma_2\sqrt{1-\rho^2}} \\
&\cdot \exp\left[\frac{-1}{2(1-\rho^2)}\left(\frac{(x_1-\mu_1)^2}{\sigma_1^2} - 2\rho\frac{(x_1-\mu_1)(x_2-\mu_2)}{\sigma_1\sigma_2} + \frac{(x_2-\mu_2)^2}{\sigma_2^2}\right)\right].\blacksquare
\end{aligned}
$$
$$(4.52)$$

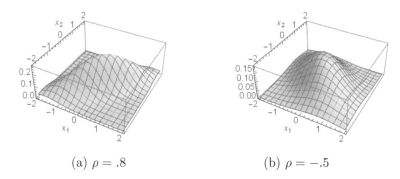

(a) $\rho = .8$ (b) $\rho = -.5$

FIGURE 4.12 Bivariate normal density, means 0, variances 1.

Two particular cases of the bivariate normal density are sketched in Figure 4.12, one with a positive value of the correlation $\rho = .8$, and the other with a negative value $\rho = -.5$. In both cases, the means were fixed at 0 and the standard deviations were taken to be 1. Notice that in the $\rho > 0$ case, the variables tend to be positive together or negative together, and in the $\rho < 0$ case the variables tend to have opposite sign. The magnitude of ρ indicates the strength of the relationship. Also, the cross-hatches on the surface graph suggest that for each fixed value of x_1, X_2 has a bell-shaped distribution (not necessarily properly scaled to be a density), and for each fixed value of x_2 the distribution of X_1 is also normal-like. The visual evidence therefore leads to an hypothesis that the conditional distributions of each variable given the other are normal. As we will see shortly, the marginal densities are also normal.

Does the correlation parameter ρ really represent the correlation between the two variables? The answer is yes, and we will not derive this here, but the special case in which there are two random variables with means $\mu_1 = \mu_2 = 0$, and variances $\sigma_1^2 = \sigma_2^2 = 1$ is carried out in Appendix A for the interested reader.

The following important theorem about marginal and conditional densities is also proved in Appendix A. The proof is based on completing the square in x_2 in the exponent of the bivariate normal density.

Theorem 1. If X_1 and X_2 have the bivariate normal distribution with parameters $\mu_1, \mu_2, \sigma_1, \sigma_2$, and ρ, then:

(a) The marginal density of X_1 is $\mathcal{N}\left(\mu_1, \sigma_1^2\right)$;

(b) The conditional density of X_2 given $X_1 = x_1$ is $\mathcal{N}\left(\mu_{2|1}, \sigma_{2|1}^2\right)$, where:

$$\mu_{2|1} = \mu_2 + \frac{\rho\sigma_2}{\sigma_1}\left(x_1 - \mu_1\right), \quad \sigma_{2|1}^2 = \sigma_2^2\left(1 - \rho^2\right). \blacksquare \qquad (4.53)$$

Remark. By symmetry, it must also be true that X_2 is $\mathcal{N}\left(\mu_2, \sigma_2^2\right)$-distributed, and that, given $X_2 = x_2$, X_1 is conditionally normal, with conditional mean and variance:

$$\mu_{1|2} = \mu_1 + \frac{\rho\sigma_1}{\sigma_2}\left(x_2 - \mu_2\right), \quad \sigma_{1|2}^2 = \sigma_1^2\left(1 - \rho^2\right). \qquad (4.54)$$

It is intuitively reasonable that the higher the correlation between the two variables, and the farther away the observed value of one from its mean, the farther the conditional mean from the marginal mean. But it is rather strange that the conditional variance does not depend on the observed value. It is interesting, however, that as the correlation approaches 1 (or -1), the conditional variance approaches zero. The more correlated the two random variables are, the less variability one has given the value of the other.

This theorem allows us to make predictions about financial assets like those in the next example.

Example 11. Suppose that the rate of return variables for McDonald's and Yum! from Example 10 have the multivariate normal distribution. Estimate the parameters by the sample statistics that were listed in the example. Compute:

(a) the probability that the McDonald's rate of return is at least .0003;

(b) the mean and variance of the Yum! rate of return given that the McDonald's rate of return is .0005;

(c) the probability that the Yum! rate of return is less than .0003 given that the McDonald's rate of return is .0005.

Solution. The estimated parameters are repeated below. Recall that asset 1 is McDonald's, and asset 2 is Yum!.

$$\mu_1 \approx \bar{X}_1 = .000240; \quad \mu_2 \approx \bar{X}_2 = .001234;$$
$$\sigma_1 \approx S_1 = .010303; \quad \sigma_2 \approx S_2 = .015431;$$
$$\rho \approx R_{12} = .584455.$$

(a) Theorem 1 states that the McDonald's rate X_1 is normally distributed with (approximate) mean $\mu_1 = .000240$ and (approximate) standard deviation $\sigma_1 = .010303$. Standardizing appropriately and consulting a normal table or technology, we get:

$$
\begin{aligned}
P[X_1 \geq .0003] &= 1 - P[X_1 \leq .0003] \\
&= 1 - P\left[\frac{X_1 - .000240}{.010303} \leq \frac{.0003 - .000240}{.010303}\right] \\
&= P[Z \leq .0058] = .4977.
\end{aligned}
$$

(b) By part (b) of Theorem 1, the conditional mean and variance of the Yum! rate given that the McDonald's rate is .0005 are:

$$
\begin{aligned}
\mu_{2|1} &= \mu_2 + \frac{\rho\sigma_2}{\sigma_1}(x_1 - \mu_1) \\
&\approx .001234 + \frac{(.584455)(.015431)}{.010303}(.0005 - .000240) \\
&= .00146
\end{aligned}
$$
$$\sigma_{2|1}^2 = \sigma_2^2\left(1 - \rho^2\right) \approx .015431\left(1 - (.584455)^2\right) = .01016.$$

(c) With the conditional mean and variance from part (b), since we know that the Yum! rate is conditionally normal given the McDonald's value, we have:

$$
\begin{aligned}
P[X_2 < .0003|X_1 = .0005] &= P[X_2 < .0003|X_1 = .0005] \\
&= P\left[Z < \frac{.0003 - .00146}{\sqrt{.01016}}\right] \\
&= P[Z < -.0115] = .4954. \ \blacksquare
\end{aligned}
$$

Exercises 4.2

1. In Example 1, compute the individual probabilities: $P[X_1 > 2]$, $P[X_2 < 1]$, and $P[X_3 \in [2,3]]$.

2. Find the joint c.d.f. of the random variables in Example 1.

3. In Example 2, compute the joint probability $P[X_1 \leq .5, X_2 > .5]$.

4. In the section, we have established formula (4.30) for linearity of expectation for any two continuous random variables. Use this and mathematical induction to verify the linearity property for arbitrarily many continuous random variables:

$$E\left[\sum_{i=1}^{n} c_i X_i\right] = \sum_{i=1}^{n} c_i E[X_i].$$

5. Let X_1 and X_2 have the joint density below. Find the constant c and the joint c.d.f. Now compute the second order mixed partial derivative $\partial^2 F/\partial x_1 \partial x_2$ of the joint c.d.f. What do you observe?

$$f(x_1, x_2) = c/(x_1 \cdot x_2) \text{, for } x_1, x_2 \in [1, 2].$$

6. Consider two continuous random variables X and Y whose joint density is:

$$f(x, y) = c(x + y), \quad x, y \in [0, 2].$$

Find:
   (a) the marginal densities of both random variables;
   (b) both $P[X < 1]$ and $P[Y > 1]$;
   (c) the joint probability $P[X < 1, Y > 1]$;
   (d) $P[Y > 1 | X = 1]$.

7. In Example 4, verify that each of the marginal densities integrates to 1 over the interval $[0, 1]$, and find the cumulative distribution functions associated with these densities.

8. In Example 5, check that (a) the conditional density of X_1 given $X_2 = x_2$ and $X_3 = x_3$ is the same as the marginal density of X_1; and also (b) the conditional density of X_2 given $X_1 = x_1$ and $X_3 = x_3$ is the same as the marginal density of X_2.

9. Show that the defining formula (4.38) for independence of random variables implies the factorization of probabilities in formula (4.40).

10. Why is independence condition (4.40) a generalization of condition (4.39), as the text claims?

11. Show that if continuous random variables $X_1, X_2, ..., X_n$ are mutually independent, then:

$$E[g_1(X_1) g_2(X_2) \cdots g_n(X_n)] = E[g_1(X_1)] \cdot E[g_2(X_2)] \cdots E[g_n(X_n)].$$

12. For the joint density in Example 8, repeated below, find:
   (a) the conditional density of X_1 given X_2;
   (b) the conditional density of X_2 given X_1;
   (c) the conditional mean of X_1 given X_2;
   (d) the conditional mean of X_2 given X_1.

$$f(x_1, x_2) = c x_1 x_2, 0 \le x_2 \le x_1 \le 2$$

13. Prove formulas (4.48).

14. Prove the bilinearity formula (4.49).

15. Compute the covariance between the two random variables in Exercise 5. Write out the covariance matrix.

16. (Technology required) If X and Y have joint density $f(x, y) = cx^2y$, $x, y \in [0, 1]$, then:
   (a) produce a graph of the joint density;
   (b) find the marginal densities of X and Y;
   (c) compute $P[X \geq Y]$.

17. Suppose that two continuous random variables X_1 and X_2 have the joint density below. First find the appropriate value of the constant c, and then compute $P[X_2 > 3/4 | X_1 = 2/3]$.

$$f(x_1, x_2) = \begin{cases} c & \text{if } 0 \leq x \leq y \leq 1; \\ 0 & \text{otherwise.} \end{cases}$$

18. Let random variables X_1 and X_2 have joint density equal to 1 on the set of points (x_1, x_2) such that $x_1 \in [-1, 1]$ and $0 \leq x_2 \leq 1 - |x_1|$. Find:
   (a) the marginal densities of both X_1 and X_2;
   (b) the conditional density of X_2 given $X_1 = x_1$;
   (c) $E\left[X_2^2 | X_1 = x_1\right]$;
   (d) the covariance between X_1 and X_2.

19. Why is the covariance of a random variable with itself the same as its variance? Why is the correlation between a random variable and itself equal to 1?

20. Return to the two assets, McDonald's and Yum!, from Examples 10 and 11. Assume that the rates of return follow the bivariate normal distribution with parameters as estimated in the example. These are repeated below for convenience. Compute:
   (a) the probability that the Yum! rate of return is less than .0002;
   (b) the conditional distribution of the McDonald's rate of return given that the Yum! rate is .0002;
   (c) the probability that the McDonald's rate of return is less than .0001 given that the Yum! rate of return is .0002;

$$\bar{X}_1 = .000240; \quad \bar{X}_2 = .001234; \quad S_1 = .010303; S_2 = .015431; R_{12} = .584455.$$

21. Two risky assets have rates of return that follow the bivariate normal distribution with means .04 and .06, standard deviations .03 and .05, and some correlation ρ. How large does ρ have to be in order that the conditional mean rate of return for asset 2 is at least .07, given that the asset 1 rate of return

is .05?

22. Let X_1 and X_2 be independent, normally distributed random variables with means μ_1 and μ_2, and standard deviations σ_1 and σ_2, respectively. Define new random variables $Y_1 = X_1$ and $Y_2 = X_1 + X_2$. Argue informally that the pair (Y_1, Y_2) has the bivariate normal distribution. Find the five parameters (means, variances, and correlation) for this distribution.

4.3 Measurability and Conditional Expectation

Next, we would like to carry over the ideas and results from Sections 3.2 and 3.3 on measurability of discrete random variables and conditional expectation to the continuous world. The development is parallel to those sections, and we will find that this does not require too much more work.

4.3.1 Sigma Algebras

As we move to the continuous world, we note immediately that finite objects are not sufficient. So we start with an analog of the idea of an algebra of events.

Definition 1. A collection of events $\mathcal{H}$ in a continuous sample space Ω is called a **σ-algebra** if it includes Ω itself and is closed under complementation and countable union, that is:

(a) $\Omega \in \mathcal{H}$;
(b) if $A \in \mathcal{H}$ then $A^c \in \mathcal{H}$;
(c) if $A_1, A_2, A_3, \ldots \in \mathcal{H}$ then $\cup_{i=1}^{\infty} A_i \in \mathcal{H}$. ∎

Remark. Parts (a) and (b) of the definition imply that the empty event $\emptyset$ must lie in any σ-algebra. Definition 1(c) does not preclude finite unions; for instance if A_1 and A_2 are both in a σ-algebra, then $A_1 \cup A_2$ can be expressed as a countable union $A_1 \cup A_2 \cup \emptyset \cup \emptyset \cup \cdots$, which belongs to the σ-algebra by condition (c). Also, as with algebras, the set identity:

$$\cap_{i=1}^{\infty} A_i = \left(\cup_{i=1}^{\infty} (A_i)^c \right)^c \tag{4.55}$$

together with conditions (b) and (c) shows that σ-algebras are closed under countable (and finite) intersections (see Exercise 1).

The idea of an atom is not relevant here in the non-finite world, but the idea of a refinement of a σ-algebra still makes sense. Once more, we are trying to represent the concept that the larger σ-algebra contains more information about the random phenomenon of interest than the smaller σ-algebra.

Definition 2. A σ-algebra $\mathcal{H}$ is said to be a **refinement** of a σ-algebra $\mathcal{G}$ if every event in $\mathcal{G}$ is also in $\mathcal{H}$, that is $\mathcal{G} \subset \mathcal{H}$. ∎

Example 1. Consider a sample space Ω that is appropriate to model the failure of a device that is meant to operate during continuously measured time. We might think of outcomes as being functions $\omega = \omega(t)$, such as the one sketched in Figure 4.13, in which the device functions through a time interval $[0, T)$, during which ω takes the value 1, and then fails at time T after which the value of ω drops to 0. The function ω is then totally determined by the continuous random variable $T = T(\omega)$ that gives the time of failure. It is not necessary at this point to put a probability measure on Ω.

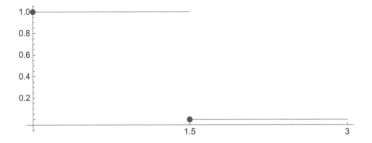

FIGURE 4.13 Typical outcome ω of a device failure process.

Define, for each fixed t, a collection of events $\mathcal{H}_t$ in which the value of path ω is known from time 0 through time t. Because of the special nature of the outcomes, the only information contained in $\mathcal{H}_t$ would be whether $T(\omega) \leq s$ or $T(\omega) > s$ for each $0 \leq s \leq t$. In view of this intuition, formally define $\mathcal{H}_t$ as the smallest σ-algebra containing all events $\{\omega | T(\omega) \leq s\}$ for each $0 \leq s \leq t$. If time t_2 is larger than time t_1, then clearly $\mathcal{H}_{t_1} \subset \mathcal{H}_{t_2}$, since $\mathcal{H}_{t_2}$ contains all events $\{T \leq s\}$ particularly for times $s \in [0, t_1] \subset [0, t_2]$. Therefore $\mathcal{H}_{t_2}$ is a refinement of $\mathcal{H}_{t_1}$. ∎

Most of our examples of interest in continuous models involve time dependent asset price processes taking values on the real line, or possibly in n-dimensional Euclidean space. Therefore we pay some particular attention to a σ-algebra on $\mathbb{R}$ called the Borel σ-algebra.

Definition 3. The *Borel σ-algebra* $\mathcal{B}$ on $\mathbb{R}$ is the smallest σ-algebra containing all rays $(-\infty, b]$ for all real numbers b. ∎

A word should be said about the idea of the "smallest σ-algebra" containing a group of sets. By this we mean the intersection of all σ-algebras containing those sets. This intersection is well-defined, because the set of all subsets (that is, the power set) of the real line, is such a σ-algebra. And an intersection of arbitrarily many (not necessarily countably many) σ-algebras is a σ-algebra, as Exercise 3 asks you to verify. So if we consider the collection of all σ-algebras containing the rays $(-\infty, b]$ and intersect those, we get a σ-algebra that also contains the rays, and must be no larger in terms of set containment than the σ-algebras that have been intersected.

Example 2. The Borel σ-algebra $\mathcal{B}$ on $\mathbb{R}$ is sufficiently broad to contain all sets of primary interest, including all intervals and therefore all finite or countable unions of intervals. To see this, let us show that all of the following are in $\mathcal{B}$:
   (a) all rays of the form (a, ∞);
   (b) all half open, half closed bounded intervals $(a, b]$;
   (c) all open intervals (a, b);
   (d) all closed rays $[a, \infty)$.
(A few other cases are dealt with in Exercise 4.)

Solution. (a) Since the closed ray $(-\infty, a]$ is in $\mathcal{B}$ and (a, ∞) is the complement of this ray, $(a, \infty) \in \mathcal{B}$ by closure of $\mathcal{B}$ under complementation.

(b) By definition and by part (a), both $(-\infty, b]$ and (a, ∞) are Borel sets. The interval $(a, b]$ is the intersection of these two; hence by closure of the σ-algebra $\mathcal{B}$ under intersection, $(a, b] \in \mathcal{B}$.

(c) By part (b), we know that all intervals of the form $\left(a, b - \frac{1}{n}\right]$ are Borel sets for integers $n = 1, 2, \dots$ (as long as $b - \frac{1}{n} > a$). Consider the countable union $B = \cup_n \left(a, b - \frac{1}{n}\right]$. By closure of $\mathcal{B}$ under countable unions, $B \in \mathcal{B}$. Since each set in the union is contained in (a, b), the union B is also contained in (a, b). But also, given $x \in (a, b)$, since x is strictly less than b, there is an integer n large enough that $b - \frac{1}{n} > x$; hence $x \in \left(a, b - \frac{1}{n}\right]$. Thus, x is in B, which implies that $B = (a, b)$, so that (a, b) is a Borel set.

(d) By part (a), all open rays $\left(a - \frac{1}{n}, \infty\right)$ are in $\mathcal{B}$ for integers $n = 1, 2, \dots$. But we can express the closed ray $[a, \infty)$ as the countable intersection

$$[a, \infty) = \bigcap_{n=1}^{\infty} \left(a - \frac{1}{n}, \infty\right), \tag{4.56}$$

since a itself and all real numbers $x > a$ are in all of the intervals $\left(a - \frac{1}{n}, \infty\right)$, and no number x that is less than a is in the intersection. The latter statement is true because if $x < a$, then there is an integer n large enough that $x < a - \frac{1}{n}$, so that x is not in the interval $\left(a - \frac{1}{n}, \infty\right)$. Because σ-algebras are closed under countable intersection, $[a, \infty)$ is in the Borel σ-algebra. ∎

The reason that we are interested in the Borel σ-algebra on $\mathbb{R}$ is that, as shown in advanced courses on measure theory, we can integrate a large class of functions $f(x)$ over Borel sets, that is:

$$\int_B f(x)\, dx$$

is well-defined, which we need for continuous probability models that are defined in terms of density functions.

We can also talk about the **restriction of the Borel sets** $\mathcal{B}$ to an interval such as $[a, b]$ on the real line, which would consist of all sets $B \in \mathcal{B}$ intersected with $[a, b]$. So when a random variable X has a state space E that is a subset of the real line, we may refer in this way to the σ-algebra $\mathcal{E}$ of Borel sets on E.

Example 3. The Borel σ-algebra $\mathcal{B}^n$ on a higher-dimensional space $\mathbb{R}^n$ can be defined analogously to the single-dimensional case. For example, in $\mathbb{R}^2$, quarter-planes of the form:

$$(-\infty, b_1] \times (-\infty, b_2] = \{(x, y) \,|\, x \leq b_1 \,, y \leq b_2\}$$

can be taken as the generating sets of $\mathcal{B}^2$. The Borel sets can be defined as the smallest σ-algebra containing all such closed quarter-planes. Then, for instance, any closed half-plane bounded on the right by a vertical line $x = b_1$ is in $\mathcal{B}^2$, because it is a countable union:

$$\{(x, y) \,|\, x \leq b_1\} = \bigcup_{n=0}^{\infty} \left((-\infty, b_1] \times (-\infty, n] \right).$$

The complementary open half-plane $\{(x, y) \,|\, x > b_1\}$ bounded on the left by $x = b_1$ is therefore also a Borel subset of $\mathbb{R}^2$. It is not hard to use a limiting process similar to part (d) of Example 2 to show that closed half-planes $\{(x, y) \,|\, x \geq b_1\}$ are also Borel. Similarly, the half-plane bounded above by the horizontal line $y = b_2$ is a countable union:

$$\{(x, y) \,|\, y \leq b_2\} = \bigcup_{n=0}^{\infty} \left((-\infty, n] \times (-\infty, b_2] \right),$$

which means that the complementary open half-plane $\{(x, y) \,|\, y > b_2\}$ is a two-dimensional Borel set. As in Exercise 5, open half-planes bounded above by horizontal lines, that is sets of the form $\{(x, y) \,|\, y < b_2\}$, can be shown to be Borel; in fact any closed or open half-plane bounded by a vertical or horizontal line, open or closed, would be in $\mathcal{B}^2$. We leave the details to the reader. Therefore, since rectangles of all forms can be expressed as intersections of four half-planes, all forms of rectangle can be shown to be in $\mathcal{B}^2$. Since many subsets of $\mathbb{R}^2$ can be constructed as finite or countable unions of rectangles, the Borel σ-algebra on $\mathbb{R}^2$ contains a rich variety of sets. ∎

4.3.2   Random Variables and Measurability

As we know, a random variable X is a function from the sample space Ω of a random phenomenon to a set E called its state space, often some subset of the real line. But we need to place an extra layer of structure on this relatively simple idea. Suppose that the sample space has a σ-algebra of events $\mathcal{H}$ defined on it, and the state space has its own σ-algebra of sets $\mathcal{E}$ (which could be the Borel σ-algebra). Then we can make the following definition.

Definition 4. A random variable X that maps the space $(\Omega, \mathcal{H})$ to the space $(E, \mathcal{E})$ is called **$\mathcal{H}/\mathcal{E}$-measurable** (or just **$\mathcal{H}$-measurable** for short) if for any set $B \in \mathcal{E}$, its inverse image

$$X^{-1}(B) = \{\omega \in \Omega | X(\omega) \in B\} \tag{4.57}$$

lies in $\mathcal{H}$. ∎

In other words, measurability means that X pulls back sets in the σ-algebra $\mathcal{E}$ to events in $\mathcal{H}$. If the sample space has a probability P on it, which works on sets in $\mathcal{H}$, this would imply that P can apply to the set $\{\omega \in \Omega | X(\omega) \in B\}$. Loosely speaking, the expression $P[X \in B]$ makes sense, which is one of the main reasons for considering the concept of measurability. In more advanced texts, the measurability condition is part of the definition of random variables; a function from Ω to E is not considered to be a valid random variable without the condition.

But more is true; "knowledge" of events in the σ-algebra $\mathcal{H}$ will imply "knowledge" about the value that X takes on. $\mathcal{H}$ may contain more information, such as information about some other random variable, but it must at least contain the information in the following related σ-algebra.

Definition 5. The **σ-algebra $\sigma(X)$ generated by a random variable X** is the smallest σ-algebra on Ω with respect to which X is measurable. ∎

Theorem 1. Let X be a random variable whose state space is $(E, \mathcal{E})$. Then $\sigma(X)$ is equal to the collection of all inverse images $X^{-1}(B) = \{\omega \in \Omega | X(\omega) \in B\}$ as B ranges through all sets in $\mathcal{E}$.

Proof. By the measurability condition, any σ-algebra $\mathcal{H}$ on Ω with respect to which X is measurable must contain all of these sets. Thus, if we can show that this family of inverse images is itself a σ-algebra, then it is therefore the smallest such, and so the family is $\sigma(X)$.

To do this, first notice that since X maps into E and E must be in $\mathcal{E}$, we have $\Omega = \{\omega | X(\omega) \in E\}$. Thus, the full sample space Ω is an inverse image, and so it belongs to our family. If event A belongs to the family, then there is a set B in $\mathcal{E}$ such that:

$$A = \{\omega \in \Omega | X(\omega) \in B\}.$$

But then:

$$A^c = \{\omega \in \Omega | X(\omega) \in B\}^c = \{\omega \in \Omega | X(\omega) \in B^c\},$$

and since B^c must be in the σ-algebra $\mathcal{E}$, A^c is in the family of inverse images. Thus, this family is closed under complementation. You are asked in Exercise

6 to check that the family is closed under countable union, which implies that it is a σ-algebra as desired. ∎

Example 4. Suppose that a sample space Ω is the square $[0,1] \times [0,1] = \{(x,y)|0 \le x \le 1, 0 \le y \le 1\}$, and assume that we put the σ-algebra $\mathcal{H} = \mathcal{B}^2$ of two-dimensional Borel sets on it (restricted to the square). Define the two component random variables:

$$X(\omega) = X(x,y) = x; \ \ Y(\omega) = Y(x,y) = y.$$

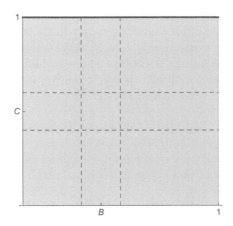

FIGURE 4.14 Events in $\sigma(X)$ and $\sigma(Y)$.

Then X has state space $E = [0,1]$ with σ-algebra $\mathcal{E}$ equal to the Borel subsets of $[0,1]$, and similarly for Y. The σ-algebra $\sigma(X)$ consists of all events of the form $\{\omega = (x,y)|X(\omega) \in B\}$, which is the strip $B \times [0,1]$ in Ω. This strip is displayed in Figure 4.14, bounded by vertical dashed lines. This is the same as $X^{-1}(B)$, and since B is a one-dimensional Borel set, $B \times [0,1] = X^{-1}(B) \in \mathcal{H}$. Therefore we know that X is $\mathcal{H}$-measurable, and that $\sigma(X) \subset \mathcal{H}$. Similarly, $\sigma(Y)$ is the set of all events of the form $\{\omega = (x,y)|Y(\omega) \in C\} = Y^{-1}(C) = [0,1] \times C$, sketched as the horizontal strip in Figure 4.14. It follows that Y is $\mathcal{H}$-measurable and $\sigma(Y) \subset \mathcal{H}$. But X is not $\sigma(Y)$-measurable, because a typical $X^{-1}(B) = B \times [0,1]$ is not in $\sigma(Y)$, whose sets are of the form $[0,1] \times C$, except in the very special case that B itself is $[0,1]$. By the same reasoning, Y is not $\sigma(X)$-measurable. ∎

4.3.3   Continuous Conditional Expectation

This subsection builds on the work we did in Section 3.3 in the discrete world to put continuous conditional expectation on solid footing. But a word or two should be said first about the general theory of expectation of real-valued random variables.

If a random variable X takes a general sample space $(\Omega, \mathcal{H})$ into a subset E of the real line together with the σ-algebra $\mathcal{E} = \mathcal{B}$ of Borel sets, then the familiar way of defining a continuous probability distribution for X is the density model, so that:

$$P[\{\omega | X(\omega) \in B\}] = \int_B f(x) dx. \tag{4.58}$$

Since $\sigma(X)$ consists of all such sets $X^{-1}(B) = \{\omega | X(\omega) \in B\}$, equation (4.58) defines a probability measure on Ω, restricted to events in $\sigma(X)$. Expectations can be defined and computed through integrals in the usual way.

But there are other ways to deal with continuous probability situations other than the density model. In more advanced measure-theoretic probability, the attention would shift back from the state space to the original probability measure P on the sample space. We would begin by introducing **simple random variables**, which are of the form:

$$S = \sum_{i=1}^{n} c_i \cdot I_{B_i}, \tag{4.59}$$

where the c_i's are real constants and the events B_i form a partition of Ω. Therefore these random variables are finite-valued, discrete, but defined on a continuous sample space. The expectation of such a simple random variable would be defined by:

$$E[S] = \sum_{i=1}^{n} c_i \cdot P[B_i]. \tag{4.60}$$

This falls into line with our previous idea of discrete expectation as a weighted sum. But then it can be shown that an arbitrary non-negative random variable X can be approached as an increasing limit of a sequence of simple random variables $S_1, S_2, S_3, \ldots$. Because of this, the expectation of X can be defined by:

$$E[X] = \lim_{n \to \infty} E[S_n]. \tag{4.61}$$

In order for this to be a sensible definition, we would also have to make sure that the limit didn't depend on which particular sequence of simple functions we might use to approximate X. Arbitrary random variables, not necessarily non-negative, can be treated by the device in Exercise 14.

This should be sufficient background for now. To move to conditional expectation, we take our cue from Theorem 3 of Section 3.3. The intuition is that the conditional expectation of X is a random variable whose average is the same as that of X over any event in the information base being conditioned on.

Definition 6. Let X be a random variable that takes a sample space $(\Omega, \mathcal{H})$ to a state space $(E, \mathcal{E})$, and let $\mathcal{A}$ be a sub-σ-algebra of $\mathcal{H}$; that is, all events in $\mathcal{A}$ are also in $\mathcal{H}$, and $\mathcal{A}$ is a σ-algebra. Then the conditional expectation of X given $\mathcal{A}$ is an $\mathcal{A}$-measurable random variable $Y(\omega) = E[X|\mathcal{A}](\omega)$ such that for all sets $A \in \mathcal{A}$:

$$E\left[I_A \cdot Y\right] = E\left[I_A \cdot X\right]. \blacksquare \tag{4.62}$$

Remark. We used the language "an $\mathcal{A}$-measurable random variable" in the definition of conditional expectation purposely; if Y is one random variable satisfying the conditions, then Y could be changed on a set of probability 0 without affecting the expectation equality $E\left[I_A \cdot Y\right] = E\left[I_A \cdot X\right]$. But with this exception, the conditional expectation can be shown to be unique up to sets of probability zero. We will adopt this proviso without further mention when we write equations of conditional expectations.

To gain intuition, it might help to reconsider an old example (Example 2 from Section 3.3) from a different perspective.

Example 5. Let $\Omega = [0, 1]$, and instead of defining an algebra $\mathcal{A}$ by its atoms as in the earlier example, let $\mathcal{H}$ be the σ-algebra of Borel sets on Ω. As before, define a random variable $X(\omega) = 2\omega$ for $\omega \in \Omega$, and let the uniform distribution characterized by a constant density $f(\omega) = 1$ be the probability measure on Ω. Figure 3.13 from the old example is repeated as Figure 4.15 below, which shows the action of random variable X. Consider the smallest σ-algebra $\mathcal{A}$ containing all of three disjoint intervals $A_1 = [0, 1/3)$, $A_2 = [1/3, 2/3)$, and $A_3 = [2/3, 1]$. A moment's thought should convince you that this σ-algebra is the same as the algebra of Example 3.3-2. Use the notation I_A for the indicator variable of event A, which equals 1 for $\omega \in A$ and equals 0 otherwise. We propose that the following simple random variable Y is the conditional expectation $E[X|\mathcal{A}]$:

$$
\begin{aligned}
Y(\omega) \;&=\; \tfrac{1}{3}I_{[0,1/3)}(\omega) + I_{[1/3,2/3)}(\omega) + \tfrac{5}{3}I_{[2/3,1]}(\omega) \\[2mm]
&=\; \begin{cases} 1/3 & \text{if } \omega \in [0, 1/3); \\ 1 & \text{if } \omega \in [1/3, 2/3); \\ 5/3 & \text{if } \omega \in [2/3, 1]. \end{cases}
\end{aligned} \tag{4.63}
$$

To show this, we need to show that Y is $\mathcal{A}$-measurable, and that the expectation equation (4.62) holds for all events $A \in \mathcal{A}$. For the former, there are three events of main concern: $\{\omega \in \Omega | Y(\omega) = 1/3\} = Y^{-1}(1/3)$, $\{\omega \in \Omega | Y(\omega) = 1\} = Y^{-1}(1)$, and $\{\omega \in \Omega | Y(\omega) = 5/3\} = Y^{-1}(5/3)$. These inverse images are, respectively, $[0, 1/3)$, $[1/3, 2/3)$, and $[2/3, 1]$, which belong to $\mathcal{A}$. The inverse image under Y of any union, such as $Y^{-1}(\{1/3, 1\})$ is a union of inverse images, namely $[0, 2/3)$ in this example, which by the σ-algebra property will belong to $\mathcal{A}$. Any set in the state space $\{1/3, 1, 5/3\}$ of Y will

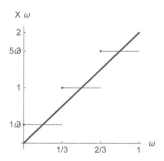

FIGURE 4.15 $Y = E[X|\mathcal{A}]$ is an average value of X over each event in $\mathcal{A}$.

clearly be representable as such a finite union, and therefore will have an inverse image in $\mathcal{A}$, by similar reasoning. Thus, Y is $\mathcal{A}$-measurable.

Proceeding to the expectation equation, the following computation checks it for the generating set A_1:

$$E\left[X \cdot I_{A_1}\right] = \int_0^{1/3} 2\omega \cdot 1 d\omega = \omega^2\big|_0^{1/3} = \frac{1}{9};$$

$$
\begin{aligned}
E\left[Y \cdot I_{A_1}\right] &= E\left[\left(\tfrac{1}{3}I_{[0,1/3)} + I_{[1/3,2/3)} + \tfrac{5}{3}I_{[2/3,1]}\right) \cdot I_{A_1}\right] \\
&= E\left[\tfrac{1}{3}I_{[0,1/3)} \cdot I_{A_1} + I_{[1/3,2/3)} \cdot I_{A_1} + \tfrac{5}{3}I_{[2/3,1]} \cdot I_{A_1}\right] \\
&= E\left[\tfrac{1}{3}\left(I_{[0,1/3)}\right)^2 + 0 + 0\right] \\
&= \tfrac{1}{3}E\left[I_{[0,1/3)}\right] \\
&= \tfrac{1}{3} \cdot P\big[[0,1/3)\big] = \tfrac{1}{3} \cdot \tfrac{1}{3} = \tfrac{1}{9}.
\end{aligned}
$$

In the third line, since $A_1 = [0,1/3)$ is disjoint from the other two intervals, the product of indicators is zero. In line 4, since indicator random variables only take on the value 0 or 1, the square of the indicator is the same as the indicator itself. In Exercise 9, you will repeat these computations for the other two events A_2 and A_3. Other events in $\mathcal{A}$ are simply disjoint unions of the generating events, and the equation can be checked easily by linearity. For example, for $A = A_1 \cup A_2 = [0,2/3)$ we have $I_A = I_{A_1} + I_{A_2}$, and therefore since the expectation equation holds for each individual A_i, we can compute:

$$
\begin{aligned}
E\left[Y \cdot I_A\right] &= E\left[Y \cdot \left(I_{A_1 \cup A_2}\right)\right] \\
&= E\left[Y \cdot \left(I_{A_1} + I_{A_2}\right)\right] \\
&= E\left[Y \cdot I_{A_1}\right] + E\left[Y \cdot I_{A_2}\right] \\
&= E\left[X \cdot I_{A_1}\right] + E\left[X \cdot I_{A_2}\right] \\
&= E\left[X \cdot \left(I_{A_1} + I_{A_2}\right)\right] \\
&= E\left[X \cdot \left(I_{A_1 \cup A_2}\right)\right] \\
&= E\left[X \cdot I_A\right].
\end{aligned}
$$

Notice that changes to Y on sets of probability 0 will not affect any of these computations. For instance if we chose instead to define Y as having the value $1/3$ on the closed interval $[0, 1/3]$ instead of the interval $[0, 1/3)$ the integrals would work out the same as would the computations involving Y, because $P[[0, 1/3]] = P[[0, 1/3)]$. ∎

Next, let us verify that conditional expectation is linear.

Theorem 2. Let X_1 and X_2 be random variables defined on a sample space $(\Omega, \mathcal{H})$, let c_1 and c_2 be constants, and let $\mathcal{A}$ be another σ-algebra on Ω. Then:

$$E\left[c_1 X_1 + c_2 X_2 \,|\, \mathcal{A}\right] = c_1 E\left[X_1 \,|\, \mathcal{A}\right] + c_2 E\left[X_2 \,|\, \mathcal{A}\right]. \qquad (4.64)$$

Proof. Let $Y_1 = E\left[X_1 \,|\, \mathcal{A}\right]$ and let $Y_2 = E\left[X_2 \,|\, \mathcal{A}\right]$. Then both Y_1 and Y_2 are $\mathcal{A}$-measurable, and the fact that a linear combination such as $c_1 Y_1 + c_2 Y_2$ is therefore also $\mathcal{A}$-measurable is a well-known result in measure theoretic probability whose proof we will omit (see for example Chapter 3 of [5]). To prove formula (4.64), we must show that the random variable on the right side of the equation, which is $c_1 Y_1 + c_2 Y_2$, has the same expectation as $c_1 X_1 + c_2 X_2$ when both are multiplied by I_A, the indicator of an event in $\mathcal{A}$. This is easy to check, since Y_1 and Y_2 are both conditional expectations in their own right:

$$
\begin{aligned}
E\left[(c_1 Y_1 + c_2 Y_2)\, I_A\right] &= c_1 E\left[Y_1 \cdot I_A\right] + c_2 E\left[Y_2 \cdot I_A\right] \\
&= c_1 E\left[X_1 \cdot I_A\right] + c_2 E\left[X_2 \cdot I_A\right] \\
&= E\left[(c_1 X_1 + c_2 X_2)\, I_A\right]. \quad \blacksquare
\end{aligned}
$$

The next sequence of results, which were useful in the discrete world, are equally useful in the continuous world.

To set up Theorem 3, recall that families of events $\mathcal{A}$ and $\mathcal{B}$ are said to be independent if any pair of events chosen from them, $A \in \mathcal{A}$ and $B \in \mathcal{B}$, are independent in the usual probabilistic sense. Then we can call a random variable X **independent** of a σ-algebra $\mathcal{A}$ if the σ-algebra $\sigma(X)$ generated by X is independent of $\mathcal{A}$.

Theorem 3. If X is a random variable that is independent of a σ-algebra $\mathcal{A}$ then:

$$E[X|\mathcal{A}] = E[X]. \qquad (4.65)$$

Proof. We would like to first establish that under our assumptions, X and the indicator I_A of any event $A \in \mathcal{A}$ are independent as random variables. This is checked in the computations below.

$$
\begin{aligned}
P\left[X \in B, I_A = 1\right] &= P\left[\{X \in B\} \cap A\right] \\
&= P\left[X \in B\right] \cdot P\left[A\right] \\
&= P\left[X \in B\right] \cdot P\left[I_A = 1\right].
\end{aligned}
$$

$$
\begin{aligned}
P\left[X \in B, I_A = 0\right] &= P\left[\{X \in B\} \cap A^c\right] \\
&= P\left[X \in B\right] \cdot P\left[A^c\right] \\
&= P\left[X \in B\right] \cdot P\left[I_A = 0\right].
\end{aligned}
$$

Line 2 in each of these follows from the fact that generic events in $\sigma(X)$ are of the form $\{X \in B\}$, and these are being assumed to be independent of events in $\mathcal{A}$.

The constant random variable $Y = E[X]$ is $\mathcal{A}$-measurable (see Exercise 12). So we must verify the expectation equation:

$$E[X \cdot I_A] = E[Y \cdot I_A] = E[E[X] \cdot I_A], \text{for all } A \in \mathcal{A}.$$

By independence, the left side is $E[X] \cdot E[I_A] = E[X] \cdot P[A]$. But factoring the constant $E[X]$ out of the right side also gives $E[X] \cdot E[I_A] = E[X] \cdot P[A]$; hence the equation follows. ∎

Theorem 4. If X is $\mathcal{A}$-measurable, then $E[X|\mathcal{A}] = X$.

We leave the proof of Theorem 4 to the reader as Exercise 11.

Theorem 5. If X is measurable with respect to a σ-algebra $\mathcal{A}$, and Z is another random variable, then $E[X \cdot Z|\mathcal{A}] = X \cdot E[Z|\mathcal{A}]$.

Proof. We give an outline of a proof, based on the fact that was mentioned above that a general random variable X may be approached as a limit of an increasing sequence of simple random variables. As a first step, for any indicator variable I_B with $B \in \mathcal{A}$, and any other event $A \in \mathcal{A}$, we have from the definition of conditional expectation:

$$E[I_B \cdot Z \cdot I_A] = E[Z \cdot I_{B \cap A}] = E[E[Z|\mathcal{A}] \cdot I_{B \cap A}] = E[I_B \cdot E[Z|\mathcal{A}] \cdot I_A].$$
$$(4.66)$$

Thus, $E[I_B \cdot Z|\mathcal{A}] = I_B \cdot E[Z|\mathcal{A}]$. That is, the theorem is true when $X = I_B, B \in \mathcal{A}$. In a similar way, by linearity, the theorem is also true when X is a simple random variable $S = \sum_{i=1}^{n} c_i \cdot I_{B_i}$ with all partition sets B_i in $\mathcal{A}$, since:

$$
\begin{aligned}
E[S \cdot Z|\mathcal{A}] &= E[(\sum_{i=1}^{n} c_i \cdot I_{B_i}) \cdot Z|\mathcal{A}] \\
&= \sum_{i=1}^{n} c_i \cdot E[I_{B_i} \cdot Z|\mathcal{A}] \\
&= \sum_{i=1}^{n} c_i \cdot I_{B_i} E[Z|\mathcal{A}] \\
&= S \cdot E[Z|\mathcal{A}].
\end{aligned}
\quad (4.67)
$$

Now, let X be the monotone increasing limit of a sequence $S_1, S_2, S_3, \dots$ of simple random variables of the form above. Write, for the kth member of this sequence:

$$S_k = \sum_{i=1}^{n_k} c_{i,k} \cdot I_{B_{i,k}}.$$

Then for any event $A \in \mathcal{A}$ we have:

$$
\begin{aligned}
\cdot \quad E\left[X \cdot Z \cdot I_A\right] &= E\left[\lim_k S_k \cdot Z \cdot I_A\right] \\
&= \lim_k E\left[S_k \cdot Z \cdot I_A\right] \\
&= \lim_k \sum_{i=1}^{n_k} c_{i,k} \cdot E\left[I_{B_{i,k}} \cdot Z \cdot I_A\right] \\
&= \lim_k \sum_{i=1}^{n_k} c_{i,k} \cdot E\left[I_{B_{i,k}} \cdot E[Z|\mathcal{A}] \cdot I_A\right] \\
&= \lim_k E\left[\sum_{i=1}^{n_k} c_{i,k} I_{B_{i,k}} \cdot E[Z|\mathcal{A}] \cdot I_A\right] \\
&= \lim_k E\left[S_k \cdot E[Z|\mathcal{A}] \cdot I_A\right] \\
&= E\left[\lim_k S_k \cdot E[Z|\mathcal{A}] \cdot I_A\right] \\
&= E\left[X \cdot E[Z|\mathcal{A}] \cdot I_A\right].
\end{aligned}
\tag{4.68}
$$

The real work is in the second line, where we are applying the monotone convergence theorem for expectation to pull the limit out of expectation. We have not shown this result. Once that is done, however, linearity gives the third line, the definition of $E[Z|\mathcal{A}]$ yields the fourth, and linearity again is applied in the fifth line. Lines 6, 7, and 8 essentially play out lines 1, 2, and 3 in reverse. The first and last lines together say that $X \cdot E[Z|\mathcal{A}]$ satisfies the condition to be $E[X \cdot Z|\mathcal{A}]$. ∎

The Tower Laws for simplifying double conditional expectations still hold in this context, as Theorem 6 asserts.

Theorem 6. (Tower Laws) If $\mathcal{A}$ and $\mathcal{B}$ are two σ-algebras on a sample space Ω such that $\mathcal{A} \subset \mathcal{B}$, and if X is a random variable on Ω, then:

$$
E[E[X|\mathcal{A}]|\mathcal{B}] = E[X|\mathcal{A}] \text{ and } E[E[X|\mathcal{B}]|\mathcal{A}] = E[X|\mathcal{A}]. \quad ∎
\tag{4.69}
$$

The proof is exactly the same as the proof of Theorem 8 of Section 3.3, and so it will not be repeated.

We finish this section with an example in which some of the main results are brought in to simplify a conditional expectation.

Example 6. Suppose that there are three random variables, X, Y, and Z, defined on the same sample space, such that Y is $\sigma(X)$-measurable and Z is independent of $\sigma(X)$. Simplify as far as possible:

$$
E[E[4XY + Z|\sigma(Y)]|\sigma(X)].
$$

Solution. First, note that since we are assuming that Y is $\sigma(X)$-measurable, all events of the form $\{Y \in B\}$ belong to $\sigma(X)$. But these events are exactly those of $\sigma(Y)$, so that $\sigma(Y) \subset \sigma(X)$. By the Tower Law and linearity, the double conditional expectation compresses to a single conditional expectation:

$$E[E[4XY+Z|\sigma(Y)]|\sigma(X)] = E[4XY+Z|\sigma(Y)] = 4E[XY|\sigma(Y)]+E[Z|\sigma(Y)].$$

Since Z is independent of $\sigma(X)$, events of the form $\{Z \in C\}$ are independent of all events in $\sigma(X)$, including in particular those in $\sigma(Y)$. In other words, Z is also independent of $\sigma(Y)$. Therefore, appealing to Theorems 4 and 5, we have:

$$4E[XY|\sigma(Y)] + E[Z|\sigma(Y)] = 4YE[X|\sigma(Y)] + E[Z].$$

This is as far as we can go with the information given. ∎

Exercises 4.3

1. Prove identity (4.55) and explain why a σ-algebra is closed under both countable and finite intersections.

2. In Example 1, argue that the events $\{r < T \leq s\}$ and $\{T = s\}$ are in $\mathcal{H}_t$ for all $r < s \leq t$.

3. Show, as mentioned in the section, that if $\{\mathcal{H}_\alpha\}_{\alpha \in A}$ is an arbitrary collection of σ-algebras on the same sample space indexed by a (not necessarily countable) index set A, then the collection of events common to all of the $\mathcal{H}_\alpha$, written as $\bigcap_{\alpha \in A} \mathcal{H}_\alpha$, is a σ-algebra.

4. Show that all intervals of the following types are in the Borel σ-algebra of subsets of $\mathbb{R}$:
    (a) closed intervals $[a, b]$;
    (b) half-closed, half-open intervals $[a, b)$;
    (c) open rays $(-\infty, b)$.

5. Show carefully that each of the following are in the Borel σ-algebra on $\mathbb{R}^2$:
    (a) half-planes $(-\infty, a_1) \times (-\infty, \infty)$ and $(-\infty, \infty) \times (-\infty, a_2)$;
    (b) the open rectangle $(a_1, b_1) \times (a_2, b_2)$;
    (c) the closed rectangle $[a_1, b_1] \times [a_2, b_2]$;
    (d) the half plane $P = \{(x, y)|x < y\}$.

6. For random variable X as in Theorem 1, verify that the collection of all inverse images $X^{-1}(B) = \{\omega \in \Omega | X(\omega) \in B\}$ as B ranges through all sets in $\mathcal{E}$ is closed under countable union.

7. Let a sample space Ω be the interval $[-1, 1]$, and let $\mathcal{H}$ be the σ-algebra of Borel sets restricted to this interval. Define two random variables $X(\omega) = \omega$, $Y(\omega) = |\omega|$. Argue that $\mathcal{H} = \sigma(X)$, that Y is $\sigma(X)$-measurable, but X is not $\sigma(Y)$-measurable. Thus, knowledge of the value of X tells you about Y, but

knowledge of Y does not give the value of X.

8. In Example 4 we did not introduce a probability measure on Ω, so we could not talk about independence of X and Y. But now suppose that we define P on the σ-algebra $\mathcal{H} = \mathcal{B}^2$ by $P[A] = $ area of A. (We assume without proof that Borel subsets of $\mathbb{R}^2$ actually have well-defined area.) Show that the σ-algebras $\sigma(X)$ and $\sigma(Y)$ are independent (that is, all events in one are independent of all events in the other).

9. In Example 5, carry out the computations to verify that $E[I_A \cdot Y] = E[I_A \cdot X]$ for the generating events $A_2 = [1/3, 2/3)$ and $A_3 = [2/3, 1]$. Also, verify by direct computation of the expectations that $E[I_\Omega \cdot Y] = E[I_\Omega \cdot X]$ (in other words, $E[Y] = E[X]$).

10. For the sample space of Example 5, with uniform probability measure, write out all events in the σ-algebra $\mathcal{A}$ generated by the intervals $[0, 1/2)$, $[1/2, 3/4)$, $[3/4, 1]$. If we define a random variable $Z(\omega) = \omega^2$, find the conditional expectation $Y = E[Z|\mathcal{A}]$, using Definition 6 to prove that your proposed Y is correct.

11. Show that if $\mathcal{A}$ is a σ-algebra and if X is $\mathcal{A}$-measurable, then $E[X|\mathcal{A}] = X$.

12. Why is a constant random variable $X(\omega) = c$ measurable with respect to any σ-algebra $\mathcal{A}$?

13. Prove the Law of Total Probability for Expectation; that is, for any σ-algebra $\mathcal{A}$ and any random variable X:

$$E[X] = E[E[X|\mathcal{A}]].$$

14. Show that any real-valued random variable X can be expressed as a difference of two non-negative random variables X^+ and X^-, that is, $X = X^+ - X^-$. Therefore, $E[X]$ can be defined as $E[X^+] - E[X^-]$, at least if both expectations are finite. What can be said if one or the other or both are infinite?

15. Suppose that we have three random variables X, Y, and Z such that X is independent of both Y and Z, and Z is $\sigma(Y)$-measurable. Simplify:

$$E[E[3XY - XZ|\sigma(Y)]|\sigma(Z)].$$

16. If two random variables X and Y are such that X is independent of $\sigma(Y)$, is it also true that Y is independent of $\sigma(X)$? Why or why not?

4.4 Brownian Motion and Geometric Brownian Motion

We now want to move from static probability models to the dynamic aspects of the motion of risky asset prices through continuous time. In so doing, we will create the natural extension of the binomial branch process to the continuous-time, continuous-state setting that was hinted at in Section 3.7.

We can take a cue from the theory of interest in continuous time, which says that a risk-free asset that grows at a continuous rate of r has value equal to:

$$X_0(t) = x_0 e^{rt}, \tag{4.70}$$

where x_0 is its initial value at time 0. Then, taking logs,

$$\log(X_0(t)) - \log(x_0) = rt. \tag{4.71}$$

Risky assets are not so well-behaved as this. But we might guess that they add some randomness to this deterministic linear drift exhibited by the log of the risk-free value.

Example 1. We looked at General Electric stock in Section 3.7; now let's try another company, H&R Block. Weekly prices during the calendar years 2012-2016 were gathered, and the logs of those were computed. Then the log of the initial price was subtracted. A best fitting line through the origin was fitted by linear regression, and its graph is superimposed upon the log price data in Figure 4.16. We see a great deal of wandering for large time values, but there is reason to suspect an underlying linear drift in the logged process. If you perform the same experiment with other companies, you will find similar results. The question is: what random model of the form:

$$\log(X(t)) - \log(x_0) = \mu t + B_t \tag{4.72}$$

captures this sort of behavior, for an appropriate stochastic process $(B_t)_{t \geq 0}$? ■

The theoretical concepts of the previous section, such as σ-algebras, measurability, independence, and conditional expectation will have an impact here. First, given any stochastic process $(X_t)_{t \geq 0}$, whether continuous or discrete, we will call the **history generated by the process**, denoted $(\mathcal{H}_t)_{t \geq 0}$, the increasing family of σ-algebras which is the smallest such that each random variable is measurable with respect to its corresponding σ-algebra, meaning that:

(a) For times $s \leq t$, $\mathcal{H}_s \subseteq \mathcal{H}_t$;
(b) X_t is $\mathcal{H}_t$-measurable for each t;
(c) For each t, $\mathcal{H}_t$ is the smallest σ-algebra satisfying (a) and (b).

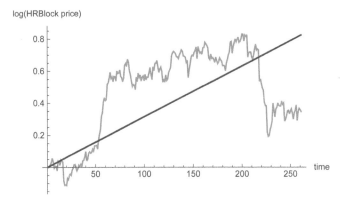

FIGURE 4.16 Time series of logs of H&R Block prices minus log(initial price).

Intuitively, $\mathcal{H}_t$ contains all information available by watching the process through time t. Events such as $\{X_{1.0} \in B_1, X_{3.2} \in B_2\}$ would be contained in $\mathcal{H}_{4.8}$, for example, where B_1 and B_2 are Borel sets. For some random variable Y to be independent of $\mathcal{H}_t$ means that all events $\{Y \in C\}$ are independent of all events $\{X_s \in B\}$ for $s \le t$.

4.4.1   Random Walk Processes

In Section 3.7 we reconsidered the binomial branch process in such a way that the logs of its price values evolved as:

$$\log(S_{t_1}) = \log(s_0) + \log(R_1);$$
$$\log(S_{t_2}) = \log(s_0) + \log(R_1) + \log(R_2);$$
$$\log(S_{t_3}) = \log(s_0) + \log(R_1) + \log(R_2) + \log(R_3),$$
$$\vdots$$

where $R_1, R_2, R_3, \ldots$ were the successive ratios of prices at adjacent times, which we assumed were independent. Thus, the logged branch process evolves as a series of independent steps taken from a starting point of $\log(s_0)$, each of which has a two-point probability distribution (value $\log(1+b)$ with probability p and $\log(1+a)$ with probability $1-p$). Such a process is called a *random walk*.

Definition 1. A process $X_0 = x_0, X_1, X_2, X_3\ldots$ is called a *random walk* if it is given by

$$X_1 = x_0 + Y_1; \ \ X_2 = X_1 + Y_2; \ \ X_3 = X_2 + Y_3; \cdots$$

where the Y_is are independent, identically distributed random variables and the initial state x_0 is known. ∎

Note that the states traversed by a random walk are just the initial state x_0 plus the accumulations of steps:

$$X_m = x_0 + \sum_{i=1}^{m} Y_i. \tag{4.73}$$

Our model will use the time scheme of Section 3.7, illustrated in Figure 4.17. The time interval $[0, T]$ is broken into n equally sized subintervals with length $\Delta t = \frac{T}{n}$, whose partition points are:

$$t_0 = 0, t_1, t_2, ..., t_{n-1}, t_n = T.$$

Then explicitly, $t_k = k \Delta t = k \frac{T}{n}$.

FIGURE 4.17 Time axis for asset motion.

Consider a random walk in which each step Y_i has the two-point distribution:

$$g(y) = \begin{cases} 1/2 & \text{if } y = \sqrt{\Delta t}; \\ 1/2 & \text{if } y = -\sqrt{\Delta t}. \end{cases} \tag{4.74}$$

In this case, the mean and variance of each step Y_i are:

$$\mu_Y = E[Y] = \frac{1}{2} \cdot \left(\sqrt{\Delta t}\right) + \frac{1}{2} \cdot \left(-\sqrt{\Delta t}\right) = 0; \tag{4.75}$$

$$\sigma_Y^2 = \text{Var}(Y) = \frac{1}{2} \cdot \left(\sqrt{\Delta t}\right)^2 + \frac{1}{2} \cdot \left(-\sqrt{\Delta t}\right)^2 = \Delta t. \tag{4.76}$$

Therefore the mean and variance of the state after m steps of the random walk are:

$$E[X_m] = E\left[x_0 + \sum_{i=1}^{m} Y_i\right] = x_0 + \sum_{i=1}^{m} E[Y_i] = x_0; \tag{4.77}$$

$$\text{Var}(X_m) = \text{Var}\left(x_0 + \sum_{i=1}^{m} Y_i\right) = \sum_{i=1}^{m} \text{Var}(Y_i) = m \cdot \Delta t. \tag{4.78}$$

This simple random walk has no tendency to drift, but instead maintains an expected value equal to the initial value x_0. Its variance after m steps, that is, at clock time $m \cdot \Delta t$, is exactly equal to the clock time. At any particular clock time $T = n \cdot \Delta t$, the state of the walk X_n is the sum of a fixed value x_0 plus a large number of independent, identically distributed random variables, and so its distribution is approximately normal, by the Central Limit Theorem. We could envision a sequence of such stochastic processes on a fixed interval $[0, T]$ of time, with numbers of subintervals n converging to infinity (and consequently time steps Δt converging to zero), and hope that there is a limiting process that inherits the properties that:

(i) the initial state is x_0;
(ii) the mean value $E[X_{t_k}] = x_0$ at each time t_k;
(iii) the variance $\mathrm{Var}(X_{t_k}) = t_k$ at each time t_k;
(iv) the distribution of X_{t_k} is normal.

But even more is true about this random walk. Consider an increment between times $t_k = k \cdot \Delta t$ and $t_{k+l} = (k+l)\Delta t$. The change in state between these two times is:

$$
\begin{aligned}
X_{t_{k+l}} - X_{t_k} &= X_{(k+l)\Delta t} - X_{k\Delta t} \\
&= \left(x_0 + \sum_{i=1}^{k+l} Y_i \right) - \left(x_0 + \sum_{i=1}^{k} Y_i \right) \qquad (4.79) \\
&= \sum_{i=k+1}^{k+l} Y_i.
\end{aligned}
$$

This increment is not a function of exactly what the initial time t_k is, and is independent of all steps Y_i from Y_1 through Y_k, and thus is independent of all states X_1 through X_{t_k} of the random walk. Also, since it is the sum of l identically distributed steps, its distribution is the same as the sum of any other disjoint collection of l of the Y_is. In particular,

$$
X_{(k+l)\Delta t} - X_{k\Delta t} \text{ has the same distribution as } X_{l\Delta t} - X_0.
$$

It is instructive to know something about the dynamic behavior of the random walk. In Figure 4.18 is a simulation of a path of the random walk with initial state $x_0 = 0$, and $n = 1000$ time points with time increment $\Delta t = 0.001$, so that the terminal time in the simulation is $T = 1$. You can observe that since the sizes of the steps are very small, the process appears to traverse a path that is nearly continuous. The path is also reminiscent of the motion of the log price process for real assets that we have looked at before. The *Mathematica* version of the text has the code that produced the simulation.

4.4.2   Standard Brownian Motion

Definition 2. A ***standard Brownian motion*** with initial state z_0 is a continuous-time stochastic process $(Z_t)_{t \geq 0}$ with state space $E = \mathbb{R}$ and history $(\mathcal{H}_t)_{t \geq 0}$, such that:

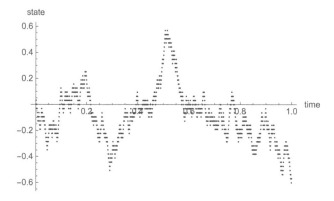

FIGURE 4.18 Simulated random walk on time interval $[0, 1]$.

(a) The function $t \to Z_t(\omega)$ is continuous at every t for all outcomes ω except possibly some in a set of probability zero;

(b) For all $t > 0$, Z_t is normally distributed with mean z_0 and variance t;

(c) For all $t, s > 0$, the distribution of $Z_{t+s} - Z_t$ does not depend on t and is independent of $\mathcal{H}_t$. ∎

This process captures all of the important properties that we discussed in the previous subsection relative to the simple random walk process. Property (c) is called the **stationary, independent increments** property, and it implies that, in concert with property (b),

$$Z_{t+s} - Z_t \sim Z_s - z_0 \sim \mathcal{N}(0, s); \tag{4.80}$$

in other words, an increment of a standard Brownian motion for a time step of s units is normally distributed with mean 0 and variance equal to the time step.

Example 2. Suppose that a process of asset rates of return $(R_t)_{t \geq 0}$ is in standard Brownian motion with initial state $r_0 = .05$. Find the probabilities that:

(a) $R_{.5}$ is between .03 and .07;

(b) the change in rate between times .5 and 1 is more than .02;

(c) the change in rate between times .5 and 1 is less than .01 given that the rate at time .5 is .04.

Solution. (a) Since the rate of return process is assumed to satisfy the standard Brownian motion conditions, $R_{.5} \sim \mathcal{N}(.05, .5)$, hence:

$$
\begin{aligned}
P\left[.03 \leq R_{.5} \leq .07\right] &= P\left[\frac{.03-.05}{\sqrt{.5}} \leq Z \leq \frac{.07-.05}{\sqrt{.5}}\right] \\
&= P[-.0283 \leq Z \leq .0283],
\end{aligned}
$$

where Z is a generic standard normal random variable. Using technology or a table, this probability comes out to about .0226.

(b) In symbols, we are asked for:

$$P[R_1 - R_{.5} \geq .02].$$

But by the Brownian conditions, the change $R_1 - R_{.5}$ has the same distribution as $R_{.5} - R_0 = R_{.5} - .05$. This distribution is $\mathcal{N}(0, .5)$. So, we can compute:

$$\begin{aligned}
P[R_1 - R_{.5} \geq .02] &= P[R_{.5} - R_0 \geq .02] \\
&= P\left[Z \geq \frac{.02 - 0}{\sqrt{.5}}\right] \\
&= P[Z \geq .0283] = .4887.
\end{aligned}$$

(c) This time, we are asked for:

$$P[R_1 - R_{.5} < .01 | R_{.5} = .04].$$

But from the stationary, independent increments condition of Brownian motion, the increment $R_1 - R_{.5}$ is independent of $R_{.5}$, and as noted in part (b) of this problem, it has the $\mathcal{N}(0, .5)$ distribution. Therefore we have:

$$\begin{aligned}
P[R_1 - R_{.5} < .01 | R_{.5} = .04] &= P[R_1 - R_{.5} < .01] \\
&= P[R_{.5} - R_0 < .01] \\
&= P\left[Z < \frac{.01 - 0}{\sqrt{.5}}\right] \\
&= P[Z < .0141] \approx .4944. \quad \blacksquare
\end{aligned}$$

You may have noticed in the computations of Example 2 that not only did the model forbid changes to the expected rate of return, but the variance was extraordinarily large in comparison to the mean return of .05. The mean and variance are enforced in the conditions for standard Brownian motion, which is a tipoff that we need something more general than this process, introduced in the next subsection.

4.4.3   Non-Standard Brownian Motion

Definition 3. A ***non-standard Brownian motion*** with initial state w_0, ***drift rate*** μ, and ***variance rate*** σ^2 is a continuous-time stochastic process $(W_t)_{t \geq 0}$ with state space $E = \mathbb{R}$ and history $(\mathcal{H}_t)_{t \geq 0}$, such that:

(a) The function $t \to W_t(\omega)$ is continuous at every t for all outcomes ω except possibly some in a set of probability zero;

(b) For all $t > 0$, W_t is normally distributed with mean $w_0 + \mu t$ and variance $\sigma^2 t$;

(c) For all $t, s > 0$, the distribution of $W_{t+s} - W_t$ does not depend on t and is independent of $\mathcal{H}_t$. $\quad \blacksquare$

Thus, a non-standard Brownian motion shares the path continuity and stationary, independent increments properties of the standard version, but it allows a drift in mean value with slope μ, and a rescaling of variance, both of which changes give us more general modeling capabilities.

One way of thinking of a non-standard Brownian motion process W is to construct it as a transformation of a standard process Z with initial state 0, in the following way:

$$W_t = w_0 + \mu t + \sigma \cdot Z_t. \tag{4.81}$$

Then since W_t is a deterministic value plus a constant times a normally distributed random variable, W_t has the normal distribution. The histories generated by the two processes would be the same. At time $t = 0$ the state is w_0. The W process inherits continuity of paths from the Z process. The mean and variance of the time t value W_t are:

$$
\begin{aligned}
E\left[W_t\right] &= E\left[w_0 + \mu t + \sigma \cdot Z_t\right] \\
&= w_0 + \mu t + \sigma \cdot 0 \\
&= w_0 + \mu t;
\end{aligned}
$$

$$
\begin{aligned}
\mathrm{Var}\left(W_t\right) &= \mathrm{Var}\left(w_0 + \mu t + \sigma \cdot Z_t\right) \\
&= \mathrm{Var}\left(\sigma \cdot Z_t\right) \\
&= \sigma^2 \mathrm{Var}\left(Z_t\right) \\
&= \sigma^2 t.
\end{aligned}
$$

Also, the W process satisfies the stationary, independent increments condition because the Z process does (see Exercise 7). Hence all of the defining conditions for non-standard Brownian motion are in place.

Example 3. Suppose that a random asset price process has week t value X_t. Assume that the logged price process, minus the log of the initial price x_0, can be modeled as a non-standard Brownian motion process (W_t) with initial state 0, that is:

$$\log\left(X_t\right) - \log\left(x_0\right) = W_t.$$

If the parameters of the process are $\mu = .0009$ and $\sigma = .005$, and $x_0 = 52$, write the price X_t using construction (4.81) as a function of standard Brownian motion. Find the probability that the price at time 3.5 weeks is (a) at least 53; (b) less than 52.

Solution. (a) From formula (4.81), since the initial state of the Brownian motion is 0, we have:

$$
\begin{aligned}
\log\left(X_t\right) - \log\left(x_0\right) &= W_t = \mu t + \sigma \cdot Z_t \\
&\implies e^{\log(X_t) - \log(x_0)} = e^{\mu t + \sigma \cdot Z_t} \\
&\implies \frac{X_t}{x_0} = e^{\mu t + \sigma \cdot Z_t} \\
&\implies X_t = x_0 e^{\mu t + \sigma \cdot Z_t} = 52 e^{.0009t + .005 Z_t}.
\end{aligned}
\tag{4.82}
$$

This derivation is quite general until the last equation in which the given parameter values were used. The next subsection explores this asset price model further. For question (a), we can compute as follows:

$$
\begin{aligned}
P[X_{3.5} \geq 53] &= P\left[52e^{.0009(3.5)+.005Z_{3.5}} \geq 53\right] \\
&= P\left[.0009(3.5)+.005Z_{3.5} \geq \log(53/52)\right] \\
&= P\left[Z_{3.5} \geq \tfrac{\log(53/52)-.0009(3.5)}{.005}\right] \\
&= P[Z_{3.5} \geq 3.17964].
\end{aligned}
$$

Since $Z_{3.5} \sim \mathcal{N}(0,3.5)$, standardization by dividing both sides of the last inequality by $\sqrt{3.5}$ finishes the computation.

$$
\begin{aligned}
P[X_{3.5} \geq 53] &= P[Z_{3.5} \geq 3.17964] \\
&= P\left[Z \geq \tfrac{3.17964}{\sqrt{3.5}}\right] \\
&= 1 - P\left[Z \leq \tfrac{3.17964}{\sqrt{3.5}} = 1.70\right] \\
&= .045.
\end{aligned}
$$

(b) For part (b) we derive similarly:

$$
\begin{aligned}
P[X_{3.5} < 52] &= P\left[52e^{.0009(3.5)+.005Z_{3.5}} < 52\right] \\
&= P\left[.0009(3.5)+.005Z_{3.5} < \log(52/52) = 0\right] \\
&= P\left[Z_{3.5} < \tfrac{-.0009(3.5)}{.005} = -.63\right] \\
&= P\left[Z < \tfrac{-.63}{\sqrt{3.5}} = -.337\right] = .368. \quad \blacksquare
\end{aligned}
$$

Example 3 begs the question: how does an analyst estimate the drift and variance parameters for a real asset price process? Because of the stationary, independent increments hypothesis, the successive increments of the Brownian motion on time intervals of length 1:

$$
W_1 - w_0, W_2 - W_1, W_3 - W_2, ..., W_n - W_{n-1}
$$

form a set of n independent and identically distributed random variables having the $\mathcal{N}\left(\mu \cdot 1, \sigma^2 \cdot 1\right) = \mathcal{N}\left(\mu, \sigma^2\right)$ distribution. Then μ and σ^2 may be estimated by the sample mean and sample variance of this list of data. We will show how this works in the next subsection in the context of the geometric Brownian motion process. Two words of caution, however. As noted earlier in analysis of real rate of return data, the higher than expected frequency of extreme rates of return tends to argue against normality, and, in addition, it is not reasonable to expect the mean and variance parameters to remain constant for a very long period of time, so that the sample size n should not be taken to be very large. Statistical methods for detection of trends may be applied, but we will not discuss those here, preferring instead the simplest approach.

4.4.4 Geometric Brownian Motion

At the beginning of Example 3, we looked at a model for the logged prices of a financial asset:

$$\log(X_t) - \log(x_0) = W_t = \mu t + \sigma \cdot Z_t, \tag{4.83}$$

where W_t was a non-standard Brownian motion and Z_t was a standard Brownian motion. Removing the logarithm, we found that:

$$X_t = x_0 e^{\mu t + \sigma \cdot Z_t} = x_0 e^{W_t}. \tag{4.84}$$

This leads us to the following definition.

Definition 4. A stochastic process $(X_t)_{t \geq 0}$ is called a **geometric Brownian motion** with parameters μ and σ^2 if it is of the form $X_t = x_0 e^{W_t}$, where W_t is a non-standard Brownian motion with these parameters and initial state 0. Equivalently, $(X_t)_{t \geq 0}$ is a geometric Brownian motion if:

$$\log\left(\frac{X_t}{x_0}\right) = W_t, \tag{4.85}$$

where $(W_t)_{t \geq 0}$ is non-standard Brownian motion with drift rate μ and variance rate σ^2. ∎

Remark. Notice from formula (4.84) that a geometric Brownian motion $X_t = x_0 e^{\mu t + \sigma \cdot Z_t}$ generalizes a deterministic exponential growth process $x_t = x_0 e^{rt}$ by including the random term $\sigma \cdot Z_t$, which is normally distributed with mean 0 and variance $\sigma^2 t$. Thus, the average tendency of the random process appears to be exponential growth at a rate μ (but we need to be careful: see the next example), and variability is greater as σ increases, and also as time increases, which is intuitively reasonable.

Example 4. Use the well-known formula for the **moment-generating function** $M(s) = E\left[e^{sY}\right] = e^{\mu s + \sigma^2 s^2 / 2}$ of a $N\left(\mu, \sigma^2\right)$ random variable Y to find the mean and variance of the time t value X_t of a geometric Brownian motion with parameters μ and σ^2.

Solution. We represent X_t as $X_t = x_0 e^{\mu t + \sigma \cdot Z_t}$, where Z_t has the $\mathcal{N}(0, t)$ distribution. For the mean value of X_t:

$$
\begin{aligned}
E[X_t] &= E\left[x_0 e^{\mu t + \sigma \cdot Z_t}\right] \\
&= x_0 e^{\mu t} E\left[e^{\sigma \cdot Z_t}\right] \\
&= x_0 e^{\mu t} \cdot e^{0 \cdot \sigma + t \cdot \sigma^2 / 2} \\
&= x_0 e^{\left(\mu + \sigma^2 / 2\right)t}.
\end{aligned}
\tag{4.86}
$$

The crucial third line of this computation follows by using the m.g.f. formula with $\mu = 0$, $\sigma^2 = t$ (which are the mean and variance of Z_t), and parameter $s = \sigma$. So notice the surprising result that the variance parameter also contributes to the expected exponential growth rate.

To compute the variance of X_t, we first compute the second moment similarly to the mean:

$$
\begin{aligned}
E\left[X_t^2\right] &= E\left[x_0^2 e^{2\mu t + 2\sigma \cdot Z_t}\right] \\
&= x_0^2 e^{2\mu t} E\left[e^{2\sigma \cdot Z_t}\right] \\
&= x_0^2 e^{2\mu t} \cdot e^{0 \cdot 2\sigma + t \cdot (2\sigma)^2/2} \\
&= x_0^2 e^{\left(2\mu + 2\sigma^2\right)t}.
\end{aligned}
\tag{4.87}
$$

Hence the variance of the state at time t is as below. Notice that the variance depends on μ as well as σ^2.

$$
\begin{aligned}
\mathrm{Var}\,(X_t) &= E\left[X_t^2\right] - (E\left[X_t\right])^2 \\
&= x_0^2 e^{\left(2\mu + 2\sigma^2\right)t} - \left(x_0 e^{\left(\mu + \sigma^2/2\right)t}\right)^2 \\
&= x_0^2\left(e^{\left(2\mu + 2\sigma^2\right)t} - e^{\left(2\mu + \sigma^2\right)t}\right) \\
&= x_0^2 e^{\left(2\mu + \sigma^2\right)t}\left(e^{\sigma^2 t} - 1\right).\ \blacksquare
\end{aligned}
\tag{4.88}
$$

Example 5. In the geometric Brownian motion model of financial asset price processes, the price X_t at time t is related to a Brownian motion process $(W_t)_{t \geq 0}$ by the formula: $\log(X_t) - \log(x_0) = W_t$. We can compute the logged prices at each of a sequence of equally spaced times $t = 1, 2, 3, ..., n$, and then as noted above, the random variables

$$
\begin{aligned}
&W_1 - w_0 = \log(X_1) - \log(x_0),\ W_2 - W_1 = \log(X_2) - \log(X_1), \\
&W_3 - W_2 = \log(X_3) - \log(X_2), ...
\end{aligned}
\tag{4.89}
$$

form a random sample from the $\mathcal{N}\left(\mu, \sigma^2\right)$ distribution. Then the sample mean and standard deviation of this list of differences of logs can be used to estimate the parameters μ and σ of the geometric Brownian motion.

For example, consider the daily price values of Twitter (ticker symbol TWTR) on the New York Stock Exchange for April 1 through July 31 of 2017, listed below.

14.84, 14.69, 14.53, 14.39, 14.29, 14.36, 14.31, 14.42, 14.3, 14.4, 14.44, 14.54,
14.65, 14.63, 14.71, 14.66, 15.82, 16.61, 16.48, 17.54, 18.24, 18.57, 18.48,
18.69, 18.31, 18.37, 18.54, 18.39, 18.61, 19.23, 19.49, 18.28, 18.51, 18.35, 18.43,
18.15, 17.98, 17.95, 18.23, 18.43, 18.32, 18.53, 18.31, 18.23, 17.57, 17.44, 17.59,
16.9, 17.04, 16.97, 16.76, 16.83, 16.67, 17.06, 16.91, 17.78, 18.15, 18.5, 18.29,
18.12, 17.95, 17.65, 17.87, 17.65, 17.82, 17.92, 18.02, 18.08, 18.64, 19.25, 19.32,
19.64, 19.94, 19.98, 20.12, 20.53, 20.11, 20., 19.97, 19.61, 16.84, 16.75, 16.09.

The logs of these prices can be taken; the first and last few are:

$$2.69733, 2.68717, 2.67622, 2.66653, ..., 2.97604, 2.82376, 2.8184, 2.7782.$$

Prices show a generally increasing trend, as Figure 4.19 illustrates, with a lot of variability.

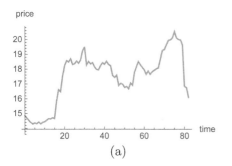
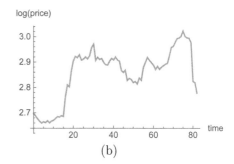

FIGURE 4.19 (a) April-July Twitter prices, 2017; (b) logged prices.

The differences in the logged prices are easy to compute. Here are the first and last few:

$$-0.0101593, -0.0109515, -0.00968191, ..., -0.152283, -0.00535876, -0.0402003.$$

The sample mean of this list of data, which we take to be our estimate of the parameter μ, is about .000986. The sample standard deviation is about .027, which estimates σ. With these values, and the initial price of $x_0 = 14.84$, the probability that the price at time 100 trading days after April 1 is at least 16 is:

$$P\left[X_{100} \geq 16\right] = P\left[14.84e^{\mu t + \sigma \cdot Z_t} \geq 16\right] = P\left[14.84e^{.000986(100) + .027Z_{100}} \geq 16\right],$$

where Z_{100} is distributed as $\mathcal{N}(0, 100)$. Unwinding the inequality, the probability that we want is:

$$P\left[14.84e^{.000986(100) + .027Z_{100}} \geq 16\right]$$
$$= P\left[Z_{100} \geq \frac{\log(16/14.84) - .000986(100)}{.027} = -.864\right].$$

Standardizing by dividing both sides by $\sqrt{100} = 10$, the probability is:

$$P\left[Z_{100} \geq -.864\right] = P\left[Z \geq \frac{-.864}{10} = -.0864\right] \approx .5344.$$

The purpose of this example was to show how estimation of parameters and price prediction works in the ideal model. But we should always be aware that

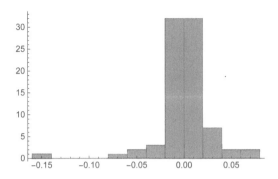

FIGURE 4.20 Histogram of differences of log Twitter prices, April-July 2017.

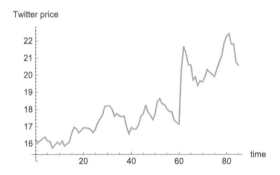

FIGURE 4.21 Twitter prices, August-November 2017.

real models may depart from assumptions. In Figure 4.20 is the histogram of the differences of logged prices, which are assumed to be normally distributed. The same phenomenon occurs here as in earlier analyses: there is a stronger than expected tendency for data to cluster near the mean, and there is a clear outlier. Parameter estimates can also be unstable over time. The Twitter price process during the months August-November of the same year was analyzed in the same way, and the time series for these four months is displayed in Figure 4.21. The asset showed stronger performance than earlier, and the mean of the differences in log prices comes out to .00281, as contrasted with .000986 in the previous four months. The standard deviation of the log difference data was about .026, very similar to the earlier data set, however. ∎

4.4.5   Brownian Motion and Binomial Branch Processes

We close this section by firming up the connection between binomial branch processes in discrete time and Brownian motion processes in continuous time.

This connection has been hinted at earlier in this section, as well as in Section 3.7.

Using a time scheme in which transitions of a binomial branch process occur at integer multiples $k\Delta t$ of a time increment Δt, in Section 3.7 we described the process of states $S_0 = s_0, S_1, S_2, \ldots$ of a binomial branch process as:

$$S_m = S(m\Delta t) = s_0 \cdot \frac{S_1}{s_0} \cdot \frac{S_2}{S_1} \cdot \frac{S_3}{S_2} \cdots \frac{S_m}{S_{m-1}} = s_0 \cdot R_1 \cdot R_2 \cdot R_3 \cdots R_m. \quad (4.90)$$

The random multipliers R_i are independent and identically distributed, with values $1 + b$ (with probability p) or $1 + a$ (with probability $1 - p$). Hence the logarithms of the prices satisfy:

$$\log\left(S_m\right) = \log\left(s_0 \cdot R_1 \cdot R_2 \cdot R_3 \cdots R_m\right) = \log\left(s_0\right) + \sum_{l=1}^{m} \log\left(R_l\right). \quad (4.91)$$

In other words, the logged price process is a random walk with initial state $\log\left(s_0\right)$ and steps of sizes $\log(1 + b)$ or $\log(1 + a)$, with probabilities p and $1 - p$, respectively. Now if these steps happened to be such that:

$$\log(1 + b) = \sqrt{\Delta t} \text{ and } \log(1 + a) = -\sqrt{\Delta t}$$
$$\Longrightarrow b = e^{\sqrt{\Delta t}} - 1 \text{ and } a = e^{-\sqrt{\Delta t}} - 1,$$

and $p = 1/2$, then the random walk is the simple symmetric walk of formulas (4.73)-(4.74). Although we are not showing it rigorously, this discrete process converges in some sense to a standard Brownian motion.

The binomial branch process has three parameters, p, b, and a, to control, whereas a Brownian motion only has two, μ and σ. So we have some flexibility in choosing the binomial branch parameters in such a way that the logged price process converges to non-standard Brownian motion. Let us take advantage of that and consider only branch processes that have $p = 1/2$. The analysis of Section 3.7 suggests the following relationship between parameters:

$$\log(1 + b) = \mu\Delta t + \sigma\sqrt{\Delta t} \text{ and } \log(1 + a) = \mu\Delta t - \sigma\sqrt{\Delta t}$$
$$\Longrightarrow b = e^{\mu\Delta t + \sigma\sqrt{\Delta t}} - 1 \text{ and } a = e^{\mu\Delta t - \sigma\sqrt{\Delta t}} - 1. \quad (4.92)$$

Inverting the relationships by adding and then subtracting the equations for $\log(1 + b)$ and $\log(1 + a)$, we derive the following expressions for μ and σ in terms of b and a:

$$\begin{cases} \log(1+b) &= \mu\Delta t + \sigma\sqrt{\Delta t} \\ \log(1+a) &= \mu\Delta t - \sigma\sqrt{\Delta t} \end{cases}$$
$$\implies \begin{cases} \log(1+b) + \log(1+a) &= 2\mu\Delta t \\ \log(1+b) - \log(1+a) &= 2\sigma\sqrt{\Delta t} \end{cases} \qquad (4.93)$$
$$\implies \begin{cases} \mu\Delta t &= (\log(1+b) + \log(1+a))/2 \\ \sigma\sqrt{\Delta t} &= (\log(1+b) - \log(1+a))/2. \end{cases}$$

Each step of the random walk has mean:

$$E\left[\log\left(R_i\right)\right] = \frac{1}{2}\log(1+b) + \frac{1}{2}\log(1+a) = \mu\Delta t,$$

from the top formula in the second system in derivation (4.93). The variance of each step is:

$$\begin{aligned} \mathrm{Var}\left(\log\left(R_i\right)\right) &= E\left[\left(\log\left(R_i\right)\right)^2\right] - \left(E\left[\log\left(R_i\right)\right]\right)^2 \\ &= \frac{1}{2}(\log(1+b))^2 + \frac{1}{2}(\log(1+a))^2 - (\mu\Delta t)^2 \\ &= \frac{1}{2}\left(\mu\Delta t + \sigma\sqrt{\Delta t}\right)^2 + \frac{1}{2}\left(\mu\Delta t - \sigma\sqrt{\Delta t}\right)^2 - (\mu\Delta t)^2 \\ &= \frac{1}{2}\left((\mu\Delta t)^2 + 2\mu\Delta t \cdot \sigma\sqrt{\Delta t} + \sigma^2\Delta t\right) \\ &\quad + \frac{1}{2}\left((\mu\Delta t)^2 - 2\mu\Delta t \cdot \sigma\sqrt{\Delta t} + \sigma^2\Delta t\right) \\ &\quad - (\mu\Delta t)^2 \\ &= \sigma^2\Delta t. \end{aligned}$$

Thus, in n steps, the logged binomial branch process reaches a state $\log\left(S_n\right) = \log(S(n\Delta t))$ which has mean $\mu \cdot n\Delta t$ and variance $\sigma^2 n\Delta t$, as does a non-standard Brownian motion at time $t = n\Delta t$. And, as n approaches infinity and Δt approaches 0, the distribution of $\log(S(t)) = \log(S(n\Delta t))$ approaches normality. Hence we have motivated that the logged branch process with b and a as in formula (4.92) and $p = 1/2$ approaches non-standard Brownian motion with parameters μ and σ as in formula (4.93). Therefore, the binomial branch process itself approaches a geometric Brownian motion.

Exercises 4.4

1. Consider a random walk (X_m) with initial state 0, in which the step random variables Y_i have the continuous uniform distribution on $[0, 1]$. Find formulas for the mean and variance of X_m.

2. Let a random walk $(X_m)_{m=0,1,2,\dots}$ be constructed so that its initial state is 2, and its step random variables Y_i have the $\mathcal{N}(0, 1)$ distribution. Find the probability that (a) $X_3 > 3$; (b) $X_3 < 1$.

3. Referring to Example 2, compute (a) the probability that at time 3 the rate of return exceeds .08; (b) the probability that, given that the rate of return is

.06 at time 1, the change in rate of return between time 1.5 and time 2 is at least .065.

4. Suppose that the process $(Z_t)_{t\geq0}$ is a standard Brownian motion starting at zero with history $(\mathcal{H}_t)_{t\geq0}$. Find:
  (a) $P[Z_{.5} \leq 1]$;
  (b) $P[Z_1 - Z_{.5} \leq 1|\mathcal{H}_{.5}]$;
  (c) $P[Z_1 \leq 1|Z_{.5} = .5]$;

(Hint on (c): Add and subtract $Z_{.5}$.)

5. If $(W_t)_{t\geq0}$ is a non-standard Brownian motion process with initial state 0, and drift and variance parameters μ and σ^2, what kind of process is $(X_t)_{t\geq0}$, where

$$X_t = \frac{W_t - \mu t}{\sigma}.$$

6. Suppose that the value of an index fund is a stochastic process $(X_t)_{t\geq0}$ with initial value 100 such that the logged price process is a non-standard Brownian motion with parameters $\mu = .04$, $\sigma^2 = .01$.
  (a) Find the probability that the value exceeds 101 at time 4;
  (b) Find the probability that the value exceeds 101 at time 4 given that it was 100.5 at time 2.

7. Referring to formula (4.81), show that the W process has stationary, independent increments as long as the Z process is standard Brownian motion with initial state 0.

8. The **covariance function** of a stochastic process $(X_t)_{t\geq0}$ is defined by $C(s,t) = \text{Cov}(X_s, X_t)$ for $s < t$. Find the covariance function of a standard Brownian motion with initial state 0, and use it to find the covariance function of a non-standard Brownian motion with parameters μ and σ^2. What is the correlation between Z_s and Z_t for a standard Brownian motion with initial state 0?

9. Suppose that an investor holds a portfolio of $10,000 in a risk-free asset with rate of return $r = .02$, and 500 shares in a risky asset whose current price is $20 per share. The risky asset undergoes geometric Brownian motion with parameters $\mu = .001$ and $\sigma = .05$. Find the probability that the portfolio is worth at least $22,000 at time 4.

10. Consider the risky asset of Exercise 9. Suppose that an investor holds a put option on the asset with strike price equal to the initial price of $20 and expiration time 5. Find the probability that the option will end in the money.

11. For a financial asset in geometric Brownian motion $(X_t)_{t \geq 0}$ with initial state 50, drift rate $\mu = .03$ and variance rate $\sigma^2 = .05$, compute the following probabilities by expressing X_t in exponential form:

  (a) $P[X_4 \leq 51]$;
  (b) $P[50 \leq X_3 \leq 52]$.

12. Find a general formula for the nth moment $E[X_t^n]$ of the state at time t of a geometric Brownian motion with initial state x_0, and parameters μ and σ^2.

13. For a geometric Brownian motion $X_t = x_0 e^{\mu t + \sigma \cdot Z_t}$, where Z_t has history $(\mathcal{H}_t)_{t \geq 0}$, and times $s \leq t$, why is X_s measurable with respect to $\mathcal{H}_t$?

4.5   Introduction to Stochastic Differential Equations

In a deterministic world, exponential growth of a non-risky asset price can be modeled as a simple differential equation in any of the four forms below:

$$
\begin{aligned}
\frac{dS_t}{dt} = rS_t \quad &\Longleftrightarrow \quad dS_t = rS_t dt \\
&\Longleftrightarrow \quad \frac{dS_t}{S_t} = r\, dt \\
&\Longleftrightarrow \quad d\left(\ln\left(S_t\right)\right) = r\, dt.
\end{aligned}
\tag{4.94}
$$

We could write any of these in integral form; for instance the second and fourth are:

$$
S_t - S_0 = \int_0^t rS_u du; \quad \ln\left(S_t\right) - \ln\left(S_0\right) = \int_0^t r\, du
\tag{4.95}
$$

and, of course, the solution is the deterministic function $S_t = S_0 e^{rt}$.

  One way of generalizing to random processes is to add in a random term driven by a standard Brownian motion $(Z_t)_{t \geq 0}$ as in equation (4.96) below.

$$
dS_t = \mu S_t dt + \sigma S_t dZ_t \quad \Longleftrightarrow \quad \frac{dS_t}{S_t} = \mu\, dt + \sigma\, dZ_t.
\tag{4.96}
$$

Such equations are called **stochastic differential equations**. The infinitesimal change dS_t in value of the solution S_t is a function of the current value of the solution, the infinitesimal change in time dt, and the infinitesimal change in a stochastic process dZ_t. The driving random process Z_t is standard Brownian motion in the work that we will do, but it could be something else as well. Sometimes we use the abbreviation SDE in place of the full phrase "stochastic differential equation".

  The equation on the left of (4.96) can be written in the integral form:

$$
S_t - S_0 = \int_0^t \mu S_u du + \int_0^t \sigma S_u dZ_u.
\tag{4.97}
$$

Giving solid meaning to either the differential form (4.96) or the integral form (4.97) is difficult, but they can be best understood as approximate increments in value:

$$S_{t+\Delta t} - S_t \approx \mu S_t \Delta t + \sigma S_t \left(Z_{t+\Delta t} - Z_t \right). \tag{4.98}$$

Clearly one can simulate such a process using this equation, because the increment $Z_{t+\Delta t} - Z_t$ is normally distributed with mean 0 and variance Δt, and is independent of other increments whose time intervals are disjoint from this one. (See the next subsection for the general simulation program.)

It is reasonable to guess that, if an ordinary function is a solution of an ordinary deterministic differential equation, then perhaps a stochastic process is a solution of a stochastic differential equation. But what process solves our linear coefficient equation (4.96)?

4.5.1 Meaning of the General SDE

The linear coefficient stochastic differential equation is actually a special case of the more general equation:

$$dS_t = \mu\left(t, S_t\right) dt + \sigma\left(t, S_t\right) dZ_t, \tag{4.99}$$

which in integral form is:

$$S_t - S_0 = \int_0^t \mu\left(s, S_s\right) du + \int_0^t \sigma\left(s, S_s\right) dZ_s. \tag{4.100}$$

The functions μ and σ are called the **drift rate** and **volatility rate** functions, respectively. In equation (4.96) we have linear functions: $\mu\left(t, S_t\right) = \mu \cdot S_t$ and $\sigma\left(t, S_t\right) = \sigma \cdot S_t$.

The most straightforward interpretation of the general stochastic differential equation (4.99) is the limit (in some manner) as the time increment Δt approaches 0 of the finite difference equations:

$$S(t + \Delta t) - S(t) \approx \mu(t, S(t))\Delta t + \sigma(t, S(t))(Z(t + \Delta t) - Z(t)). \tag{4.101}$$

What makes these stochastic differential equations very hard to rigorously define and work with is the strange nature of Brownian motion; it can be shown that a path of Brownian motion is non-differentiable almost everywhere and has infinite path length for all outcomes except perhaps for some in a set of probability zero. So the derivative dZ/dt doesn't exist, and the Riemann-Stieltjes sum that normally defines something like $\int_0^T \sigma\left(t, S_t\right) dZ_t$ can't converge in the usual sense. This difficulty can be sidestepped, but the technicalities of doing so are beyond the scope of this text, and do not prevent us from proceeding.

But equation (4.101) gives a very good intuitive sense of how the solution process evolves, and as well, shows us how we might simulate paths of a solution process. Consider breaking the time interval $[0, T]$ over which you want to simulate into n equally sized subintervals with partition points $t_0 = 0, t_1, t_2, ..., t_n = T$, where the spacing between points is $\Delta t = T/n$. Then equation (4.101) implies that:

$$S(t_i) \approx S(t_{i-1}) + \mu(t_{i-1}, S(t_{i-1})) \Delta t + \sigma(t_{i-1}, S(t_{i-1})) Z_i, \qquad (4.102)$$

where $Z_i = Z(t_i) - Z(t_{i-1})$, which by the properties of standard Brownian motion would be normally distributed with mean 0 and variance $t_i - t_{i-1} = \Delta t$.

An algorithm for simulating paths of a solution process to the SDE based upon the finite difference approximation (4.102) is below. Given the explanation of the previous paragraph, in should be self-explanatory. In particular, focus on the key step 3(b) in which the next state of the process is simulated given the current state and a normal increment with mean 0 and variance Δt, representing the change in value of the standard Brownian motion that gives the process its randomness.

SDE Simulation Algorithm

Input: initial value $s_0 = S(0)$, drift rate function $\mu(t, s)$, volatility rate function $\sigma(t, s)$, final time T, number of steps n.

Output: list of simulated states $\{s_0, s_1, s_2, ..., s_n\}$ for the SDE.

1. Initialize the list of states to $\{s_0\}$, and time t_0 to 0;

2. Compute $\Delta t = T/n$;

3. Do the following steps for $i = 1$ to n:
    (a) Simulate a new normal increment Z_i with the $\mathcal{N}(0, \Delta t)$ distribution;
    (b) Let the next state s_i equal the previous state s_{i-1} plus $\mu(t_{i-1}, s_{i-1}) \Delta t + \sigma(t_{i-1}, s_{i-1}) Z_i$;
    (c) Append s_i to the list of states;
    (d) Increment clock time by Δt;

4. Return the list of states.

The algorithm is easy to implement in any high-level computer language.

In Figure 4.22 is an example run of a simulation program (implemented on *Mathematica*) on time interval $[0, 1]$, with 50 steps, initial state 0, constant drift rate function $\mu = .05$ and constant volatility rate function $\sigma = .1$. (See Exercise 2, which asks what the solution process must be for these parameters.)

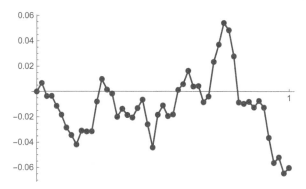

FIGURE 4.22 Simulated path of a solution to a stochastic differential equation.

It is possible to extend the idea of a stochastic differential equation and its solution to the higher dimensions. To illustrate briefly, we look at the two-variable case.

Let $\boldsymbol{Z} = (Z^1, Z^2)$ be a two-dimensional Brownian motion; that is, each Z^i is itself a Brownian motion, and the two processes are independent. A **two-dimensional stochastic differential equation** for a joint process $\boldsymbol{S} = (S^1, S^2)$ has the form of a system:

$$\begin{cases} dS_t^1 = \mu_1 \left(t, S_t^1, S_t^2\right) dt + \sigma_{11} \left(t, S_t^1, S_t^2\right) dZ_t^1 + \sigma_{12} \left(t, S_t^1, S_t^2\right) dZ_t^2; \\ dS_t^2 = \mu_2 \left(t, S_t^1, S_t^2\right) dt + \sigma_{21} \left(t, S_t^1, S_t^2\right) dZ_t^1 + \sigma_{22} \left(t, S_t^1, S_t^2\right) dZ_t^2. \end{cases}$$

(4.103)

The two-dimensional process is still reasonably easy to simulate. (See Exercise 6.) It is also easy to extend the approach to many dimensions. This is a rather general version; in more special cases the drift functions μ_1 and μ_2 may only depend on one process at a time. They may take on linear forms as well, for instance $\mu_1 \left(t, S_t^1, S_t^2\right) = \mu_1 \cdot S_t^1$ and $\mu_2 \left(t, S_t^1, S_t^2\right) = \mu_2 \cdot S_t^2$. But the presence of stochastic terms involving changes in both the Z^1 and Z^2 processes allows for correlation in the motions of the two price processes.

4.5.2 Ito's Formula

We cannot be too rigorous at this point, but on a surface level we can still understand and work with these equations and their solutions, especially with the aid of a theorem called **Ito's Theorem** (or Ito's Formula, or Ito's Lemma, depending on whom you talk to).

Return to the general SDE:

$$dS_t = \mu\left(t, S_t\right) dt + \sigma\left(t, S_t\right) dZ_t, \text{ or}$$
$$S_t - S_0 = \int_0^t \mu\left(s, S_s\right) ds + \int_0^t \sigma\left(s, S_s\right) dZ_s.$$

The value of a derivative on a risky asset that satisfies such an equation is a function of the asset value: $V = V(t, S_t)$. To value the derivative, we can set up a differential equation for V, utilizing known behavior at the boundary, in order to solve the equation (for example, a terminal boundary $V(T, S_T) = \max\{S_T - E, 0\}$ is appropriate for a European call with strike price E and exercise time T). Solving the differential equation can be difficult in any but the simplest models.

Ito's Formula, which helps us carry through this plan, is basically a stochastic version of the chain rule. In the deterministic world, the two-variable chain rule would say

$$
\begin{aligned}
\frac{df(t,x(t))}{dt} &= f_t(t, x(t)) + f_x(t, x(t)) \cdot \frac{dx(t)}{dt} \\
\implies df(t, x(t)) &= f_t(t, x(t))dt + f_x(t, x(t))dx(t).
\end{aligned}
\tag{4.104}
$$

The subscripts indicate the variable or variables with respect to which we differentiate. In other words, the infinitesimal change in the function $f(t, x(t))$ that results from an infinitesimal time increment dt is the sum of two terms: (1) the partial derivative with respect to time of f evaluated at $x(t)$ in its second component times the change in time, plus (2) the partial of f with respect to the state variable x evaluated at $x(t)$ times the infinitesimal change in the value of $x(t)$. In the stochastic world, if $X = (X_t)$ is a process satisfying the general SDE, then an additional, second order term is added to the dt part, which is:

$$
\frac{1}{2} f_{xx}(t, X_t)\, \sigma^2(t, X_t)\, dt.
$$

This term comes up because of the fact that square changes in Z, $(Z_{t+\Delta t} - Z_t)^2$ are of the order of Δt and cannot be ignored. We will see this shortly. Ito's Formula becomes in full:

$$
\begin{aligned}
df(t, X_t) &= f_t(t, X_t)\, dt + f_x(t, X_t)\, dX_t + \tfrac{1}{2} f_{xx}(t, X_t)\, \sigma^2(t, X_t)\, dt \\
&= \left(f_t(t, X_t) + f_x(t, X_t) \cdot \mu(t, X_t) + \tfrac{1}{2} f_{xx}(t, X_t)\, \sigma^2(t, X_t) \right) dt \\
&\quad + f_x(t, X_t) \cdot \sigma(t, X_t)\, dZ_t.
\end{aligned}
\tag{4.105}
$$

In the second line of the derivation above, we substituted in the expression for dX_t from the general SDE that the process is assumed to satisfy. Then all of the coefficients of dt were assembled together.

Here is a heuristic argument that helps to motivate Ito's Formula. Let $Y_t = f(t, X_t)$ where (X_t) is a stochastic process satisfying:

$$
dX_t = \mu(t, X_t)\, dt + \sigma(t, X_t)\, dZ_t.
$$

By the two-dimensional version of Taylor's Theorem, we can expand as follows:

$$
\begin{aligned}
Y_{t+\Delta t} - Y_t &= f\left(t + \Delta t, X_{t+\Delta t}\right) - f\left(t, X_t\right) \\
&= f_t\left(t, X_t\right)\Delta t + f_x\left(t, X_t\right)\Delta X + \tfrac{1}{2}\cdot f_{t,t}\left(t, X_t\right)(\Delta t)^2 \\
&\quad + f_{t,x}\left(t, X_t\right)\Delta t \cdot \Delta X + f_{x,x}\left(t, X_t\right)\tfrac{1}{2}(\Delta X)^2 + \cdots .
\end{aligned}
$$
(4.106)

The ellipsis stands for higher order terms involving products of Δt's and ΔX's with total exponent 3 or more. We anticipate factoring Δt out of the expression above and sending Δt to the infinitesimal dt. Now ΔX is approximated by:

$$
\Delta X \approx \mu\left(t, X_t\right)\Delta t + \sigma\left(t, X_t\right)\Delta Z_t = \mu\left(t, X_t\right)\Delta t + \sigma\left(t, X_t\right)\left(Z_{t+\Delta t} - Z_t\right).
$$

The change in state of the standard Brownian motion process in a short interval is "of the order of" $\sqrt{\Delta t}$, roughly because of the fact that the variance is Δt. (See Exercise 4.) Therefore, the third and fourth terms, $\tfrac{1}{2}\cdot f_{t,t}\left(t, X_t\right)(\Delta t)^2$ and $f_{t,x}\left(t, X_t\right)\Delta t \cdot \Delta X$ on the right of (4.106) are of the order $(\Delta t)^2 + (\Delta t)^{3/2}$, so clearly those can be ignored, as can all third order, fourth order, etc. terms hidden by the ellipsis. But what about the $(\Delta X)^2$ term?

We can approximate:

$$
\begin{aligned}
(\Delta X)^2 &\approx \left(\mu\left(t, X_t\right)\Delta t + \sigma\left(t, X_t\right)\Delta Z_t\right)^2 \\
&= \mu^2\left(t, X_t\right)(\Delta t)^2 + 2\mu\left(t, X_t\right)\sigma\left(t, X_t\right)\Delta t\Delta Z_t + \sigma^2\left(t, X_t\right)(\Delta Z_t)^2,
\end{aligned}
$$

and the $(\Delta t)^2$ and $\Delta t\Delta Z_t$ terms are safe to ignore, but $(\Delta Z_t)^2$ is "of the order of" Δt. So $(\Delta X)^2$ in formula (4.106) may be replaced by $\sigma^2\left(t, X_t\right)\Delta t$ in the expansion. This gives (ignoring high order terms in Δt):

$$
\begin{aligned}
Y_{t+\Delta t} - Y_t &\approx f_t\left(t, X_t\right)\Delta t + f_x\left(t, X_t\right)\Delta X + \tfrac{1}{2}f_{x,x}\left(t, X_t\right)(\Delta X)^2 \\
&\approx f_t\left(t, X_t\right)\Delta t + f_x\left(t, X_t\right)\Delta X + \tfrac{1}{2}f_{x,x}\left(t, X_t\right)\sigma^2\left(t, X_t\right)\Delta t \\
&= \left(f_t\left(t, X_t\right) + \mu\left(t, X_t\right)f_x\left(t, X_t\right)\tfrac{1}{2}\sigma^2\left(t, X_t\right)f_{x,x}\left(t, X_t\right)\right)\Delta t \\
&\quad + \sigma\left(t, X_t\right)f_x\left(t, X_t\right)\Delta Z_t,
\end{aligned}
$$

which in its infinitesimal form is Ito's Formula. We summarize in the theorem below.

Theorem 1. Let $(X_t)_{t \geq 0}$ be a stochastic process satisfying the SDE:

$$
dX_t = \mu\left(t, X_t\right)dt + \sigma\left(t, X_t\right)dZ_t.
$$

Suppose that $f(t, x)$ is a function that has a continuous first partial derivative in t and continuous first and second partial derivatives in x. Then the SDE satisfied by the process defined by $Y_t = f\left(t, X_t\right)$ is:

$$
\begin{aligned}
dY_t &= \left(f_t\left(t, X_t\right) + f_x\left(t, X_t\right)\cdot \mu\left(t, X_t\right) + \tfrac{1}{2}f_{xx}\left(t, X_t\right)\sigma^2\left(t, X_t\right)\right)dt \\
&\quad + f_x\left(t, X_t\right)\cdot \sigma\left(t, X_t\right)dZ_t. \quad \blacksquare
\end{aligned}
$$
(4.107)

Example 1. If $(Z_t)_{t\geq 0}$ is a standard Brownian motion, what stochastic differential equation does the process defined by $Y_t = \sigma Z_t + \mu t$ satisfy?

Solution. Let $\mu(t, x) = 0$, $\sigma(t, x) = 1$ and $X_t = Z_t$ in the general SDE. Then we can view the process (X_t) as satisfying $dX_t = 0dt + 1dZ_t = dZ_t$. Let $f(t, x) = \sigma x + \mu t$. Then $f_t = \mu$, $f_x = \sigma$, and $f_{xx} = 0$; hence by Ito's formula:

$$
\begin{aligned}
dY_t &= df(t, X_t) \\
&= \left(f_t(t, X_t) + f_x(t, X_t)\, \mu(t, X_t) + \tfrac{1}{2} f_{xx}(t, X_t)\, \sigma^2(t, X_t) \right) dt \\
&\quad + f_x(t, X_t)\, \sigma(t, X_t)\, dZ_t \\
&= \left(\mu + \sigma \cdot 0 + \tfrac{1}{2} \cdot 0 \cdot 1^2 \right) dt + \sigma \cdot 1 dZ_t \\
&= \mu dt + \sigma dZ_t,
\end{aligned}
$$

or in integral form,

$$
Y_{t+\Delta t} - Y_t = \int_t^{t+\Delta t} \mu \, dt + \int_t^{t+\Delta t} \sigma \, dZ_t = \mu \Delta t + \sigma \left(Z_{t+\Delta t} - Z_t \right).
$$

Notice that this Y process is just non-standard Brownian motion with drift rate μ, and volatility σ. ∎

4.5.3   Geometric Brownian Motion as Solution

The next example shows that there is a close relationship between stochastic differential equation models and the geometric Brownian motion process that we have used to model the continuous-time motion of risky asset prices.

Example 2. Use Ito's formula to find the stochastic differential equation satisfied by the process defined by $Y_t = x_0 e^{\sigma Z_t + (\mu - \sigma^2/2)t}$, where $(Z_t)_{t\geq 0}$ is a standard Brownian motion.

Solution. Example 1 allows us to treat the process defined by $X_t = \sigma Z_t + (\mu - \sigma^2/2)t$ as a solution of the SDE $dX_t = (\mu - \sigma^2/2)\, dt + \sigma dZ_t$. So the drift rate and volatility rate parameters to use in Ito's formula are the constant functions:

$$
\mu(t, x) = \mu - \sigma^2/2; \ \sigma(t, x) = \sigma.
$$

The function of X_t whose derivatives we need is $Y_t = f(X_t) = x_0 e^{X_t}$. The first and second partial derivatives in x of f are just $f_x(x) = x_0 e^x$ and $f_{xx}(x) = x_0 e^x$. Also, the time partial is 0, since f does not explicitly depend on time.

By Ito's formula, the process $Y_t = f(t, X_t) = e^{\sigma Z_t + (\mu - \sigma^2/2)t}$ satisfies:

$$
\begin{aligned}
dY_t &= \left(f_t(t, X_t) + f_x(t, X_t)\mu(t, X_t) + \tfrac{1}{2}f_{xx}(t, X_t)\sigma^2(t, X_t) \right) dt \\
&\quad + f_x(t, X_t)\,\sigma(t, X_t)\,dZ_t \\
&= \left(x_0 e^{X_t}\left(\mu - \sigma^2/2\right) + \tfrac{1}{2}\cdot x_0 e^{X_t}\cdot \sigma^2 \right) dt + \sigma \cdot x_0 e^{X_t}\,dZ_t \\
&= \mu x_0 e^{X_t}\,dt + \sigma \cdot x_0 e^{X_t}\,dZ_t \\
&= \mu Y_t\,dt + \sigma \cdot Y_t\,dZ_t.
\end{aligned}
$$

(4.108)

Thus, a geometric Brownian motion process with parameters $\mu - \sigma^2/2$ and σ is a solution of the stochastic differential equation $dY_t = \mu Y_t\,dt + \sigma Y_t\,dZ_t$. ∎

Remark. In Example 4 of Section 4.4 we showed that the expected value of a geometric Brownian motion with initial value x_0 and parameters μ and σ at time t was:

$$
E[X_t] = x_0 e^{(\mu + \sigma^2/2)t}.
$$

Let r be the continuous risk-free rate. For the particular process in the last example, with mean parameter $r - \sigma^2/2$ in place of μ, the expected value becomes:

$$
E[X_t] = x_0 e^{(r - \sigma^2/2 + \sigma^2/2)t} = x_0 e^{rt} \implies E\left[e^{-rt}X_t\right] = x_0. \qquad (4.109)
$$

This strongly suggests that if a measure can be found on the sample space of the geometric Brownian motion that forces the drift rate parameter to be $r - \sigma^2/2$, then the discounted asset process defined by $\bar{X}_t = e^{-rt}X_t$ is a martingale. Continuous versions of the Fundamental Theorems of Asset Pricing would therefore allow us to apply martingale valuation $V_0 = E_Q\left[e^{-rT}V_T\right]$ to derivatives expiring at time T.

Exercises 4.5

1. Consider the stochastic differential equation $dS_t = \mu S_t dt + \sigma S_t dZ_t$, where the drift rate $\mu = .05$ and the volatility is $\sigma = .15$. Suppose that the solution process is observed at time t to have value 50. Use formula (4.98) to approximate the probability that $S_{t+.1} > 50.5$.

2. What is the solution to the SDE that was simulated in Figure 4.22?

3. Let X be the process satisfying the stochastic differential equation

$$
dX_t = \left(\mu + \sigma^2/2\right)X_t dt + \sigma X_t dZ_t.
$$

Use Ito's Formula to find the differential equation satisfied by $Y_t = f(t, X_t) = \ln(X_t)$.

4. The heuristic argument in favor of Ito's Formula rested to a large extent on the fact that ΔZ itself could not be too large, and on estimation of the rough order, as a function of Δt, of the powers $(\Delta Z)^n = (Z_{t+\Delta t} - Z_t)^n$, $n \geq 2$.

(a) Show that for any fixed $h > 0$, $P[|\Delta Z| > h]$ is no greater than a constant times Δt.

(b) Using the known general formula below for the even moments of a $N(0, \sigma^2)$ random variable X, explain intuitively why it made sense for us to ignore terms involving $(\Delta Z)^i \cdot (\Delta t)^j$, for $i + j \geq 3$.

$$E\left[X^{2n}\right] = \frac{(2n)!}{2^n \cdot n!} \cdot \sigma^{2n}$$

5. In the Remark following Example 2, if the history $(\mathcal{H}_t)_{t \geq 0}$ is the one generated by the geometric Brownian motion with parameters x_0, $\mu = r - \sigma^2/2$ and σ, argue that the martingale condition $E\left[\bar{X}_{t+s} | \mathcal{H}_t\right] = \bar{X}_t$ holds. (Hint: multiply and divide by X_t in the expectation and use the independence of increments.)

6. Write a simulation algorithm similar to the single-dimensional case for a two-dimensional s.d.e. as described by formula (4.103).

7. Let $(Y_t)_{t \geq 0}$ be a process defined by $Y_t = (Z_t)^2$, where $(Z_t)_{t \geq 0}$ is a standard Brownian motion. Find the stochastic differential equation satisfied by the Y process. (Note that you are deriving a kind of extended power rule for "differentiation" of the square of the Z process.)

8. Use Ito's formula to show a form of a product rule for deterministic functions of time and solutions of stochastic differential equations:

$$d\left(g(t) \cdot X_t\right) = g'(t) \cdot X_t dt + g(t) dX_t.$$

9. A so-called **Ornstein-Uhlenbeck process** $(X_t)_{t \geq 0}$ is a process satisfying the stochastic differential equation

$$dX_t = \alpha\left(\mu - X_t\right) dt + \sigma dZ_t.$$

The fact that the drift is negative when X_t exceeds the constant μ and positive when it is less than μ suggests that this process tries to revert to a mean of μ, while being disturbed by the random variability of the stochastic term σdZ_t. The constant α would control the rate of the reversion. This process has been used in models of the fluctuation of interest rates, and in more sophisticated models of asset price motion where the volatility σ moves randomly according to such a process. Show that the process below is a solution of the Ornstein-Uhlenbeck s.d.e.

$$X_t = \mu + (x_0 - \mu) e^{-\alpha t} + \sigma \int_0^t e^{-\alpha(t-s)} dZ_s.$$

(Hint: First apply the transformation $Y_t = X_t - \mu$ to obtain an s.d.e. for the Y process; then let $W_t = e^{\alpha t} \cdot Y_t$ and use the result of Exercise 8 to find the s.d.e. satisfied by the W process and its solution. Then, reverse the transformations to derive the form of X_t.)

In our discussion of stochastic differential equations we sidestepped the question of the meaning of Ito integral expressions such as $\int_0^T \sigma(t, X_t) \, dZ_t$. In the next two exercises we give a first step toward an understanding of such expressions.

To set up these problems, as usual let $(\mathcal{H}_t)_{t \geq 0}$ denote the history generated by a standard Brownian motion process (Z_t). A process $(Y_t)_{t \in [0,T]}$ is called **elementary** if there is a time partition $0 = t_0 < t_1 < \cdots < t_n = T$ and a sequence of random variables $W_0, W_1, ..., W_{n-1}$ such that each W_i is $\mathcal{H}_{t_i}$-measurable and

$$Y_t = \sum_{i=0}^{n-2} W_i \cdot I_{[t_i, t_{i+1})} + W_{n-1} \cdot I_{[t_{n-1}, t_n]}.$$

Intuitively, an elementary process is a step process whose values at the left endpoints and throughout the subintervals $[t_i, t_{i+1})$ are determined by these random variables W_i, whose values are known at the beginning of the subinterval. An appropriate correction is made so that the process is defined at the rightmost endpoint $t_n = T$.

10. Define the (Ito) **stochastic integral** of an elementary process by:

$$\int_0^T Y_t \, dZ_t = \sum_{i=0}^{n-1} W_i \left(Z_{t_{i+1}} - Z_{t_i} \right).$$

Show that $E\left[\int_0^T Y_t \, dZ_t \right] = 0$.

11. In reference to the previous problem, show that:

$$E\left[\left(\int_0^T Y_t \, dZ_t \right)^2 \right] = E\left[\left(\int_0^T (Y_t)^2 \, dt \right) \right].$$

5

Derivative Valuation in Continuous Time

5.1 Black-Scholes Via Limits

5.1.1 Black-Scholes Formula

In this section we state the famous Black-Scholes Theorem[2], which gives the formula for the value of a simple European call option with strike price E and exercise time T in the continuous-time case. There are several ways of deriving the formula; here we motivate it as the limit of the binomial branch discrete-time option value as the number of time steps goes to infinity.

The market price at time t of the asset on which the option is based is denoted by either S_t or $S(t)$. In the Black-Scholes world, we assume that the process $(S_t)_{t \geq 0}$ satisfies the stochastic differential equation:

$$S_t = \mu S_t dt + \sigma S_t dZ_t, S_0 = s_0. \tag{5.1}$$

The Black-Scholes formula for the call option is as follows.

Theorem 1. (Black-Scholes call option) The initial value C at time 0 of the European call option with strike price E and termination time T is:

$$
\begin{aligned}
C \;=\; & s_0 \mathcal{N}\left(\frac{\log(s_0/E) + (r + \sigma^2/2)T)}{\sigma\sqrt{T}} \right) \\
& - E \cdot e^{-rT} \mathcal{N}\left(\frac{\log(s_0/E) + (r - \sigma^2/2)T}{\sigma\sqrt{T}} \right),
\end{aligned}
$$

where σ is the volatility, r is the continuous risk-free rate, and $\mathcal{N}$ is the standard normal cumulative distribution function. ∎

The formula can be written more briefly by introducing constants:

$$d_1 = \frac{\log(s_0/E) + (r + \sigma^2/2)T}{\sigma\sqrt{T}}, \quad d_2 = \frac{\log(s_0/E) + (r - \sigma^2/2)T}{\sigma\sqrt{T}}. \tag{5.2}$$

Then, the Black-Scholes formula becomes:

$$
\begin{aligned}
C &= s_0 \mathcal{N}\left(\frac{\log(s_0/E)+\left(r+\sigma^2/2\right)T}{\sigma\sqrt{T}}\right) - E \cdot e^{-rT} \mathcal{N}\left(\frac{\log(s_0/E)+\left(r-\sigma^2/2\right)T}{\sigma\sqrt{T}}\right) \\
&= s_0 \mathcal{N}\left(d_1\right) - E \cdot e^{-rT} \mathcal{N}\left(d_2\right).
\end{aligned}
$$

(5.3)

The development in Section 4.5 yielded that the Black-Scholes assumption about the price process is equivalent to the assumption that it is a geometric Brownian motion:

$$
S_t = s_0 e^{\left(\mu-\sigma^2/2\right)t+\sigma Z_t},
$$

whose drift parameter is $\mu - \sigma^2/2$ and whose volatility parameter is σ. But notice that μ doesn't even figure in to the Black-Scholes formula. This odd result reminds us of the corresponding result in the discrete-time world that the binomial branch up probability p does not matter in valuing the option. We will see the connection in the subsection below.

Remark. With very slight alterations, the formula can be shown to also give the value at time t, $C(t)$, where all of the T's in formulas (5.2) and (5.3) are replaced by the remaining time $T - t$, and the initial asset price s_0 is replaced by s_t, the price at time t.

5.1.2 Limiting Approach

In Sections 3.7 and 4.4 we set up the machinery through which we can derive the Black-Scholes formula as a limit of option values in the discrete binomial branch process model. As we have done several times, break up the time interval $[0,T]$ into n equally sized subintervals with length $\Delta t = \frac{T}{n}$, whose partition points are $t_k = k\Delta t = k\frac{T}{n}$.

FIGURE 5.1 Time axis for asset motion.

We assume that the continuous interest rate is known to be r per year. (For loans and for risk-free investment growth). We write $r\Delta t$ for the interest rate per discrete period, and we know that $(1 + r\Delta t)^n \to e^{r \cdot n\Delta t} = e^{rT}$ as $n \longrightarrow \infty$ and $\Delta t \to 0$.

As we have done before, multiplicatively telescope the binomial branch process as:

$$
S_n = S(n\Delta t) = s_0 \cdot \frac{S_1}{s_0} \cdot \frac{S_2}{S_1} \cdot \frac{S_3}{S_2} \cdots \frac{S_n}{S_{n-1}} = s_0 \cdot R_1 \cdot R_2 \cdot R_3 \cdots R_n. \quad (5.4)
$$

The logarithm of the price value after n transitions satisfies:

$$\log(S_n) = \log(s_0) + \sum_{i=1}^{n} \log(R_i). \tag{5.5}$$

Thus, the logged price process is a random walk with initial state $\log(s_0)$ and steps of sizes $\log(1+b)$ or $\log(1+a)$, with probabilities p and $1-p$, respectively.

Reparameterize the problem by introducing parameters μ and σ such that:

$$\log(1+b) = \mu\Delta t + \sigma\sqrt{\Delta t} \text{ and } \log(1+a) = \mu\Delta t - \sigma\sqrt{\Delta t}$$
$$\implies b = e^{\mu\Delta t + \sigma\sqrt{\Delta t}} - 1 \text{ and } a = e^{\mu\Delta t - \sigma\sqrt{\Delta t}} - 1. \tag{5.6}$$

For very small values of the time step Δt, because $e^x \approx 1 + x$ for x near 0, we would have the approximations:

$$1 + b = e^{\mu\Delta t + \sigma\sqrt{\Delta t}} \approx 1 + \mu\Delta t + \sigma\sqrt{\Delta t}$$
$$\text{and } 1 + a = e^{\mu\Delta t - \sigma\sqrt{\Delta t}} \approx 1 + \mu\Delta t - \sigma\sqrt{\Delta t}. \tag{5.7}$$

Let us compute the mean of the logged price process, given an up probability p. We have:

$$
\begin{aligned}
E[\log(S_n)] &= E\left[\log(s_0) + \sum_{i=1}^{n}\log(R_i)\right] \\
&= \log(s_0) + n \cdot E[\log(R_i)] \\
&= \log(s_0) + n\left(p \cdot \left(\mu\Delta t + \sigma\sqrt{\Delta t}\right) + (1-p) \cdot \left(\mu\Delta t - \sigma\sqrt{\Delta t}\right)\right) \\
&= \log(s_0) + n\left(\mu\Delta t + p \cdot \sigma\sqrt{\Delta t} - (1-p)\sigma\sqrt{\Delta t}\right) \\
&= \log(s_0) + \mu \cdot n \cdot \Delta t + n(2p-1)\sigma\sqrt{\Delta t} \\
&= \log(s_0) + \mu T + n(2p-1)\sigma\sqrt{\Delta t}.
\end{aligned}
\tag{5.8}
$$

We would like our branch process to converge to a geometric Brownian motion whose log would have mean equal to a linear function of T, but what can we do with the extra term on the right: $n(2p-1)\sigma\sqrt{\Delta t}$?

In the discrete world of binomial branch processes, in valuing derivatives, the only probability p that mattered was the risk-neutral probability q, which in its raw form was:

$$q = \frac{S_0(1+r) - S_d}{S_u - S_d} = \frac{S_0(1+r) - S_0(1+a)}{S_0(1+b) - S_0(1+a)}.$$

In our revised model, this would become:

$$
\begin{aligned}
q &\approx \frac{S_0(1+r\Delta t) - S_0\left(1 + \mu\Delta t - \sigma\sqrt{\Delta t}\right)}{S_0\left(1 + \mu\Delta t + \sigma\sqrt{\Delta t}\right) - S_0\left(1 + \mu\Delta t - \sigma\sqrt{\Delta t}\right)} \\
&= \frac{(r-\mu)\Delta t + \sigma\sqrt{\Delta t}}{2\sigma\sqrt{\Delta t}} \\
&= \frac{1}{2} + \frac{(r-\mu)\sqrt{\Delta t}}{2\sigma}.
\end{aligned}
\tag{5.9}
$$

Then:

$$2q - 1 \approx 2\left(\frac{1}{2} + \frac{(r-\mu)\sqrt{\Delta t}}{2\sigma}\right) - 1 = \frac{(r-\mu)\sqrt{\Delta t}}{\sigma}, \tag{5.10}$$

and so for $p = q$, the extra term in formula (5.8) is:

$$n(2q-1)\sigma\sqrt{\Delta t} = n\left(\frac{(r-\mu)\sqrt{\Delta t}}{\sigma}\right)\sigma\sqrt{\Delta t} = (r-\mu)n \cdot \Delta t = (r-\mu)T.$$

Formula (5.8) now implies that:

$$E_q\left[\log\left(S_n\right)\right] \approx \log\left(s_0\right) + \mu T + (r-\mu)T = \log\left(s_0\right) + rT.$$

Notice how the parameter μ has dropped out of the formula for the mean of the logged price process. Not only does this give us the linear relationship of mean logged price to time, it says in addition that under the risk-neutral probability q, the risky asset moves in expectation like a continuous risk-free asset S^0 with rate r:

$$S^0(T) = s_0^0 e^{rT} \iff \log\left(S^0(T)\right) = \log\left(s_0^0\right) + rT.$$

To compute the approximate variance of the logged price after n steps, we first need a formula for the variance of a particular step $\log\left(R_i\right)$. For general up probability p, a straightforward calculation (see Exercise 2) shows that:

$$\text{Var}\left(\log\left(R_i\right)\right) = p(1-p) \cdot 4\sigma^2\Delta t. \tag{5.11}$$

Therefore:

$$\begin{aligned}
\text{Var}\left(\log\left(S_n\right)\right) &= \text{Var}\left(\log\left(s_0\right) + \sum_{i=1}^n \log\left(R_i\right)\right) \\
&= n \cdot \text{Var}\left(\log\left(R_i\right)\right) \\
&\approx np(1-p)4\sigma^2\Delta t \\
&= 4p(1-p)\sigma^2 \cdot n\Delta t \\
&= 4p(1-p)\sigma^2 \cdot T.
\end{aligned} \tag{5.12}$$

When $p = q$, formula (5.9) implies that $q \longrightarrow \frac{1}{2}$ as $\Delta t \longrightarrow 0$, and so under the risk-neutral probability the variance of the state after n steps approaches

$$\text{Var}\left(\log\left(S_n\right)\right) \longrightarrow 4 \cdot \frac{1}{2}\left(1 - \frac{1}{2}\right)\sigma^2 \cdot T = \sigma^2 \cdot T.$$

This suggests that the limiting logged price process converges to Brownian motion with drift rate r and volatility σ, under q.

In addition, Example 2 of Section 4.5 showed that the process defined by:

$$S_t = s_0 e^{\sigma Z_t + (r - \sigma^2/2)t} \tag{5.13}$$

satisfies the stochastic differential equation:

$$dS_t = rS_t dt + \sigma S_t dZ_t \iff d\left(\ln\left(S_t\right)\right) = \frac{dS_t}{S_t} = rdt + \sigma dZ_t. \qquad (5.14)$$

So, the logged continuous process is Brownian motion with drift r and volatility σ. Thus, under the risk-neutral q for the binomial branch process, the price converges to solution of the Black-Scholes SDE (5.1), with $\mu = r$.

Returning to the valuation of the n-step European call option, martingale valuation says that the initial value V_0 of a discrete-time European call expiring at time T with strike price E is:

$$V_0 = E_q\left[(1 + r\Delta t)^{-n} \cdot \max\left(S_n - E, 0\right)\right]. \qquad (5.15)$$

By our arguments above, as $n \to \infty$, this expectation converges to:

$$C = E\left[e^{-rT} \cdot \max\left(S_T - E, 0\right)\right], \text{where } S_T = s_0 e^{(r - \sigma^2/2)T + \sigma Z_T}. \qquad (5.16)$$

It remains only to use calculus to compute the expectation.

We know that Z_T is normally distributed with mean 0 and variance T. The expected value is therefore:

$$C = e^{-rT} \int_{-\infty}^{\infty} \max\left(s_0 e^{(r - \frac{1}{2}\sigma^2)T + \sigma x} - E, 0\right) \frac{1}{\sqrt{2\pi T}} e^{-\frac{x^2}{2T}} dx. \qquad (5.17)$$

Substitute $u = \frac{x}{\sqrt{T}}$, $du = \frac{1}{\sqrt{T}}dx$. Then the integral transforms to:

$$C = e^{-rT} \int_{-\infty}^{\infty} \max\left(s_0 e^{(r - \frac{1}{2}\sigma^2)T + \sigma u \sqrt{T}} - E, 0\right) \frac{1}{\sqrt{2\pi}} e^{-\frac{u^2}{2}} du. \qquad (5.18)$$

The integrand is greater than zero when:

$$s_0 e^{(r - \frac{1}{2}\sigma^2)T + \sigma u \sqrt{T}} - E > 0 \iff \left(r - \tfrac{1}{2}\sigma^2\right) T + \sigma u \sqrt{T} > \log\left(\frac{E}{s_0}\right)$$

$$\iff u > \frac{\log\left(\frac{E}{s_0}\right) - \left(r - \frac{1}{2}\sigma^2\right)T}{\sigma\sqrt{T}} \equiv a.$$

$$(5.19)$$

Thus,

$$
\begin{aligned}
C &= e^{-rT} \int_a^\infty \left(s_0 e^{(r - \frac{1}{2}\sigma^2)T + \sigma u\sqrt{T}} - E\right) \frac{1}{\sqrt{2\pi}} e^{-\frac{u^2}{2}} du \\
&= e^{-rT} \int_a^\infty \left(s_0 e^{(r - \frac{1}{2}\sigma^2)T + \sigma u\sqrt{T}}\right) \frac{1}{\sqrt{2\pi}} e^{-\frac{u^2}{2}} du \\
&\quad - e^{-rT} \int_a^\infty E \frac{1}{\sqrt{2\pi}} e^{-\frac{u^2}{2}} du \\
&= e^{-rT} e^{rT} s_0 \int_a^\infty \frac{1}{\sqrt{2\pi}} e^{-\frac{u^2}{2} + \sigma u\sqrt{T} - \frac{1}{2}\sigma^2 T} du \\
&\quad - e^{-rT} E \int_a^\infty \frac{1}{\sqrt{2\pi}} e^{-\frac{u^2}{2}} du \\
&= s_0 \int_a^\infty \frac{1}{\sqrt{2\pi}} e^{-\frac{1}{2}\left(u - \sigma\sqrt{T}\right)^2} du - e^{-rT} E \cdot (1 - \mathcal{N}(a)).
\end{aligned}
\qquad (5.20)
$$

Here, $\mathcal{N}(a)$ is the standard normal c.d.f. By symmetry, we can replace $1-\mathcal{N}(a)$ by $\mathcal{N}(-a)$. Apply another substitution:

$$z = u - \sigma\sqrt{T}, \quad du = dz.$$

Then,

$$
\begin{aligned}
C &= s_0 \int_{a-\sigma\sqrt{T}}^{\infty} \frac{1}{\sqrt{2\pi}} e^{-\frac{z^2}{2}} dz - e^{-rT} E \cdot \mathcal{N}(-a) \\
&= s_0 \left(1 - \mathcal{N}\left(a - \sigma\sqrt{T} \right) \right) - e^{-rT} E \cdot \mathcal{N}(-a) \\
&= s_0 \cdot \mathcal{N}\left(-\left(a - \sigma\sqrt{T} \right) \right) - e^{-rT} E \cdot \mathcal{N}(-a) \\
&= s_0 \cdot \mathcal{N}\left(-a + \sigma\sqrt{T} \right) - e^{-rT} E \cdot \mathcal{N}(-a).
\end{aligned}
\tag{5.21}
$$

We have almost completed the derivation. Looking back to the Black-Scholes formula (5.3), we need only check that:

$$
\begin{aligned}
-a + \sigma\sqrt{T} &= \frac{\log(s_0/E) + \left(r + \sigma^2/2\right)T}{\sigma\sqrt{T}} \\
\text{and} - a &= \frac{\log(s_0/E) + \left(r - \sigma^2/2\right)T}{\sigma\sqrt{T}}
\end{aligned}
\tag{5.22}
$$

where from (5.19), we had defined the constant a by:

$$a = \frac{\log\left(E/s_0\right) - \left(r - \frac{1}{2}\sigma^2\right)T}{\sigma\sqrt{T}}.$$

These are routine simplifications that we leave to the reader as Exercise 1. The derivation of Black-Scholes' theorem by limits is finished.

Example 1. Compute the value at time 0 of a continuous-time European call option, given the following problem parameters: initial price $S_0 = 50$, strike price $E = 60$, risk-free rate $r = .04$, exercise time $T = 1$ year, volatility $\sigma = .25$.

Solution. We can use the Black-Scholes formula:

$$C = s_0 N\left(d_1\right) - Ee^{-rT} N\left(d_2\right),$$

where

$$d_1 = \frac{\log\left(s_0/E\right) + \left(r + \sigma^2/2\right)T}{\sqrt{T}\sigma}, \quad d_2 = \frac{\log\left(s_0/E\right) + \left(r - \sigma^2/2\right)T}{\sqrt{T}\sigma}.$$

Using the given parameters we get:

$$
\begin{aligned}
d_1 &= \frac{\log(50/60) + \left(.04 + .25^2/2\right)\cdot 1}{\sqrt{1}\cdot.25} = -.444286; \\
d_2 &= \frac{\log(50/60) + \left(.04 - .25^2/2\right)\cdot 1}{\sqrt{1}\cdot.25} = -.694286.
\end{aligned}
$$

Hence,

$$C = 50N(-.444286) - 60e^{-.04 \cdot 1}N(-.694286) = 2.36927. \blacksquare$$

With technological support, we can investigate the dependence of the call option value on the input parameters. For instance, let us see how the value depends on the exercise time T. Using the asset parameters of the last example, we would have:

$$d_1 = \frac{\log(50/60) + (.04 + .25^2/2)T}{\sqrt{T} \cdot .25};$$
$$d_2 = \frac{\log(50/60) + (.04 - .25^2/2)T}{\sqrt{T} \cdot .25};$$
$$C = 50N(d_1) - 60e^{-.04T}N(d_2).$$

The graph is shown in Figure 5.2.

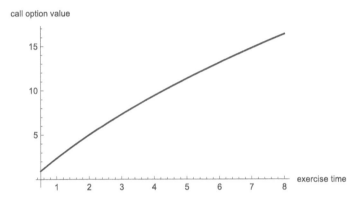

FIGURE 5.2 Call value as a function of T.

The initial value of the call increases as the time horizon for execution of the option increases. This makes perfect sense, since the passage of more time makes it more likely for the asset to rise in value enough to bring the option to "in-the-money" status, and to increase the potential claim value when it is. Exercise 10 explores the relationship between the call value and the strike price E and between the call value and the volatility parameter σ.

5.1.3 Put Options and Put-Call Parity

In both the continuous and the discrete world, there is a surprisingly simple relationship between the value of a European call option and that of a European put option. This relationship is called the **put-call parity formula**, and it is expressed as follows. If P is the initial value of a put on an asset S expiring at time T with strike price E, and C is the initial value of a call on the same asset with the same expiration and strike price, then:

$$P = C - S_0 + e^{-rT}E. \tag{5.23}$$

For a discrete-time option problem in which the expiration of both options occurs after n steps, the discount factor e^{-rT} is replaced by $(1+r)^{-n}$.

To see where the formula comes from, consider a portfolio that is long one share of stock and one put, and short one call. The values of the three assets at the exercise time T can be denoted by S_T, P_T, and C_T, so that the portfolio value at that time is $S_T + P_T - C_T$. If $S_T \leq E$, then the call is worthless, and the put can be executed by selling the share of stock that is already owned in exchange for amount E. In other words, $C_T = 0$, $P_T = E - S_T$ and the overall value of the portfolio is E. On the other hand, if $S_T > E$, then the put is worthless but the call is in the money. It will be executed by the call partner, but the share is already available to turn over for the price of E. Then, $C_T = S_T - E$ and $P_T = 0$ in the formula $S_T + P_T - C_T$. Hence, in either case, the final value of the portfolio is constantly E. By the Law of One Price, to avoid arbitrage, the initial value of the portfolio must be the same as the initial value of the cash amount of E, so that:

$$S_0 + P - C = e^{-rT}E. \tag{5.24}$$

Solving for P yields the put-call parity formula (5.23).

We can apply put-call parity together with the Black-Scholes formula in order to derive a corresponding formula for the European put option. The result is in the next theorem, and we leave the proof for the exercises. Again the underlying asset is presumed to satisfy the stochastic differential equation (5.1).

Theorem 2. (Black-Scholes put option) Suppose that the underlying asset follows the Black-Scholes model of Theorem 1, where s_0 is the initial price, σ is the volatility, r is the continuous risk-free rate, and $\mathcal{N}$ is the standard normal cumulative distribution function. The initial value P at time 0 of the European put option with strike price E and termination time T is:

$$
\begin{aligned}
P \;=\; & E \cdot e^{-rT}\mathcal{N}\left(\frac{\log(E/s_0)-\left(r-\sigma^2/2\right)T}{\sigma\sqrt{T}}\right) \\
& - s_0\mathcal{N}\left(\frac{\log(E/s_0)-\left(r+\sigma^2/2\right)T}{\sigma\sqrt{T}}\right).
\end{aligned}
\tag{5.25}
$$

Proof. Exercise 9. ∎

Example 2. As a quick calculation, consider an underlying asset whose initial price s_0 is 50, and for which the volatility constant is $\sigma = .3$. Let the risk-free continuous rate be $r = .04$. Let us compute the value of a European put option with expiration time $T = 4$ and strike price $E = 48$.

Solution. Let the first quantity inside the normal c.d.f. be denoted by c_1, and the second by c_2. Then:

$$
\begin{aligned}
c_1 &= \frac{\log(E/s_0) - (r - \sigma^2/2)T}{\sigma\sqrt{T}} \\
&= \frac{\log(48/50) - (.04 - .3^2/2)4}{.3\sqrt{4}} = -.0347033; \\
c_2 &= \frac{\log(E/s_0) - (r + \sigma^2/2)T}{\sigma\sqrt{T}} \\
&= \frac{\log(48/50) - (.04 + .3^2/2)4}{.3\sqrt{4}} = -.634703.
\end{aligned}
$$

Thus, the initial put option value is:

$$
\begin{aligned}
P &= E \cdot e^{-rT} \mathcal{N}(c_1) - s_0 \mathcal{N}(c_2) \\
&= 48 \cdot e^{-.04(4)} \mathcal{N}(-.0347033) - 50 \cdot \mathcal{N}(-.634703) \\
&= 6.74473. \ \blacksquare
\end{aligned}
$$

The reader can do Exercise 11 to study the relationships between the put value and the underlying parameters.

Exercises 5.1

1. Verify identities (5.22).

2. Verify formula (5.11) for the variance of $\log(R_i)$.

3. (Technology required) Compute the value of a call option on an underlying asset whose initial price is 20, and whose volatility is $\sigma = .2$. The strike price of the option is 21, and it expires at time $T = .5$ year. The continuous risk-free rate is $r = .01$.

4. (Technology required) A put option on an underlying asset is offered with strike price 50 and expiration time 1 year. The risk-free rate is .005. The asset has initial price 55 and volatility $\sigma = .4$. Find the initial value of the put.

5. For an alternative derivation of the put-call parity formula, compare the final values of two portfolios: portfolio A consisting of a share of the underlying asset and a put with strike price E expiring at time T, and portfolio B consisting of a call on the same asset with the same strike and expiration time plus $e^{-rT}E$ in the risk-free asset. Draw a conclusion.

6. A risky asset currently has price 50. Both a call and a put option on that asset are available, both with strike prices 50 and expiration times 2. The call is worth .5 monetary units more than the put. What must be the risk-free rate?

7. There is a common method for calibrating binomial branch processes to continuous-time processes such as those we are studying, called the **Hull-White**

model. The approach is to reparameterize the branch process by setting $p = 1/2$, and using up and down rates:

$$b = \mu \Delta t + \sigma \sqrt{\Delta t} \text{ and } a = \mu \Delta t - \sigma \sqrt{\Delta t}.$$

Find $\lim_{n \to \infty} E[\log(S(T))]$ in the Hull-White model without using linear approximation, in the case that $p = 1/2$. (Hint: express the difference of logs as the log of a quotient, transform variables to $m = 1/n$, and use L'Hopital's Rule to find the limit of the resulting expression as $m \to 0$.) Simplify to the form $\log(s_0) + \left(\mu - \frac{1}{2}\sigma^2\right) T$. Conclude that the limit of the branch process in the Hull-White model is consistent with the mean of the Black-Scholes asset process in formula (5.1).

8. This exercise follows up on Exercise 7 on the Hull-White model for the binomial branch process. First recall that the second order Taylor approximation for $\log(1 + x)$ is $x - \frac{x^2}{2}$, which fits the function very nicely near zero, as the graph below shows ($\log(1 + x)$ is in black, the quadratic function is in gray).

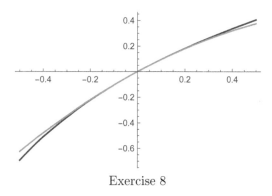

Exercise 8

Show that for a Hull-White branch process, formula (5.12) for the limit of the variance $\text{Var}(\log(S(T))) = 4p(1 - p)\sigma^2 \cdot T$ still holds, even for general p, by doing the following steps:

(a) Show that for any two-point probability mass function with $P[X = c] = p$, $P[X = d] = 1 - p$,

$$\text{Var}(X) = p(1 - p)(c - d)^2;$$

(b) Use part (a) and a second order Taylor approximation of $\log(1 + x)$ to establish the formula:

$$\text{Var}\left(\log(S(T))\right) = p(1 - p) \cdot 4\sigma^2(1 - 2\mu \Delta t)^2 T;$$

(c) Conclude that in the limit, the desired form for $\text{Var}(\log(S(T)))$ holds.

9. Use put-call parity to derive formula (5.25) for the value of a European put option in the Black-Scholes world.

10. (Technology required) For the call option in the situation of Example 1 with $T = 1$, graph the option value as a function of: (a) the strike price E; and (b) the parameter σ. Explain intuitively why you see the behavior that you see.

11. (Technology required) For the put option in the situation of Example 2, graph the option value as a function of: (a) the exercise time T; (b) the strike price E; and (c) the parameter σ. Explain intuitively why you see the behavior that you see.

5.2 Black-Scholes Via Martingales

Here we take an alternative approach to derive valuation formulas, not only for the European call and put, but for other kinds of derivatives in continuous time. The development will be analogous to the discrete-time case, in which the concepts of replication of a derivative by an admissible trading strategy, no arbitrage, and martingale measure are crucial. Therefore, we need continuous-time versions of the Fundamental Theorems of Asset Pricing, but in the Black-Scholes world the full derivations involve technical details that we prefer to avoid, particularly in the multiple asset case. But we can at least motivate the existence of a martingale measure, after which we will go forward with the confidence that in an arbitrage-free market, all derivatives that are replicable can be valued using the martingale approach.

To illustrate the usefulness of the ideas, we will look at two simple classes of derivatives: the **bond and asset binary derivatives**. These will then be used to value simple European calls and puts, and a few other interesting derivatives.

The subsection on martingale valuation draws from the fine books by Baxter and Rennie [1] and Williams [24], while the material on valuing binary derivatives and other non-vanilla derivatives relies heavily on the interesting book by Buchen [3].

5.2.1 Trading Strategies and Martingale Valuation

We will be studying derivatives based on a single risky asset whose price at time t is S_t, and we assume that we are in the Black-Scholes world in which the dynamics of the asset process satisfy:

$$dS_t = \mu S_t dt + \sigma S_t dZ_t, S_0 = s_0, \tag{5.26}$$

where (Z_t) is a standard Brownian motion under a given probability P. There is a risk-free asset B growing exponentially at continuous rate r, so that (assuming without loss of generality that it is worth one monetary unit at time 0):

$$\frac{dB_t}{dt} = rB, B_0 = 1 \Longleftrightarrow B_t = e^{rt}. \qquad (5.27)$$

To make progress, we first need to define a few terms. Probability measures P and Q are called **equivalent** if they have the same events of probability 0, that is:

$$P[A] = 0 \Longleftrightarrow Q[A] = 0. \qquad (5.28)$$

Note that this is a fairly weak condition; we are not supposing that the two measures give the same probability to every event. The **history** generated by the process (S_t) is the family of σ-algebras $(\mathcal{F}_t)_{t \geq 0}$ such that each $\mathcal{F}_t$ is the smallest σ-algebra with respect to which all $S_r, r \leq t$ are measurable. Intuitively, each $\mathcal{F}_t$ contains the information obtained by observing the S process up through time t.

We will consider derivatives expiring at a time T. A **trading strategy** over time interval $[0, T]$ is a family $(\alpha_t, \beta_t)_{t \geq 0}$ of pairs of stochastic processes that describes the numbers of units α_t of the risky asset S and the number of units β_t of the non-risky asset B that are held at each time t. We assume that the trading strategy cannot look ahead, but can only be based on the information available at the time, so that α_t and β_t are assumed to be measurable with respect to $\mathcal{F}_t$. Then the value of a trading strategy at time t is:

$$V_t = \alpha_t S_t + \beta_t B_t, \qquad (5.29)$$

which is also $\mathcal{F}_t$-measurable.

The concept of a **self-financing trading strategy** is best understood by approximation. The idea is to state a condition that says that instantaneous changes in value of a portfolio can only occur via changes in value of the assets that the portfolio is based on, not by adding or taking away funds. For a short interval of length Δt beginning at time t, we would have:

$$V_{t+\Delta t} - V_t \approx \alpha_t \left(S_{t+\Delta t} - S_t\right) + \beta_t \left(B_{t+\Delta t} - B_t\right). \qquad (5.30)$$

Passing to the limit as $\Delta t \longrightarrow 0$, in the language of stochastic calculus, a trading strategy is **self-financing** if its value process (V_t) satisfies:

$$dV_t = \alpha_t dS_t + \beta_t dB_t. \qquad (5.31)$$

For technical reasons later, we say that an **admissible trading strategy** (α_t, β_t) is one that is self-financing and satisfies the growth condition:

$$E_Q \left[\int_0^T \left(\alpha_t \bar{S}_t\right)^2 dt \right] < \infty, \qquad (5.32)$$

where $\bar{S}_t = e^{-rt} S_t$ is the discounted asset price process.

A *financial derivative* is an $\mathcal{F}_T$ - measurable random variable X. So our structure allows for derivatives that are path dependent. A derivative X is **attainable**, or **replicable**, if there is a self-financing trading strategy whose value function V_T agrees with X except perhaps on a set of probability 0. As in the discrete-time case, we define an **arbitrage trading strategy** to be a self-financing strategy such that its initial value V_0 is 0, but there is a time t such that $V_t \geq 0$ for all outcomes, and with positive probability, $V_t > 0$.

5.2.2   Martingale Measures

As in the discrete-time case, define a **martingale measure** Q on the probability space on which the process resides as a probability measure under which the discounted asset process $\bar{S}_t = e^{-rt} S_t$ is an $(\mathcal{F}_t)$-martingale. An **equivalent martingale measure** Q is a martingale measure that is equivalent to P.

How do we know that there is such a thing as an equivalent martingale measure Q in our Black-Scholes world? An important theorem of probability called the **Cameron-Martin-Girsanov Theorem** points the way (see for example [24], p. 142). We state the special case that is of most interest to us.

Theorem 1. (C-M-G Theorem) Let P be the probability measure on the sample space Ω on which the asset process is defined, and let $(Y_t)_{t \geq 0}$ be the process defined by $Y_t = e^{-\theta Z_t - \frac{1}{2}\theta^2 t}$, where θ is a constant. Then (Y_t) is a martingale under P. Furthermore, define a new probability measure Q on Ω by:

$$Q(A) = E_P\left[I_A \cdot Y_T\right], A \subset \Omega. \tag{5.33}$$

Then Q is equivalent to P, and also, the process defined by:

$$\tilde{Z}_t = Z_t + \theta t \tag{5.34}$$

is standard Brownian motion under Q.

Proof. We can show part of this theorem. The fact that (Y_t) in the theorem is a P-martingale is a routine expectation calculation and is left for Exercise 6. We need to show that Q actually is a probability measure. Since $I_A \cdot Y_T$ must be non-negative, so must be its expectation; hence Q does satisfy the non-negativity axiom of probability. Q is easily shown to satisfy the additivity axiom for disjoint events, which leaves the axiom $Q(\Omega) = 1$ to be shown. For this, we can derive:

$$
\begin{aligned}
Q(\Omega) &= E_P\left[I_\Omega \cdot Y_T\right] \\
&= E_P\left[Y_T\right] \\
&= E_P\left[e^{-\theta Z_T - \frac{1}{2}\theta^2 T}\right] \\
&= e^{-\frac{1}{2}\theta^2 T} E_P\left[e^{-\theta Z_T}\right] \\
&= e^{-\frac{1}{2}\theta^2 T} e^{\frac{1}{2}\theta^2 T} = 1.
\end{aligned}
$$

Line 2 is correct because $I_\Omega = 1$. In line 3 we substitute the expression for Y_T, and the non-random factor is factored out of expectation in line 4. The final line follows from the known moment-generating function of the $N\left(0, \sigma^2\right)$ distribution, $E\left[e^{sZ}\right] = e^{\frac{1}{2}s^2\sigma^2}$, since the variance of our Z_T is T.

Exercise 7 asks you to show that Q and P are equivalent. We omit the rest of the proof. ∎

As we have seen previously, in a Black-Scholes world, the price of a risky asset in continuous time is modeled by a general geometric Brownian motion:

$$S_t = s_0 e^{\left(\mu - \sigma^2/2\right)t + \sigma Z_t}. \tag{5.35}$$

Recalling that $E\left[e^{\sigma Z_t}\right] = e^{\left(\sigma^2/2\right)t}$, we get:

$$E\left[S_t\right] = s_0 E\left[e^{\left(\mu - \sigma^2/2\right)t + \sigma Z_t}\right] = s_0 e^{\left(\mu - \sigma^2/2\right)t} e^{\left(\sigma^2/2\right)t} = S_0 e^{\mu t}, \tag{5.36}$$

so that in expectation we get a process that on average experiences exponential growth.

Consider the case $\mu = r$, and again define the σ-algebras $\mathcal{F}_t = \sigma\left(S_u; u \le t\right)$ to be the histories of information through time t. Then the discounted process $\left(\bar{S}_t = e^{-rt}S_t\right)_{t \ge 0}$ satisfies:

$$
\begin{aligned}
E\left[\bar{S}_{t+s} | \mathcal{F}_t\right] &= E\left[e^{-r(t+s)}S_{t+s} | \mathcal{F}_t\right] \\
&= E\left[e^{-r(t+s)}s_0 e^{\left(r - \sigma^2/2\right)(t+s) + \sigma Z_{t+s}} | \mathcal{F}_t\right] \\
&= e^{-r(t+s)}e^{+\left(r - \sigma^2/2\right)(t+s)}s_0 E\left[e^{\sigma(Z_{t+s} - Z_t) + \sigma Z_t} | \mathcal{F}_t\right] \\
&= e^{-\left(\sigma^2/2\right)(t+s)}s_0 e^{\sigma Z_t} E\left[e^{\sigma(Z_{t+s} - Z_t)}\right] \\
&= e^{-\left(\sigma^2/2\right)(t+s)}s_0 e^{\sigma Z_t} E\left[e^{\sigma Z_s}\right] \\
&= e^{-\left(\sigma^2/2\right)(t+s)}s_0 e^{\sigma Z_t} e^{\left(\sigma^2/2\right)s} \\
&= e^{-rt}s_0 e^{\left(r - \sigma^2/2\right)t + \sigma Z_t} \\
&= \bar{S}_t.
\end{aligned}
\tag{5.37}
$$

In line 2 of derivation (5.37), we substituted the proper expression for S_{t+s}. Then in line 3 we pulled all non-random factors out of the expectation and added and subtracted σZ_t in the exponent. This was done so that in line 4 we could take advantage of the measurability of Z_t with respect to $\mathcal{F}_t$ and pull out the $e^{\sigma Z_t}$ factor as well. In the same line we used the fact that $Z_{t+s} - Z_t$ is independent of $\mathcal{F}_t$ to remove the conditional. Stationary increments allowed us to replace $Z_{t+s} - Z_t$ by Z_s in line 5. The sixth line used the formula mentioned above for $E\left[e^{\sigma Z_s}\right]$, and finally we multiplied and divided by e^{rt} and then simplified to get the last line.

Derivation (5.37) shows that the discounted price process is a continuous-time martingale if a measure can be found such that the drift rate is r. This is given to us by the Cameron-Martin-Girsanov Theorem.

To see this, let Q be the equivalent measure to P guaranteed by the theorem, and let $\tilde{Z}_t = Z_t + \theta t$ be the new standard Brownian motion under Q, where $\theta = \frac{\mu - r}{\sigma}$. Then the differential equation satisfied by the asset process under Q becomes

$$
\begin{aligned}
dS_t &= \mu S_t dt + \sigma S_t dZ_t \\
&= \mu S_t dt + \sigma S_t \left(d\tilde{Z}_t - \theta\, dt \right) \\
&= (\mu - \sigma \cdot \theta) S_t dt + \sigma S_t d\tilde{Z}_t \\
&= r S_t dt + \sigma S_t d\tilde{Z}_t.
\end{aligned}
\tag{5.38}
$$

The combination of derivations (5.37) and (5.38) shows that the discounted asset process is a martingale, and therefore Q is an equivalent martingale measure.

Next, we would like to show the following theorem.

Theorem 2. Let Q be a martingale measure, let (α_t, β_t) be an admissible self-financing trading strategy, and let $V_t = \alpha_t S_t + \beta_t B_t$ be the value function of the strategy. Then the discounted value process defined by $\bar{V}_t = e^{-rt} V_t$ is a Q-martingale.

Proof. Under Q, the dynamics of the discounted asset value process become:

$$
\begin{aligned}
d\bar{S}_t = d\left(e^{-rt} S_t\right) &= -re^{-rt} S_t dt + e^{-rt} dS_t \\
&= -re^{-rt} S_t dt + e^{-rt} \left(r \cdot S_t dt + \sigma S_t d\tilde{Z}_t \right) \\
&= \sigma \bar{S}_t d\tilde{Z}_t.
\end{aligned}
$$

Recall the self-financing condition $dV_t = \alpha_t dS_t + \beta_t dB_t$ and the fact that B_t is just the non-risky asset satisfying $B_t = e^{rt}$, $dB_t = r \cdot e^{rt} dt = r \cdot B_t dt$. Then the stochastic differential satisfied by the discounted value process $\bar{V}_t = e^{-rt} V_t$ under the martingale measure Q is:

$$
\begin{aligned}
d\bar{V}_t &= -re^{-rt} V_t dt + e^{-rt} dV_t \\
&= -re^{-rt} V_t dt + e^{-rt} \left(\alpha_t dS_t + \beta_t dB_t \right) \\
&= -re^{-rt} V_t dt + e^{-rt} \left(\alpha_t dS_t + \beta_t \cdot r \cdot B_t dt \right) \\
&= -re^{-rt} V_t dt + e^{-rt} \left(\alpha_t \left(r \cdot S_t dt + \sigma \cdot S_t d\tilde{Z}_t \right) + \beta_t \cdot r \cdot B_t dt \right) \\
&= -r\bar{V}_t dt + e^{-rt} \left(r\left(\alpha_t S_t + \beta_t \cdot B_t \right) dt + \alpha_t \sigma \cdot S_t d\tilde{Z}_t \right) \\
&= -r\bar{V}_t dt + e^{-rt} \left(rV_t dt + \alpha_t \sigma \cdot S_t d\tilde{Z}_t \right) \\
&= -r\bar{V}_t dt + r\bar{V}_t dt + \alpha_t \sigma \cdot \bar{S}_t d\tilde{Z}_t \\
&= \alpha_t \sigma \cdot \bar{S}_t d\tilde{Z}_t.
\end{aligned}
$$

The product rule is invoked in the first line, followed by substitution of the self-financing condition in line 2. Then in the third and fourth lines we take advantage of the expressions above for dB_t and dS_t (under Q). Regrouping the dt terms together in parenthesis, identifying the linear combination of S_t and B_t as V_t, distributing the discount factor through and simplifying in lines 5-8 yields the final expression.

We notice that there is no dt drift term in the dynamics of the $(\bar{V}_t)$ process. To finish the proof of Theorem 2, notice that:

$$E_Q\left[\bar{V}_{t+s} - \bar{V}_t | \mathcal{F}_t\right] = E\left[\int_t^{t+s} \alpha_u \sigma \cdot \bar{S}_u d\tilde{Z}_u | \mathcal{F}_t\right]$$
$$= 0. \tag{5.39}$$

The fact that the expression is 0 is a well known result in the theory of stochastic integrals, as long as the integrand satisfies a growth condition on its expectation (see [8], Theorem 2.8, p. 65):

$$E_Q\left[\int_0^T \left(\alpha_t \sigma \cdot \bar{S}_t\right)^2 dt\right] < \infty. \tag{5.40}$$

Up to the constant σ, this is precisely the technical admissibility assumption for trading strategies. Hence the value process $(\bar{V}_t)$ of an admissible trading strategy forms a martingale under Q. ■

Williams ([24], Theorem 4.4.2, p. 64) is able to exhibit an admissible trading strategy $\phi = (\alpha_t, \beta_t)$ that replicates a derivative X, as long as $E_Q[\|X\|]$ is finite. Following our reasoning in the discrete world, we therefore have the following martingale valuation result.

Theorem 3. If X is a financial derivative such that $E_Q[\|X\|] < \infty$, and Q is a martingale measure, then the initial value of the derivative is $e^{-rT} E_Q[X]$.

Proof. Let ϕ be the admissible trading strategy mentioned above, and let $(V_t)_{t\geq 0}$ be its value process. Replication says that $V_T = X$ for all outcomes except perhaps those in a set of P probability (and hence Q-probability) 0. Since the discounted value process is a martingale under Q, we have:

$$V_0 = \bar{V}_0 = E_Q\left[\bar{V}_T\right] = e^{-rT} E_Q\left[V_T\right] = e^{-rT} E[X]. \tag{5.41}$$

To avoid arbitrage, the initial value V_0 of the replicating portfolio must also be the initial value of the derivative. ■

For example, the martingale method would lead us to value the continuous-time European call option based on an asset obeying the geometric Brownian motion $S_t = s_0 e^{(r-\sigma^2/2)t+\sigma Z_t}$ as:

$$E_Q\left[e^{-rT} \max\left(s_0 e^{(r-\sigma^2/2)t+\sigma Z_T} - E, 0\right)\right], \tag{5.42}$$

where $Z_T \sim N(0, T)$. But this is exactly the same expectation as in the limit derivation of Section 5.1 (see formula (5.16)). Thus we have a second derivation of the Black-Scholes formula for the European call option starting at time 0, which in its final form was:

$$
\begin{aligned}
C &= s_0 \mathcal{N}\left(\frac{\log(s_0/E) + (r + \sigma^2/2)T)}{\sigma\sqrt{T}}\right) - E \cdot e^{-rT}\mathcal{N}\left(\frac{\log(s_0/E) + (r - \sigma^2/2)T}{\sigma\sqrt{T}}\right) \\
&= s_0 \mathcal{N}(d_1) - E \cdot e^{-rT}\mathcal{N}(d_2).
\end{aligned}
\tag{5.43}
$$

5.2.3 Asset and Bond Binaries

A third derivation of the Black-Scholes formula is made possible by considering two simple building block derivatives called **asset** and **bond binaries**. These will serve as tools for pricing certain exotic options as well.

We have seen that, as in the discrete world, the idea of valuing derivatives as the expectation of their discounted final claim value under a martingale measure still is applicable in the continuous world. We can use this approach to value the following two elementary derivatives.

Definition 1. (a) An **up-type bond binary derivative** pays one monetary unit (for instance a zero coupon bond with face value 1 and yield rate r) if its underlying asset exceeds a strike price E at T. Then the final claim value of such a derivative is denoted and represented as:

$$
B_T^+ \equiv 1 \cdot I_{\{S_T \geq E\}}.
\tag{5.44}
$$

(b) Similarly, a **down-type bond binary derivative** pays one monetary unit at time T if the underlying asset falls below the strike price E at T. The final claim value of the down-type bond binary is:

$$
B_T^- \equiv 1 \cdot I_{\{S_T \leq E\}}.
\tag{5.45}
$$

Definition 2. (a) An **up-type asset binary derivative** pays one unit of its underlying asset at expiration time T if that asset exceeds a strike price E at T. The final claim value of the up-type asset binary can be written as:

$$
A_T^+ \equiv S_T \cdot I_{\{S_T \geq E\}}.
\tag{5.46}
$$

(b) A **down-type asset binary derivative** pays one unit of its underlying asset at time T if the underlying asset falls below E at time T. The final claim value of the down-type asset binary is therefore:

$$
A_T^- \equiv S_T \cdot I_{\{S_T \leq E\}}.
\tag{5.47}
$$

To value bond and asset binary derivatives, it is useful to have a quick algebraic lemma, whose proof is requested in Exercise 1.

Lemma 1. For all real numbers z, the following inequalities are equivalent:

$$s_0 e^{(r-\frac{1}{2}\sigma^2)T+\sigma z\sqrt{T}} - E \geq (\text{or} <=\ )0$$

$$\Longleftrightarrow z \geq (\text{or} <=)\frac{\log\left(\frac{E}{s_0}\right)-(r-\frac{1}{2}\sigma^2)T}{\sigma\sqrt{T}}. \quad \blacksquare$$

(5.48)

Example 1. Apply martingale valuation to value the up-type bond binary in Definition 1(a).

Solution. Observe that at time T, the value of the driving Brownian motion in the geometric Brownian motion model Z_T is a random variable with the same distribution as $\sqrt{T} \cdot Z$, where Z is standard normal. Then by martingale valuation and Lemma 1:

$$
\begin{aligned}
B_0^+ &= E_Q\left[e^{-rT}B_T^+\right] \\
&= E_Q\left[e^{-rT}1 \cdot I_{\{S_T \geq E\}}\right] \\
&= e^{-rT}E\left[I_{\{S_T \geq E\}}\right] \\
&= e^{-rT}P\left[S_T \geq E\right] \\
&= e^{-rT}P\left[s_0 e^{(r-\sigma^2/2)T+\sigma Z_T} \geq E\right] \\
&= e^{-rT}P\left[s_0 e^{(r-\sigma^2/2)T+\sigma\sqrt{T}Z} \geq E\right] \\
&= e^{-rT}P\left[Z \geq \frac{\log\left(\frac{E}{s_0}\right)-(r-\frac{1}{2}\sigma^2)T}{\sigma\sqrt{T}}\right] \\
&= e^{-rT}\left(1-\mathcal{N}\left(\frac{\log\left(\frac{E}{s_0}\right)-(r-\frac{1}{2}\sigma^2)T}{\sigma\sqrt{T}}\right)\right) \\
&= e^{-rT}\mathcal{N}\left(-\frac{\log\left(\frac{E}{s_0}\right)-(r-\frac{1}{2}\sigma^2)T}{\sigma\sqrt{T}}\right) \\
&= e^{-rT}\mathcal{N}\left(\frac{\log\left(\frac{s_0}{E}\right)+(r-\frac{1}{2}\sigma^2)T}{\sigma\sqrt{T}}\right) = e^{-rT}\mathcal{N}(d_2),
\end{aligned}
$$

(5.49)

where d_2 is the constant in the Black-Scholes call option value formula (5.2). Notice also that E units of such a derivative would be valued at $E \cdot e^{-rT}\mathcal{N}(d_2)$, matching the second term in the call option formula. ∎

Exercise 3 asks you to derive in a similar way the formula below for the value of a down-type bond binary.

$$B_0^- = e^{-rT}\mathcal{N}\left(-\frac{\log\left(\frac{s_0}{E}\right)+(r-\frac{1}{2}\sigma^2)T}{\sigma\sqrt{T}}\right) = e^{-rT}\mathcal{N}(-d_2).$$

(5.50)

For the asset derivatives we will need the following lemma, called the **Gaussian Shift Theorem**.

Lemma 2. Let Z be a random variable with the standard normal distribution, let c be a constant, and let B be a subset of the real line. Denote by $B - c$ the set of all numbers in B minus the value c, i.e. the set B translated to the left by c units. Then:

$$E\left[e^{cZ} \cdot I_B(Z)\right] = e^{c^2/2} \cdot P[Z \in B - c]. \tag{5.51}$$

Proof. The expectation on the left of formula (5.51) is:

$$E\left[e^{cZ} \cdot I_B(Z)\right] = \int_{-\infty}^{+\infty} e^{cz} I_B(z) \cdot \frac{1}{\sqrt{2\pi}} e^{-z^2/2} dz = \int_B e^{cz} \cdot \frac{1}{\sqrt{2\pi}} e^{-z^2/2} dz.$$

Bring the two exponential factors together and complete the square in the exponent to get:

$$
\begin{aligned}
E\left[e^{cZ} \cdot I_B(Z)\right] &= \int_B \frac{1}{\sqrt{2\pi}} e^{(z^2 - 2cz)/2} dz \\
&= \int_B \frac{1}{\sqrt{2\pi}} e^{(z^2 - 2cz + c^2)/2 + c^2/2} dz \\
&= e^{c^2/2} \cdot \int_B \frac{1}{\sqrt{2\pi}} e^{-(z-c)^2/2} dz.
\end{aligned}
$$

Finally, substitute $u = z - c, du = dz$ in the integral. This yields:

$$
\begin{aligned}
E\left[e^{cZ} \cdot I_B(Z)\right] &= e^{c^2/2} \cdot \int_B \frac{1}{\sqrt{2\pi}} e^{-(z-c)^2/2} dz \\
&= e^{c^2/2} \cdot \int_{B-c} \frac{1}{\sqrt{2\pi}} e^{-u^2/2} du \\
&= e^{c^2/2} \cdot P[Z \in B - c]. \ \blacksquare
\end{aligned}
$$

Example 2. Derive the initial value of the down-type asset binary, whose claim value is $A_T^- = S_T \cdot I_{\{S_T \le E\}}$.

Solution. Notice that from Lemma 1, since the distribution of Z_T is the same as that of $\sqrt{T} \cdot Z$, where Z is standard normal, the events $\{S_T \le E\}$ and $\{Z \le -d_2\}$ are equivalent. Using Lemma 2 with $c = \sigma\sqrt{T}$, we can write the initial value of the derivative as the expectation under the martingale measure as follows:

$$
\begin{aligned}
A_0^- &= E_Q\left[e^{-rT}A_T^-\right] \\
&= e^{-rT}E_Q\left[S_T \cdot I_{\{S_T \le E\}}\right] \\
&= e^{-rT}E_Q\left[s_0 e^{(r-\sigma^2/2)T+\sigma Z_T} \cdot I_{\{S_T \le E\}}\right] \\
&= s_0 e^{-rT}e^{(r-\sigma^2/2)T}E_Q\left[e^{\sigma\sqrt{T}Z} \cdot I_{\{Z \le -d_2\}}\right] \\
&= s_0 e^{(-\sigma^2/2)T} \cdot e^{\left(\sigma\sqrt{T}\right)^2/2} \cdot P\left[Z \le -d_2 - \sigma\sqrt{T}\right] \\
&= s_0 \cdot P\left[Z \le -\frac{\log\left(\frac{s_0}{E}\right)+\left(r-\frac{1}{2}\sigma^2\right)T}{\sigma\sqrt{T}} - \sigma\sqrt{T}\right] \qquad (5.52) \\
&= s_0 \cdot P\left[Z \le \frac{\log\left(\frac{E}{s_0}\right)-\left(r-\frac{1}{2}\sigma^2\right)T}{\sigma\sqrt{T}} - \frac{\left(\sigma\sqrt{T}\right)^2}{\sigma\sqrt{T}}\right] \\
&= s_0 \cdot P\left[Z \le \frac{\log(E/s_0)-\left(r+\sigma^2/2\right)T)}{\sigma\sqrt{T}}\right] \\
&= s_0 \cdot \mathcal{N}\left(\frac{\log(E/s_0)-\left(r+\sigma^2/2\right)T)}{\sigma\sqrt{T}}\right). \ \blacksquare
\end{aligned}
$$

The derivation in formula (5.52) can be adapted to give the initial value of the up-type asset binary as:

$$
A_0^+ = s_0 \cdot \mathcal{N}\left(\frac{\log\left(s_0/E\right)+\left(r+\sigma^2/2\right)T)}{\sigma\sqrt{T}}\right). \qquad (5.53)
$$

(See Exercise 4.)

Consider formulas (5.49) and (5.53) for B_0^+ and A_0^+. We can easily use them to derive the Black-Scholes formula for the initial value of a European call option. Since the call claim value is $(S_T - E) \cdot I_{\{S_T \ge E\}}$, we have:

$$
C_T = S_T \cdot I_{\{S_T \ge E\}} - E \cdot I_{\{S_T \ge E\}} = A_T^+ - E \cdot B_T^+.
$$

In other words, the final value of the call is the same as the final value of a portfolio that is long one up-type asset binary and short E units of an up-type bond binary. To avoid arbitrage, the initial value of the call must match the initial value of the portfolio; hence:

$$
\begin{aligned}
C_0 &= A_0^+ - E \cdot B_0^+ \\
&= s_0 \cdot \mathcal{N}\left(\frac{\log\left(s_0/E\right)+\left(r+\sigma^2/2\right)T)}{\sigma\sqrt{T}}\right) \\
&\quad - E \cdot e^{-rT}\mathcal{N}\left(\frac{\log\left(\frac{s_0}{E}\right)+\left(r-\frac{1}{2}\sigma^2\right)T}{\sigma\sqrt{T}}\right).
\end{aligned}
$$

This is identical to the Black-Scholes formula.

So we have derived the Black-Scholes result using only the minimal integration necessary to prove the Gaussian Shift Theorem. In Exercise 5, you will make a similar argument for the initial value of the European put option.

5.2.4 Other Binary Derivatives

Let us now take further advantage of knowing the initial values of up and down asset and bond binaries to value a few more exotic options expiring at a single time T.

The three examples that we are about to present are all variations of more general first order up- and down-type binary derivatives, with time T payoffs of:

$$V_T^+ = f(S_T) \cdot I_{\{S_T \geq E\}}, V_T^- = f(S_T) \cdot I_{\{S_T \leq E\}}. \qquad (5.54)$$

For our simple asset binary, the payoff function is $f(x) = x$, and for the bond binary the payoff is the constant $f(x) = 1$.

Example 3. In a ***gap option***, the event of whether the option is in-the-money or not is still determined by the strike price E, but the payoff is no longer just $S_T - E$ for a call or $E - S_T$ for a put. A different constant k is used for the amount paid, which would be $S_T - k$ for a gap call option, or $k - S_T$ for a gap put option. To ensure non-negative payoffs, we would have to take $k \leq E$ for the call and $k \geq E$ for the put. Derive a formula for the initial value of the gap call option in terms of the initial values of the up-type bond and asset binaries.

Solution. The payoff for this derivative is:

$$V_T = (S_T - k) I_{\{S_T \geq E\}} = S_T \cdot I_{\{S_T \geq E\}} - k \cdot I_{\{S_T \geq E\}}.$$

This is the same as the payoff for a portfolio that is long one asset binary and short k units of a bond binary. To avoid arbitrage, the initial value of the gap call must also agree with the initial value of the portfolio. Therefore the gap call has the following time 0 value:

$$V_0 = A_0^+ - k \cdot B_0^+. \qquad (5.55)$$

In light of the fact that we have the explicit formulas (5.49) and (5.53) for B_0^+ and A_0^+, this is sufficient to characterize the initial value of the gap call. ∎

Example 4. Option issuers may well be concerned about major losses, if the underlying asset doesn't behave as they expect. For call options, a cap on payoff is desirable for the issuer, which kicks in if the payoff $S_T - E$ exceeds a cap c. (See Exercise 12.) In the case of a put option, the issuer would benefit if the option could not pay more than c; in the case that $E - S_T \geq c$, in other words, if $S_T \leq E - c$, then the payoff is c. A typical payoff graph for this derivative is in Figure 5.3, with strike price 50 and cap 20. Write an expression for the initial value of this ***capped put option*** in terms of asset and bond binaries.

Solution. We can divide the claim value into three cases as follows.

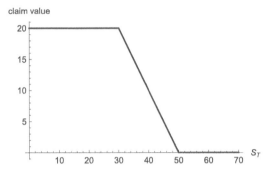

FIGURE 5.3 Final claim value of capped put option, $E = 50, c = 20$.

$$V_T\left(S_T\right) = \begin{cases} 0 & \text{if } S_T > E; \\ E - S_T & \text{if } E - c \leq S_T \leq E; \\ c & \text{if } S_T < E - c. \end{cases}$$

We need to write this piecewise function appropriately in terms of indicators whose sets involve comparisons between S_T and constants E and $E - c$. Since this derivative has no value whatsoever if $S_T > E$, we think of multiplying some expression by $I_{\{S_T \leq E\}}$. That being determined, the claim value depends on whether furthermore $S_T < E - c$ or not. The following function captures the payoff correctly:

$$I_{\{S_T \leq E\}} \left((E - S_T)\, I_{\{S_T \geq E-c\}} + c \cdot I_{\{S_T < E-c\}} \right).$$

Multiplying through the sum by the first indicator $I_{\{S_T \leq E\}}$, note that $I_{\{S_T \leq E\}} \cdot I_{\{S_T < E-c\}} = I_{\{S_T < E-c\}}$, since both equal 1 only when $S_T < E - c$. Also, we observe that

$$I_{\{S_T \leq E\}} \cdot I_{\{S_T \geq E-c\}} = I_{\{S_T \leq E\}} - I_{\{S_T < E-c\}}$$

because both are 1 only when $E - c \leq S_T \leq E$. Now substituting into the payoff and simplifying, we find:

$$\begin{aligned} V_T\left(S_T\right) &= I_{\{S_T \leq E\}} \left((E - S_T)\, I_{\{S_T \geq E-c\}} + c \cdot I_{\{S_T < E-c\}} \right) \\ &= (E - S_T) \left(I_{\{S_T \leq E\}} - I_{\{S_T < E-c\}} \right) + c \cdot I_{\{S_T < E-c\}} \\ &= E \cdot I_{\{S_T \leq E\}} - S_T \cdot I_{\{S_T \leq E\}} \\ &\quad + S_T \cdot I_{\{S_T < E-c\}} - (E - c) \cdot I_{\{S_T < E-c\}}. \end{aligned} \tag{5.56}$$

This final claim value matches that of a portfolio of E units of a down-type bond binary with exercise price E and one unit of a down type asset binary with exercise price $E - c$, together with a short position in each of a unit of down-type asset binary with exercise price E and $(E - c)$ units of a down-type bond binary with exercise price $(E - c)$. (Note also that the first two terms of the bottom line of formula (5.56) constitute a long put with strike E, and the second two terms form the claim value of a short put with strike $E - c$.) Introducing the strike price explicitly into the notation for the bond and asset binary values, it follows that the initial value of the capped put option is:

$$V_0 = E \cdot B_0^-(E) + A_0^-(E - c) - A_0^-(E) - (E - c) \cdot B_0^-(E - c). \quad \blacksquare \quad (5.57)$$

Example 5. The derivative called a **range-forward contract** behaves for the holder in a way similar to the outcome of a forward contract to the seller, earning a profit when the underlying asset goes down, but incurring a loss when it goes up. However in this derivative, there is a range of values in the middle in which the contract has no final claim value. Typically, conditions are set so that the derivative has value 0 at issuance, that is, no money changes hands at the start, much like a forward contract.

A typical picture of the payoff is in Figure 5.4. The range forward contract is defined in terms of two constants, k_1 and k_2, with $k_1 < k_2$. Its final value as a function of the final asset price is in the equation below. Let us find an expression for the initial value.

$$V_T(S_T) = \begin{cases} k_1 - S_T & \text{if } S_T < k_1; \\ 0 & \text{if } k_1 \le S_T \le k_2; \\ -(S_T - k_2) = k_2 - S_T & \text{if } S_T > k_2. \end{cases} \quad (5.58)$$

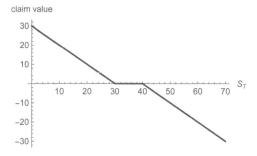

FIGURE 5.4 Final claim value of range forward contract, $k_1 = 30, k_2 = 40$.

Solution. We take a different approach here, which is to express the range forward contract as equivalent to a portfolio which is long on a put option and short on a call. This is suggested by the top and bottom formulas in the piecewise definition (5.58) of the payoff. Specifically, let's check that the same payoff is earned by holding a put option with strike price k_1 while issuing a call option with strike price k_2. If $S_T < k_1$, then the call expires worthless while the put value is $k_1 - S_T$. If $k_1 \le S_T \le k_2$, then both of the options are worthless, so the portfolio is worth 0. In the last case where $S_T > k_2$, the put is worthless but the call is in the money, so since we have issued the call, or are short in it, we must pay out $S_T - k_2$. In each case the portfolio's payoff agrees with the payoff of the contract; hence the initial values of the two assets must agree. Previously we showed that a European call option with strike price E could be viewed as a portfolio of an asset binary and a bond binary with strike price E, so that the initial value of a call is $C_0 = A_0^+(E) - E \cdot B_0^+(E)$, and similarly in Exercise 5 you will determine that a European put has initial value $P_0 = E \cdot B_0^-(E) - A_0^-(E)$. The call here has strike price k_2 and the put has strike price k_1; therefore in terms of asset and bond binaries we would have the formula below for the initial value of the range forward contract.

$$
\begin{aligned}
V_0 &= P_0(k_1) - C_0(k_2) \\
&= \left(k_1 \cdot B_0^-(k_1) - A_0^-(k_1)\right) - \left(A_0^+(k_2) - k_2 \cdot B_0^+(k_2)\right). \ \blacksquare
\end{aligned}
\tag{5.59}
$$

In Exercise 13 we reverse the range forward to pay off similarly to the buyer's position in an ordinary forward. See Exercise 9 for another interesting derivative, the **pay-at-expiry** or PAX option and Exercise 14 for the **butterfly spread**.

Exercises 5.2

1. Prove Lemma 1.

2. What change to the derivation and formula in (5.49) results if we want to compute B_t^+, that is the value of the up-type bond binary at time t conditioned on the value of the underlying asset at time t?

3. Derive formula (5.50) for the initial value of the down-type bond binary.

4. Derive formula (5.53) for the initial value of the up-type asset binary.

5. Use formulas (5.50) and (5.52) to verify the formula below from Section 5.1 for the initial value of the European put option:

$$
E \cdot e^{-rT} \mathcal{N}\left(\frac{\log(E/s_0) - (r - \sigma^2/2)T}{\sigma\sqrt{T}}\right)
$$
$$
- s_0 \cdot \mathcal{N}\left(\frac{\log(E/s_0) - (r + \sigma^2/2)T)}{\sigma\sqrt{T}}\right).
$$

6. Show that if (Z_t) is a standard Brownian motion with initial state 0 under a probability measure P, then for any constant θ, the process defined by $Y_t = e^{-\theta Z_t - \frac{1}{2}\theta^2 t}$ is a martingale with respect to P.

7. Argue that the C-M-G measure $Q(A) = E_P[I_A \cdot Y_T]$ is equivalent to P.

8. Show the two parity relations below for bond and asset binaries. (Hint: You may argue for and use the identity $I_{\{S_T \geq E\}} + I_{\{S_T \leq E\}} = 1$, since the probability that a continuous random variable such as S_T equals a value such as E identically is 0.)

$$B_0^+ + B_0^- = e^{-rT}$$
$$A_0^+ + A_0^- = S_0.$$

9. A **pay-at-expiry** (PAX) option is a derivative which does not cost initially, but for which, if it expires in-the-money, a premium p is paid by the holder. Derive the expression below for the arbitrage-free premium for a PAX call option with strike price E and expiration time T. (This option would pay off as a regular European call option pays, but no money changes hands at the start, and the premium is paid at the end if the asset exceeds the strike price.)

$$p = \frac{A_0^+}{B_0^+} - E.$$

10. Referring to Exercise 9, derive a similar formula for a PAX put option, which costs nothing at time 0 but pays off what an ordinary put option would pay less the premium p if the put is in the money.

11. Derive a formula for the initial value of the gap put option in terms of the initial values of the down-type bond and asset binaries.

12. A **capped call option** with cap c works like the capped put in Example 4, except that if the difference $S_T - E$ is more than c, then only c is paid. Find a formula for the initial value of a capped call, and show that it is the same as the difference between the value of an ordinary European call with strike price E and a European call with strike price $E + c$.

13. In Example 5 a range forward contract was defined so as to give a payoff similar to that of the seller in an ordinary forward. Modify this to define a range forward contract similar to the buyer's investment in a forward, and find a formula for its initial value.

14. A **butterfly spread** is a derivative that appeals to an investor who feels that an asset is likely to take a value near E at expiry time T and variability of the asset is likely to be low. The formula for the payoff is below, depending on a constant c as well as the strike price E, and we also show a typical graph in the Exercise 14 figure. Find a formula for the initial value of this derivative in terms of the formulas for asset and bond binaries. (Hint: argue that this payoff is equivalent to a portfolio which is long on two calls, one with strike $E - c$ and the other with strike $E + c$, and short two calls which both have strike E.)

$$V_T(S_T) = \begin{cases} 0 & \text{if } S_T < E - c; \\ S_T - (E - c) & \text{if } E - c \leq S_T \leq E; \\ E + c - S_T & \text{if } E < S_T \leq E + c \\ 0 & \text{if } S_T > E + c. \end{cases}$$

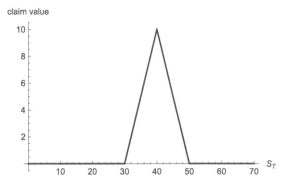

Exercise 14. Butterfly spread payoff, $E = 40, c = 10$.

5.3 Black-Scholes Via Differential Equations

We have now seen two different approaches to deriving the formula for the value of a European call option that was presented in the seminal 1972 paper by Fischer Black and Myron Scholes [2]. But neither the limiting process in Section 5.1 nor the martingale approach in Section 5.2 was the one that the original authors used. Instead, they derived and solved a second order partial differential equation, which in fact applies in general to many derivatives. It is valuable not only from the standpoint of the history of the field, but for the financial reasoning that is illustrated, to see what Black and Scholes actually did. This section serves a purpose that is more cultural than illustrative of new computational techniques; consequently the exercise list at the end is very limited.

5.3.1 Deriving the PDE

We denote the price process of the underlying asset by $(X_t)_{t \geq 0}$. Let $V(x,t)$ be value of a financial derivative at time t if the current price of the underlying asset is x. We use subscripts such as V_x, V_t, and V_{xx} to indicate the first order partial derivatives in the individual variables and the second order partial in x. The underlying asset is assumed to obey the stochastic differential equation:

$$dX_t = \mu X_t dt + \sigma X_t dZ_t, \tag{5.60}$$

which, as we know, describes a geometric Brownian motion.

Black and Scholes suggested that a portfolio that is long one share of the asset and short by $1/V_x$ shares of the derivative is fully hedged; that is, it has no risk. To see this, approximately, if the share of stock changes by Δx then the option value changes by about $V_x \Delta x$, and since the portfolio is short $1/V_x$ shares of the option, the total value of the units of the derivative changes by:

$$-(1/V_x) \cdot V_x \Delta x = -\Delta x. \qquad (5.61)$$

Therefore, the overall portfolio value change is $\Delta x - \Delta x = 0$. In other words, the value of the portfolio doesn't change with price movements of the underlying asset.

In general, the value of the portfolio at time t is:

$$X_t - \frac{V(X_t, t)}{V_x(X_t, t)}. \qquad (5.62)$$

Its approximate change in value during the time interval $(t, t+\Delta t)$ is therefore:

$$\Delta X_t - \frac{\Delta V(X_t, t)}{V_x(X_t, t)}. \qquad (5.63)$$

By Ito's Lemma:

$$\Delta V(X_t, t) \approx V_t(X_t, t) \Delta t + \frac{1}{2} V_{xx}(X_t, t) \cdot \sigma^2 X_t^2 \Delta t + V_x(X_t, t) \Delta X_t. \qquad (5.64)$$

Substituting into Equation (5.63), the approximate change in value of the hedged portfolio is:

$$
\begin{aligned}
\Delta X_t - \frac{\Delta V(X_t,t)}{V_x(X_t,t)} &\approx \Delta X_t - \frac{V_t(X_t,t)\Delta t + \frac{1}{2}V_{xx}(X_t,t)\cdot\sigma^2 X_t^2 \Delta t + V_x(X_t,t)\Delta X_t}{V_x(X_t,t)} \\
&= \Delta X_t - \frac{V_t(X_t,t)\Delta t + \frac{1}{2}V_{xx}(X_t,t)\cdot\sigma^2 X_t^2 \Delta t}{V_x(X_t,t)} - \frac{V_x(X_t,t)\Delta X_t}{V_x(X_t,t)} \\
&= \frac{-\left(V_t(X_t,t) + \frac{1}{2}V_{xx}(X_t,t)\cdot\sigma^2 X_t^2\right)\Delta t}{V_x(X_t,t)}.
\end{aligned}
$$
$$(5.65)$$

Since there is no risk, the anti-arbitrage assumption implies that this change should equal $r\Delta t$ times the current value of the portfolio. Thus:

$$\frac{-\left(V_t(X_t, t) + \frac{1}{2}V_{xx}(X_t, t) \cdot \sigma^2 X_t^2\right)\Delta t}{V_x(X_t, t)} \approx r\Delta t \cdot \left(X_t - \frac{V(X_t, t)}{V_x(X_t, t)}\right). \qquad (5.66)$$

Multiplying both sides of this equation by $V_x(X_t, t)$, dividing by Δt, and passing to the limit as $\Delta t \to 0$ (given $X_t = x$) gives us the differential equation:

$$- \left(V_t(x,t) + \tfrac{1}{2} V_{xx}(x,t) \cdot \sigma^2 x^2 \right) = r \cdot x \cdot V_x(x,t) - rV(x,t)$$
$$\Longrightarrow V_t(x,t) = rV(x,t) - rxV_x(x,t) - \tfrac{1}{2} \cdot \sigma^2 x^2 \cdot V_{xx}(x,t). \tag{5.67}$$

Formula (5.67) is the Black-Scholes differential equation for the value $V(x,t)$ of the portfolio at time t when the underlying asset has value x. Given an appropriate boundary condition and an assumption that a derivative may be replicated; this also characterizes the value of the derivative. And notice that we have not specialized the type of derivative in any way, so that the p.d.e. (5.67) should be applicable in a wide variety of settings, not just simple European calls and puts, by adjusting the boundary conditions.

The following interesting alternative approach, following Buchen [3], takes the point of view of replicating portfolios very directly. Consider a continuously rebalanced portfolio, which at time t consists of δ_t shares of the underlying asset and β_t units of the risk-free asset. For concreteness this risk-free asset may be a zero-coupon bond with face value 1 expiring at time T, whose value at time t would therefore satisfy:

$$P(t) = 1 \cdot e^{-r(T-t)} \implies \frac{dP}{dt} = r \cdot P(t) \implies dP(t) = r \cdot P(t) \cdot dt. \tag{5.68}$$

The portfolio values at time t and at a time $t + \Delta t$ shortly later are:

$$\begin{aligned} V(X(t),t) &= \delta_t X(t) + \beta_t P(t); \\ V(X(t + \Delta t), t + \Delta t) &= \delta_t X(t + \Delta t) + \beta_t P(t + \Delta t). \end{aligned} \tag{5.69}$$

The approximate change in portfolio value during $(t, t + \Delta t)$ would be:

$$\begin{aligned} V(X(t + \Delta t), t + \Delta t) - V(X(t),t) &\approx \delta_t(X(t + \Delta t) - X(t)) \\ &\quad + \beta_t(P(t + \Delta t) - P(t)) \\ &\approx \delta_t(X(t + \Delta t) - X(t)) \\ &\quad + \beta_t(r \cdot P(t)) \cdot \Delta t. \end{aligned} \tag{5.70}$$

Passing to the limit as $\Delta t \longrightarrow 0$, we obtain the infinitesimal form:

$$dV_t(X(t),t) = \delta_t dX(t) + r\beta_t P(t) dt. \tag{5.71}$$

On the other hand, Ito's Lemma yields an alternative form for dV_t:

$$dV_t(X(t),t) = V_x(X(t),t)dX(t) + \left(V_t(X(t),t) + \frac{1}{2}\sigma^2 X(t)^2 V_{xx}(X(t),t) \right) dt. \tag{5.72}$$

In order for the $dX(t)$ and dt terms to match in formulas (5.71) and (5.72), we must have:

$$\delta_t = V_x(X(t), t) \text{ and } r\beta_t P(t) = V_t(X(t), t) + \frac{1}{2}\sigma^2 X(t)^2 V_{xx}(X(t), t). \quad (5.73)$$

Solving for $\beta_t P(t)$ in equation (5.69) and substituting into the second equation in (5.73) yields:

$$r\left(V(X(t), t) - \delta_t X(t)\right) = V_t(X(t), t) + \frac{1}{2}\sigma^2 X(t)^2 V_{xx}(X(t), t)$$
$$\Longleftrightarrow r\left(V(X(t), t) - X(t) \cdot V_x(X(t), t)\right) = V_t(X(t), t) + \frac{1}{2}\sigma^2 X(t)^2 V_{xx}(X(t), t). \quad (5.74)$$

Isolating V_t and using the condition $X_t = x$ results in the Black-Scholes equation (5.67). But we have additional information from formula (5.73) (once the value function V is solved for) that characterizes the corresponding replicating portfolio (δ_t, β_t) for the derivative at time t with underlying asset price $X(t)$.

5.3.2 Boundary Conditions; Solving the PDE

The Black-Scholes partial differential equation (5.67) has the following boundary condition for a European call option with strike price E, expiring at time T:

$$V(x, T) = \max(x - E, 0).$$

Similarly, the boundary condition for a put option is:

$$V(x, T) = \max(E - x, 0).$$

You are asked in Exercise 3 for the appropriate boundary conditions for a few other derivatives depending on the final price of the underlying asset at the expiration time T.

The authors, having shown that derivatives such as the call option value satisfy the PDE, proceed to establish the value function formula for the call by proving the following result.

Theorem 1. The solution of the Black-Scholes partial differential equation $V_t(x, t) = rV(x, t) - rxV_x(x, t) - \frac{1}{2} \cdot \sigma^2 x^2 \cdot V_{xx}(x, t)$ with boundary condition $V(x, T) = \max(x - E, 0)$ is:

$$V(x, t) = x \cdot \mathcal{N}(d_1) - E \cdot e^{-r(T-t)} \mathcal{N}(d_2), \quad (5.75)$$

where

$$
\begin{aligned}
d_1 &= \frac{\log(x/E) + \left(r + \sigma^2/2\right)(T-t)}{\sigma\sqrt{T-t}}, \\
d_2 &= \frac{\log(x/E) + \left(r - \sigma^2/2\right)(T-t)}{\sigma\sqrt{T-t}}.
\end{aligned}
\quad (5.76)
$$

How do we get to the formula for the call option value? The Black-Scholes p.d.e. $V_t(x,t) = rV(x,t) - rxV_x(x,t) - \frac{1}{2} \cdot \sigma^2 x^2 \cdot V_{xx}(x,t)$ belongs to a well-known class of equations called **heat equations**, and Black and Scholes pull a complex substitution trick to simplify the equation to the form $y_t = y_{xx}$ whose solution is known. We look at this trick in the proof of the next theorem.

Theorem 2. The Black-Scholes p.d.e. (5.67) may be reexpressed in the form $Y_t(u,s) = Y_{uu}(u,s)$ by the transformation:

$$V(x,t) = e^{-r(T-t)} \cdot Y(u(x,t), s(t)),$$

where

$$u = \tfrac{2}{\sigma^2} \left(r - \tfrac{1}{2}\sigma^2\right) \left(\log\left(\tfrac{x}{E}\right) + \left(r - \tfrac{1}{2}\sigma^2\right)(T-t)\right),$$
$$s = \tfrac{2}{\sigma^2} \left(r - \tfrac{1}{2}\sigma^2\right)^2 (T-t).$$

Proof. By the product rule and the chain rule:

$$V_t = r \cdot e^{-r(T-t)} Y(u,s) + e^{-r(T-t)} Y_u(u,s) \cdot \tfrac{\partial u}{\partial t} + e^{-r(T-t)} Y_s(u,s) \cdot \tfrac{\partial s}{\partial t};$$
$$V_x = e^{-r(T-t)} Y_u(u,s) \cdot \tfrac{\partial u}{\partial x};$$
$$V_{xx} = e^{-r(T-t)} Y_{uu}(u,s) \cdot \left(\tfrac{\partial u}{\partial x}\right)^2 + e^{-r(T-t)} Y_u(u,s) \cdot \tfrac{\partial^2 u}{\partial x^2}.$$

$$(5.77)$$

All terms in the Black-Scholes equation, V, V_t, V_x, and V_{xx} have common factor $e^{-r(T-t)}$, which will henceforward be dropped as we rewrite the equation with its new parameters Y, u, and s. The partial derivatives that we need in Equations (5.77) are:

$$\tfrac{\partial u}{\partial t} = -\tfrac{2}{\sigma^2}\left(r - \tfrac{1}{2}\sigma^2\right)^2; \quad \tfrac{\partial s}{\partial t} = -\tfrac{2}{\sigma^2}\left(r - \tfrac{1}{2}\sigma^2\right)^2;$$
$$\tfrac{\partial u}{\partial x} = \tfrac{2}{\sigma^2}\left(r - \tfrac{1}{2}\sigma^2\right) \cdot \tfrac{1}{x}; \quad \tfrac{\partial^2 u}{\partial x^2} = -\tfrac{2}{\sigma^2}\left(r - \tfrac{1}{2}\sigma^2\right) \cdot \tfrac{1}{x^2}$$

$$(5.78)$$

The left side $V_t(x,t)$ of the Black-Scholes equation now becomes:

$$r \cdot Y(u,s) - \frac{2}{\sigma^2}\left(r - \frac{1}{2}\sigma^2\right)^2 Y_u(u,s) - \frac{2}{\sigma^2}\left(r - \frac{1}{2}\sigma^2\right)^2 Y_s(u,s). \quad (5.79)$$

This is equated to the right side $rV(x,t) - rxV_x(x,t) - \frac{1}{2} \cdot \sigma^2 x^2 \cdot V_{xx}(x,t)$, which in the new parameter set is:

$$r \cdot Y(u,s) - rxY_u(u,s) \cdot \tfrac{2}{\sigma^2}\left(r - \tfrac{1}{2}\sigma^2\right) \cdot \tfrac{1}{x}$$
$$- \tfrac{1}{2} \cdot \sigma^2 x^2 (Y_{uu}(u,s) \cdot \left(\tfrac{2}{\sigma^2}\left(r - \tfrac{1}{2}\sigma^2\right) \cdot \tfrac{1}{x}\right)^2 \quad (5.80)$$
$$+ Y_u(u,s) \cdot \left(-\tfrac{2}{\sigma^2}\left(r - \tfrac{1}{2}\sigma^2\right) \cdot \tfrac{1}{x^2}\right)).$$

All of the x factors divide away, and regrouping what remains gives:

$$
\begin{aligned}
& r \cdot Y(u,s) - r \cdot Y_u(u,s) \cdot \tfrac{2}{\sigma^2}\left(r - \tfrac{1}{2}\sigma^2\right) - \tfrac{1}{2} \cdot \sigma^2 (Y_{uu}(u,s) \cdot \left(\tfrac{2}{\sigma^2}\left(r - \tfrac{1}{2}\sigma^2\right)\right)^2 \\
& + Y_u(u,s) \cdot \left(-\tfrac{2}{\sigma^2}\left(r - \tfrac{1}{2}\sigma^2\right)\right) \\
& = r \cdot Y(u,s) - \left(r \cdot \tfrac{2}{\sigma^2}\left(r - \tfrac{1}{2}\sigma^2\right) + \tfrac{1}{2} \cdot \sigma^2 \cdot \left(-\tfrac{2}{\sigma^2}\left(r - \tfrac{1}{2}\sigma^2\right)\right)\right) Y_u(u,s) \\
& \quad -\tfrac{2}{\sigma^2}\left(r - \tfrac{1}{2}\sigma^2\right)^2 Y_{uu}(u,s) \\
& = r \cdot Y(u,s) - \tfrac{2}{\sigma^2}\left(r - \tfrac{1}{2}\sigma^2\right)^2 Y_u(u,s) - \tfrac{2}{\sigma^2}\left(r - \tfrac{1}{2}\sigma^2\right)^2 Y_{uu}(u,s).
\end{aligned}
\tag{5.81}
$$

We now equate (5.79) and (5.81), noticing right away that each has the term $r \cdot Y(u,s)$ which may be dropped:

$$
\begin{aligned}
& -\tfrac{2}{\sigma^2}\left(r - \tfrac{1}{2}\sigma^2\right)^2 Y_u(u,s) - \tfrac{2}{\sigma^2}\left(r - \tfrac{1}{2}\sigma^2\right)^2 Y_s(u,s) \\
& = -\tfrac{2}{\sigma^2}\left(r - \tfrac{1}{2}\sigma^2\right)^2 Y_u(u,s) - \tfrac{2}{\sigma^2}\left(r - \tfrac{1}{2}\sigma^2\right)^2 Y_{uu}(u,s).
\end{aligned}
\tag{5.82}
$$

The Y_u terms also may be subtracted away, and the common coefficients $-\tfrac{2}{\sigma^2}$ and $\left(r - \tfrac{1}{2}\sigma^2\right)^2$ may be divided away, resulting in the heat equation $Y_s = Y_{uu}$. ∎

Black and Scholes then state the boundary condition for a call option in the new parameter set Y, u, s. For your reference it is below, and the interested reader can step through Exercise 5 to verify it.

Theorem 3. The value function of the European call option with termination time T and strike price E has the following initial condition in the transformed system of Theorem 2:

$$
Y(u,0) = E \cdot \left(\exp\left(\frac{\sigma^2/2}{\left(r - \tfrac{1}{2}\sigma^2\right)} u \right) - 1 \right), u \geq 0. \ \blacksquare
$$

The last necessary step for the authors is to bring in a known result from partial differential equations for the simple heat equation $Y_s = Y_{uu}$ with boundary condition as in Theorem 3, solve for the function Y, and then invert to get the value V of the call option. We are not so much interested here in the theory of differential equations, so we will end this discussion here.

We have been mostly concerned with the mathematics of the option valuation problem. But we would be remiss if we did not mention the economic assumptions that make the Black-Scholes derivation work, above and beyond the stochastic process model of geometric Brownian motion with constant variance parameter. These include:

(1) The non-risky rate of return r remains constant, and is known;

(2) The underlying asset pays no dividends;

(3) There are no transactions costs for transacting in either the stock

or the option;

(4) One may buy or short sell any fraction of shares of a security;

(5) Short-selling incurs no penalties.

The subtle part of what we have done in this section is to work as if rebalancing transactions can take place continually and without costs, time delays, restrictions about discrete numbers of shares, or resistance of any kind, which accounts for most of these necessary assumptions. As you might imagine, subsequent work has tried to weaken some of them, and the persistent reader can track down a wealth of references. In the next section, we will focus on possible violations of the geometric Brownian motion model assumption.

Exercises 5.3

1. In the Black-Scholes solution expressed in formulas (5.75) and (5.76), show that

$$\lim_{t \to T} d_1 = \begin{cases} +\infty & \text{if } x > E \\ 0 & \text{if } x = E \\ -\infty & \text{if } x < E \end{cases} \quad \text{and} \quad \lim_{t \to T} d_2 = \begin{cases} +\infty & \text{if } x > E \\ 0 & \text{if } x = E \\ -\infty & \text{if } x < E \end{cases}$$

2. Assuming that the call option value function satisfies the Black-Scholes p.d.e. (5.67), check that the put option value also does. (Hint: use the put-call parity formula applied at time t, which is $P(X_t, t) = C(X_t, t) - X_t + e^{-r(T-t)} E$.)

3. What are the boundary conditions in the Black-Scholes equation for:
   (a) A derivative that returns a fixed amount of money C only if the underlying asset reaches a price X_T less than a strike price E at time T;
   (b) A derivative that returns a share of the underlying asset only if its price X_T exceeds a strike price E at time T;
   (c) A future expiring at time T with price E agreed upon for the buyer and seller in the future;
   (d) A binary derivative which pays an amount c_1 if $X_T \geq E_1$ or an amount c_2 if $X_T \leq E_2$ and otherwise pays 0. (Assume that $E_1 > E_2$.)

4. Refer back to the discussion of hedging using a portfolio of asset and option in Sections 1.6 and 3.1. (For instance, display (1.61)). Relate hedging in the discrete case to the Black and Scholes version of hedging expressed in (5.61) for the continuous case.

5. Verify the formula in Theorem 3 for the initial condition for the value function of the call option in the Black and Scholes pde. To do so, first invert the expressions that relate u to x and t, and s to t, to get:

$$x = E \cdot \exp\left(\frac{\sigma^2/2}{\left(r - \frac{1}{2}\sigma^2\right)} u - \frac{\sigma^2}{2}s\right); \quad t = T - \frac{\sigma^2}{2\left(r - \frac{1}{2}\sigma^2\right)^2}s$$

Then, decide what is implied about s and u by the original boundary conditions at time T.

5.4   Checking Black-Scholes Assumptions

The model for the motion of the prices of underlying assets that we have been using in this chapter is, of course, the stochastic differential equation:

$$S_t = \mu S_t dt + \sigma S_t dZ_t, S_0 = s_0. \tag{5.83}$$

The solution, we know, is geometric Brownian motion with parameters μ and σ, that is:

$$S_t = s_0 e^{(\mu - \sigma^2/2)t + \sigma Z_t}, \tag{5.84}$$

where Z_t is standard Brownian motion. If this model is correct, then the logged process whose t^{th} value is given by $Y_t = \log(S_t)$ satisfies:

$$\begin{aligned} Y_t &= \log(s_0) + \left(\mu - \sigma^2/2\right)t + \sigma Z_t \\ \implies Y_{t+s} - Y_t &= \left(\mu - \sigma^2/2\right)s + \sigma\left(Z_{t+s} - Z_t\right). \end{aligned} \tag{5.85}$$

The stationary, independent increments property of standard Brownian motion would imply that for time increments $s = 1$, the differences

$$\begin{aligned} D_1 &= Y_1 - Y_0 &= \log(S_1) - \log(s_0); \\ D_2 &= Y_2 - Y_1 &= \log(S_2) - \log(S_1); \\ D_3 &= Y_3 - Y_2 &= \log(S_3) - \log(S_2); \\ &\qquad\qquad \vdots \end{aligned} \tag{5.86}$$

would be independent, identically distributed normal random variables with mean $\mu - \sigma^2/2$ and variance $\sigma^2 \cdot 1 = \sigma^2$.

It is appropriate to close this chapter and this book by doing some sample data analysis to check (1) the normality of the differences in log prices, (2) the constancy of the mean and variance parameters, and (3) the independence of the differences in log prices. As we do so, we can make some qualitative observations about the consequences of violations of these assumptions to the financial theory that we have built. The problem set follows up with a few quantitative questions involving such violations in simple settings.

5.4.1 Normality of Rates of Return

The subsection heading above is concise and memorable, but slightly off of what we will actually do. A one period rate of return on the underlying asset is actually defined as:

$$R_t = \frac{S_t - S_{t-1}}{S_{t-1}} = \frac{S_t}{S_{t-1}} - 1,$$

so that $1 + R_t = S_t/S_{t-1}$. In the Black-Scholes price model above, we suppose normality of the differences of logs:

$$\log(S_t) - \log(S_{t-1}) = \log\left(\frac{S_t}{S_{t-1}}\right) = \log\left(1 + \frac{S_t}{S_{t-1}} - 1\right) = \log(1 + R_t).$$

But the rates of return R_t are going to be rather small magnitude numbers near zero. A first order Taylor approximation $\log(1+x) \approx x$ around 0 suggests that:

$$\log(S_t) - \log(S_{t-1}) = \log(1 + R_t) \approx R_t.$$

Therefore, we can speak informally about the normality of the differences of logs as normality of the asset rates of return.

In the rest of the section we will use as examples two assets on the New York Stock Exchange in two different time periods: Procter & Gamble (ticker symbol PG) in the four years extending from January of 2015 through December of 2018, and Target (symbol TGT) from 2018 through 2021. The price data and ensuing calculations were found using *Mathematica*'s FinancialData command. The details are in the electronic version of this text. For reference and for use in problems, we display the differences in logs.

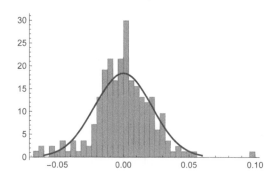

FIGURE 5.5 Procter & Gamble weekly rates, 2015-2018.

−.0021, .0110, −.0129, −.0664, .0155, .0034, −.0121, .0031,
−.0294, −.0101, .0349, −.0291, .0015, .0111, −.0099, −.0187,
−.0088, .0083, .0011, −.0137, −.0197, −.0123, .0184, .0210,
−.0150, .00741, .0127, .0158, −.0240, −.0457, −.0160, .0019,
−.0513, −.0088, −.0350, −.0050, .0220, .0383, −.0035, .0280,
.0056, .0280, −.0085, −.0107, −.0215, .0248, −.0016, .0277,
−.00064, .0045, .0210, −.00477, −.0443, −.0131, .0312, .0545,
−.0060, −.0026, .0098, −.0085, .0290, −.0211, .0170, −.0031,
.0077, −.0040, −.0109, −.0165, −.0103, .0248, −.0110, −.0150,
.0175, .0127, .0088, −.00084, −.0105, .0302, .0116, .0028,
−.0034, −.0015, .0022, .0146, .0031, .0031, .0071, −.0225,
.0208, −.0033, .0224, .0028, −.0176, −.0475, .0293, −.0205,
−.0178, −.0191, .0176, −.0128, .0236, .0037, .0033, −.0104,
.0112, −.0121, .0401, −.0084, .0079, .0064, .0349, −.00044,
−.0061, .0063, −.00077, −.0047, −.0080, −.0069, .0089, −.0158,
−.0147, −.0096, −.0036, .00058, .0116, .0152, −.0049, .0169,
−.0027, −.0257, .0057, −.0063, .0172, .0179, .0051, .0074,
.0123, .00043, .00022, .0033, .0046, −.0111, −.0138, .0147,
.0077, −.0529, −.0138, −.0053, .0181, .0031, .00023, .0214,
.00011, .0167, .0026, −.0027, −.0070, −.0180, .0162, −.0374,
−.0405, −.0528, .0330, −.0189, −.0193, .0099, −.0166, −.0395,
.0434, −.0108, −.00077, −.0601, −.0135, −.0052, .0129, .0011,
.0116, −.0116, .0495, .0026, .00065, .0081, .0159, 0, −.0080,
.0239, .0215, −.0110, .0274, −.0040, −.0049, −.0126, .0205,
.0261, −.0306, −.0131, −.0383, .0991, .0064, .0220, .0285,
.0151, −.0246, .0319, −.0220, .0443, −.0605, .0023, .0081

For the Procter & Gamble data, Figure 5.5 plots a histogram of 209 successive weekly differences of logs (which we will hereafter call the "difflogs" for short) as in equation (5.86), together with a normal density chosen to have the best fit in terms of the sample mean of about .0000777 and sample standard deviation of .0217. You can observe a generally bell-shaped, symmetric distribution of these approximate rates of return; however there is a problem that is commonly observed when these analyses are performed. The peak of the histogram in the middle near 0 is too high. This occurs because the sample standard deviation accommodated the highest and lowest values in the data so that just two points, −0.0664 and 0.0991 were outside a 3-standard deviation interval about the mean, which turns out to be (−0.0650, 0.0651). This serves to flatten the normal density in the middle in order to have longer tails.

The economic problem could be one of two things, or a combination of both. Either there are just more inactive trading weeks with little price change than a normal model can anticipate, or there are more unusually high or low rates of return than are predictable by the normal model.

Next let us look at Target in a similar way, for weekly observations extending from 2018 through 2021. There are 208 differences in the data set, a histogram of which is in Figure 5.6, with a superimposed normal density scaled to have mean and standard deviation equal to the sample mean of about .0060 and standard deviation of .0431. A 3-standard deviation interval about the mean comes out to about $(-0.1232, 0.1352)$, and this time there are a rather unusual number of five observations among the 208 that fall outside this interval: $-0.168, -0.128, 0.140, 0.143$, and 0.206. (See Exercise 1.) The peakedness phenomenon in the middle of the distribution is again apparent.

For later use, here is the difflog data for Target:

.1430, .0168, −.0148, −.0534, −.0069, .0439, −.0038, −.0035,
−.0640, .0071, −.0448, .0226, .0404, −.0107, −.0169, .0353,
−.0250, −.0113, .0779, −.0643, .0221, .0692, −.0099, −.0158,
.0011, .0088, .0120, .00052, .0300, .0163, .0154, .0040,
.0501, .0022, .0141, −.0091, −.0072, .0103, −.0427, .0011,
−.0311, −.00098, .0297, .0295, −.0872, −.1680, .0522, −.0454,
−.0095, −.0942, .0608, .0224, .0468, .0153, .0239, −.0170,
−.0041, .0273, −.0081, .0095, .0386, .0111, .0214, .0245,
.0084, −.0026, .0309, −.0766, −.0154, −.0173, −.0515, .1400,
−.0138, .0765, .0108, −.0022, −.0114, .0232, −.0126, .0055,
−.0109, −.0657, .0109, .0216, .2060, .0337, .0259, −.0171,
−.0123, −.0052, .0277, .0247, .0087, −.0342, −.0111, .0214,
.0274, .1150, −.0160, .00016, .0174, .0151, .00047, −.0350,
−.00024, −.0647, −.0225, −.0318, .0430, .0088, .0033, −.1280,
.0282, −.0476, −.0365, −.0277, −.0232, .1180, .0848, −.0506,
.0038, .0679, .0432, −.0289, .0404, −.0066, −.0388, .0308,
−.0298, .0178, −.0022, .0234, .0166, .0174, .0456, .0356,
.1180, −.0183, −.0243, .0029, .0079, .0365, .0303, .0286,
.0074, −.0289, −.0513, .0592, .0085, .0563, .0424, −.0267,
−.0192, −.0075, .0276, .0076, .0923, .0062, −.0149, −.0576,
.0416, .0135, −.0137, −.0289, −.0609, .0420, .0410, .0690,
−.0012, .0229, .0154, −.0123, .0061, .0353, −.0167, .0639,
.0081, .0193, .0069, −.0104, .0425, .0248, .0081, .0103,
.0386, .000077, −.0037, .0056, −.0316, −.0168, −.0197, .00094,
.0087, −.0214, −.0554, .0021, .0709, .0408, .0143, −.0208,
.0224, −.0365, −.0166, .00405, −.0387, −.0645, −.0104, .0461

Perhaps there are distributional models other than the normal which could help to account for the discrepancies that we tend to observe for these individual stocks. We would like to model heavier tails and a more peaked middle than in the normal distribution.

The first alternative that we can try is the t-distribution with ν degrees of freedom. Its density function is taller in the middle than the normal density and has heavier tails. But we need the generalized **non-central t-distribution**

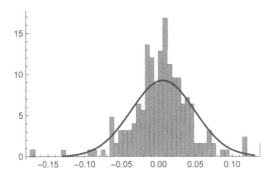

FIGURE 5.6 Target weekly rates, 2018-2021.

with location parameter μ and scale parameter σ, whose density function is below.

$$f(x; \mu, \sigma, \nu) = \frac{\left(\frac{\nu}{\nu + \frac{(x-\mu)^2}{\sigma^2}}\right)^{\frac{1+\nu}{2}}}{\sqrt{\nu}\sigma B\left(\frac{\nu}{2}, \frac{1}{2}\right)}, \tag{5.87}$$

where the beta function in the denominator is the standard mathematical function:

$$B(\alpha, \beta) = \int_0^1 t^{\alpha-1}(1-t)^{\beta-1}dt. \tag{5.88}$$

As long as the degree of freedom parameter $\nu > 1$, the mean of the distribution exists and equals μ. Also, as long as $\nu > 2$, the variance exists and is:

$$\text{Var}(X) = \frac{\nu \cdot \sigma^2}{\nu - 2}.$$

We need to calibrate the rate of return data to a t-distribution with appropriate parameters. It is not appropriate to delve deeply into the theory of statistical estimation here, but only to put into play a couple of standard techniques. Probably the simplest estimation method is the **method of moments**, where the moments of the data (sample mean and variance) are equated to the distributional mean and variance in order to solve for estimates of the parameters of the distribution. For the t-distribution model, we would have:

$$\mu \approx \bar{X}, \text{and } S^2 \approx \frac{\nu \cdot \sigma^2}{\nu - 2} \implies \sigma^2 \approx \frac{(\nu - 2) \cdot S^2}{\nu}. \tag{5.89}$$

But that still leaves us the problem of finding the best degree of freedom parameter ν. Complicated criteria can be developed, for instance finding the value of ν that minimizes a chi-square goodness-of-fit statistic using chosen class intervals for the data, but here we just experiment with a couple of graphs.

The smaller the degrees of freedom, the more peaked the density graph, so let us try two possible values of ν, 4 and 8, to see which better captures the histogram for the Procter & Gamble data. Recall that the sample mean of the difflogs was about .0000777, and the sample standard deviation was .0217. Then our parameter estimates would be:

$$\mu \approx \bar{X} \approx .0000777, \text{and } \sigma^2 \approx \frac{(\nu - 2) \cdot (.0217)^2}{\nu} \implies \sigma \approx .0217 \sqrt{\frac{(\nu - 2)}{\nu}}.$$

In Figure 5.7 we display the histogram, the normal density approximation, and the non-central t-densities with 4 and 8 degrees of freedom. Both t-distributions capture the peakedness in the middle better than normal. The $t(4)$ density may be a bit too peaked to properly account for the extreme data in the tails. Visually, the $t(8)$ distribution appears to be a good compromise. Exercise 2 asks you to do similar experiments with the Target data.

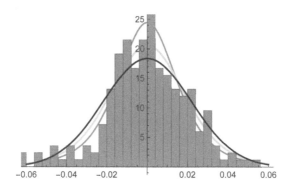

FIGURE 5.7 Procter & Gamble data and distributions (normal: black, $t(4)$: dark gray, $t(8)$: light gray).

The t-density with 8 degrees of freedom may look like a better fit to the data than the normal distribution, but is there a way of quantifying the improvement?

To check this, we can compute chi-square goodness-of-fit statistics for both distributions. There are several ways to proceed. We will divide the data set into groups according to the deciles of the respective distributions, and then tally the frequencies in those ten groups. Since there are 209 data points, the expected frequencies per group under the null hypothesis that the data came from the distribution would each be 20.9. The chi-square statistic is computed as:

$$Q = \sum_{i=1}^{10} \frac{(\text{observed frequency} - \text{expected frequency})^2}{\text{expected frequency}}. \tag{5.90}$$

This test statistic is generally used in a formal statistical test, but can also be viewed as a quantitative measure of how well the distribution giving rise to the frequencies fits the data.

The 10th, 20th, 30th, ... , 90th percentiles are listed in the table below for the $t(8)$ and normal distributions, with parameters μ and σ estimated from the data.

decile	1	2	3	4	5	6	7	8	9
$t(8)$	−.026	−.017	−.010	−.005	.000	.005	.010	.017	.026
normal	−.028	−.018	−.011	−.005	.000	.006	.011	.018	.028

The nine cutoff points divide the data into 10 categories. The percentiles are quite similar, although some differences in frequency counts do arise, as shown in the next table.

decile	1	2	3	4	5	6	7	8	9	10
$t(8)$	17	16	29	20	21	26	19	19	22	20
normal	17	13	23	24	26	27	21	23	17	18

The values of the chi-square statistic Q in formula (5.90) are 6.74 and 8.75, respectively, for the $t(8)$ and normal distributions. This indicates that the t-distribution is the better fit to the data than normal. If we were to conduct a formal hypothesis test, though, the p-values come out to be .456 for the $t(8)$ distribution and .271 for normal, so neither distribution can be rejected at reasonable levels. Perhaps using more class intervals or a different goodness-of-fit test such as the Kolmogorov-Smirnov test might provide better statistical evidence to back up the graphical evidence.

Exercises 2 and 3 ask you to do a parallel analysis for the Target data.

Another probability distribution holds promise as a model for the distribution of the differences of logs. This is the **Laplace distribution**, whose density function is in formula (5.91) and for which a typical graph is in Figure 5.8.

$$f(x; \mu, \beta) := \frac{1}{2\beta} E^{\left(\frac{-|x-\mu|}{\beta}\right)} \tag{5.91}$$

Much like the mean and standard deviation parameters for the normal distribution, the parameter μ dictates the center of symmetry of the density, and β determines the spread. In this figure, the normal(0, 1) density is superimposed on the graph of the Laplace(0, 1) density, and we see that, like the t-distribution, the Laplace distribution focuses more weight in the center, and has fatter tails than normal. This is a desirable characteristic for modeling the kind of difflog data in our situation.

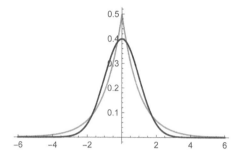

FIGURE 5.8 Laplace distribution (gray) with location parameter $\mu = 0$ and scale parameter $\beta = 1$.

The mean and variance of the Laplace(μ, β) distribution are:

$$E[X] = \mu, \ \text{Var}(X) = 2\beta^2. \tag{5.92}$$

You are asked to verify the mean in Exercise 4. Thus, the method of moments estimators of the parameters are given by:

$$\mu \approx \bar{X}, \text{and } 2\beta^2 \approx S^2 \Longrightarrow \beta \approx S \big/ \sqrt{2}.$$

Next, let us fit a Laplace distribution to the Target difflog data. The sample mean and standard deviation were .0060 and .0431 respectively, so our parameter estimates are:

$$\mu \approx .0060, \text{and } \beta \approx S \big/ \sqrt{2} = .0431 \big/ \sqrt{2} \approx .0305.$$

The Target histogram with the associated normal density and the Laplace density with these parameters are shown is Figure 5.9. Neither density is able to account for the very extreme observations on the two tails, but the Laplace density appears to fit the rest of the data far better than the normal density.

Let's look at how the chi-square statistics compare for the two distributions. Stepping through the same procedure as above with the Procter & Gamble data and the t-distribution, we first find the decile dividing points for the approximate normal and Laplace distributions:

decile	1	2	3	4	5	6	7	8	9
Laplace	−.043	−.022	−.010	−.001	.006	.013	.022	.034	.055
normal	−.049	−.030	−.017	−.005	.006	.017	.029	.042	.061

There were 208 observations in the Target data; hence with 10 equally likely categories, the expected frequency counts are each 20.8. The actual frequency counts for both distributions are in the next table.

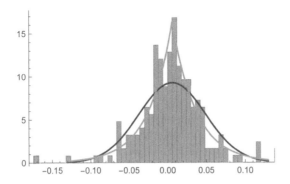

FIGURE 5.9 Target data and distributions (normal: black, Laplace: gray).

decile	1	2	3	4	5	6	7	8	9	10
Laplace	20	20	30	17	17	22	16	26	22	18
normal	17	13	22	26	26	32	25	19	13	15

Notice that the actual observations are high in the middle for the normal approximation and low on the ends. The chi-square statistics for the two distributions are 8.44 for Laplace and 17.87 for normal. So Laplace performs noticeably better than normal, and in fact a statistical test of goodness-of-fit results in p-values for Laplace of .295 and for normal of .013; hence normality would actually be rejected at reasonable levels.

All of this hard work has led to the conclusion that normality of the differences in logs of prices is in some doubt. Large-scale, deeper investigations of the normality assumption would be well worth conducting.

In terms of investor behavior, all three of these models tend to underestimate the frequency of very large positive or negative rates of return, although both the t-distribution and the Laplace distribution do a better job of predicting the frequency of small magnitude rates. A risk-preferring investor hoping to cash in on an investment in a limited time window would be disappointed by how often the asset's price stays relatively stable. A conservative investor who holds an asset for a long time and who believes that he is relatively safe from a sudden large scale decline would also be unpleasantly surprised when such a decline does occur. For derivative valuation, a normal model could overestimate the probability that the derivative ends in the money if there is not much time remaining and the normal model underestimates how often the asset price stays stable. For a longer term option, because of the occasional large bumps in the underlying asset, actual volatility may in fact be underestimated by a normal model and the probability that the option is in the money may correspondingly be underestimated. Some of these consequences are studied in narrow situations in Exercises 7 and 8.

5.4.2   Stability of Parameters

Another key assumption in the Black-Scholes model for the motion of the underlying asset is that the drift rate and volatility, μ and σ, remain constant, at least through the lifetime of the derivative. Do we have a right to expect that this is true?

Tools from elementary statistics provide one way of testing this. Consider the parameter μ first. If the model is correct, the differences in successive logs would be i.i.d normal random variables with mean $\mu - \sigma^2/2$ and variance $\sigma^2 \cdot 1 = \sigma^2$. Suppose that we split the data into groups according to the block of time in which the data values were obtained. For the Procter & Gamble example, we have four years' worth of weekly difflog data, with one extra week totalling 209 observations. We could split the data into four sets, one per year, and do an analysis of variance F-test to check for equality of means for the four groups. Assuming for the moment that the variance doesn't change, equality of the difflog means of $\mu_i - \sigma^2/2$ would be equivalent to equality of the μ_i themselves.

The manipulations and statistical calculations can be done in any statistics package. The ANOVA table used to test the null hypothesis $H_0 : \mu_1 = \mu_2 = \mu_3 = \mu_4$ is below, obtained from *Mathematica* as described in the electronic version of the text. Since the F-ratio is very small and the p-value is large, it would be a serious mistake to reject the null hypothesis. As far as this asset is concerned, for this time frame, the drift rate does not seem to change from one year to another.

	DF	SumOfSq	MeanSq	FRatio	PValue
Model	3	.000540	.000180	.377176	.7696
Error	204	.097403	.000477		
Total	207	.097843			

We could further explore the data by wondering if there are significant changes to the drift rate over shorter time periods, such as 6 months. For this, we could take the first two years of data and divide into four equal half-year data sets of size 26 each, for example. Doing this results in the ANOVA table below. Again the p-value is very large, so we do not have evidence against the hypothesis that the drift rate is the same over the first four half-year periods.

	DF	SumOfSq	MeanSq	FRatio	PValue
Model	3	.000705	.000235	.561385	.64174
Error	100	.041849	.000418		
Total	103	.042554			

Turning to the volatility, we could be content with comparing the variances of two groups using the classical F-ratio test from elementary statistics, but to parallel what we did above for means, we could carry out more advanced tests such as Bartlett's test or Levene's test for equality of several variances.

Bartlett's test, which assumes normality of the group data, is based on a complicated test statistic whose value is large when the group sample variances differ substantially from the pooled variance of the full sample. The test statistic has a χ^2-distribution under the null hypothesis of equal variances. Levene's test is recommended when the analyst is not willing to assume normality of the group data. Its test statistic is large when the absolute deviations of group data from the group mean deviation differ greatly from the analogous absolute deviation for the full pooled data set. Under the null hypothesis, the Levene statistic has an F-distribution.

As a sample analysis, we split the latter pair of years for the Procter & Gamble difflogs into four half-year data sets. We show below a table of p-values obtained from *Mathematica* for the tests described above, as well as two others that are in common use.

	Statistic	p − value
Bartlett	21, 4578	.0000846
Levene	5.31646	.0020700
Brown − Forsythe	18.0244	.0019302
Conover	5.25927	.0004348

All four tests resoundingly reject the null hypothesis of equal variance. The sample variances for the four periods come out to be about .000217, .000221, .000719, and .000972 respectively, which indicates a trend toward increasing volatility with time. This is supported by the time series graph of these difflogs shown in Figure 5.10.

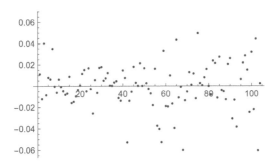

FIGURE 5.10 Time series graph of P & G difflogs, 2017-2018.

We can do a quick F-test for equality of two variances using the information we have. For the first year the sample variance of difflogs for Procter & Gamble was $S_1^2 = .000217$, and for the fourth year the variance was $S_4^2 = .000972$. The sample sizes were both 52, and the classical F-statistic is:

$$F = \frac{S_4^2}{S_1^2} = \frac{.000972}{.000217} = 4.479,$$

and the statistic has the $F(51,51)$-distribution under the null hypothesis of equal population variances (and the assumption of normality of the samples). The p-value is very tiny, on the order of 10^{-7}, so it appears very likely that the year 4 variance is significantly larger than the year 1 variance.

Of course this analysis only considers one risky asset. You are asked in Exercises 9-10 to work with the Target data set similarly. But our analysis of Procter & Gamble alerts us to the fact that for a longer time period such as two years, we can see significant changes in the parameters.

What are the consequences of such changes to the financial theory? If a European option was originally purchased with a price that assumed a certain volatility, but that volatility changed sometime during the lifetime of the option, then the probability that the option ends in the money is not what was expected. Considering the case of the issuer of a put option, if the volatility increased, then the chance that the issuer is forced to execute the agreement by the holder is greater, to his detriment. We have been talking in the context of derivative valuation, but in a portfolio problem, if either μ or σ change prior to time 1, then the mean and variance of the rate of return on the investment are not what the investor anticipated, or, stated alternatively, the optimal portfolio that was originally selected is no longer optimal. Exercises 11-13 investigate some of these issues in specific situations.

5.4.3 Independence of Rates of Return

The last issue that we would like to study relative to the assumptions in the Black-Scholes model is the independence of the series of rates of return, or differences in log prices.

A topic of longstanding debate in the finance community has been the "random walk hypothesis", which states that price changes and rates of return are simply unpredictable. Future changes are independent of the past. In our context, the Black-Scholes assumptions lead to the fact that the logged price process is a random walk, which implies that the successive differences in logs are independent of each other. This would be disappointing to analysts who would like to believe that past performance of an asset can predict future performance. So the assumption of independence of the difflogs is an important one.

Independence can be checked in a number of different ways. The theme is that independence implies a few more special conditions; if we test such a condition and the condition fails, then independence also fails. If the condition is not rejected, that does not mean that independence is true, just that we cannot reject it on the basis of that particular condition.

One condition that could be checked is whether successive difflogs D_i and D_{i+1} are significantly correlated. The idea is that if there is some form of dependence, it is likely to show up in adjacent observations rather than ones that are farther apart.

Let us use the Target data set again to illustrate. There were 208 observations, using which we can form the 207 pairs $(D_1, D_2),(D_2, D_3), \dots ,$ (D_{207}, D_{208}). Graphically, we can look at a plot of the pairs to visually assess independence of the first coordinate from the second. Analytically, we can compute the sample correlation coefficient R, and the $t(n-2)$-distributed test statistic $T = \sqrt{n-2} \cdot R / \sqrt{1 - R^2}$, rejecting the null hypothesis that the population coefficient $\rho = 0$ if T is too large and the p-value is too small.

Software tells us that the sample correlation is about .043, making the t-statistic come out to a rather small value of about .617. The hypothesis test yields a p-value of .269, which is not sufficient to reject the null hypothesis of zero correlation at any reasonable level of significance. The scattergram is shown in Figure 5.11, and it makes visually clear why the null hypothesis was not rejected.

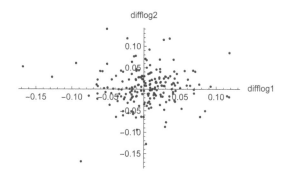

FIGURE 5.11 Scattergram of Target difflogs vs. lagged difflogs.

Once again, just because we were not able to reject the null hypothesis on this consequence of the independence assumption does not mean that we accept independence. Is there another way of looking at the data?

The active imagination might come up with a number of different conditions to check. For instance if, in contrast to independence, there was some behavior of the difflogs that was periodic by month (4 weeks), we could try a chi-square independence test on two factors: value of the time index mod 4, and the sign of the difflog.

The results are in the contingency table below. Except perhaps for week 3 in a month, the positives and negatives distribute themselves quite similarly for the other weeks, which suggests independence.

week\sign	negative	positive	total
0	22	30	52
1	21	31	52
2	20	32	52
3	25	27	52
total	88	120	208

The formal test is carried out with the chi-square statistic for factor independence:

$$Q = \sum_{\text{category}=1}^{8} \frac{(\text{observed frequency} - \text{expected frequency})^2}{\text{expected frequency}},$$

where the expected frequency in a cell is the row count times the column count for the cell divided by the overall count of 208. Since the row counts are each 52, each cell is expected to have $1/4$ of the observations in its column. In short, in each row the expected frequencies are 22 and 30. This implies that:

$$
\begin{aligned}
Q &= \frac{(22-22)^2}{22} + \frac{(21-22)^2}{22} + \frac{(20-22)^2}{22} + \frac{(25-22)^2}{22} \\
&\quad + \frac{(30-30)^2}{30} + \frac{(31-30)^2}{30} + \frac{(32-30)^2}{30} + \frac{(27-30)^2}{30} \\
&= 1.103.
\end{aligned}
$$

The degrees of freedom for the chi-square statistic are $(\#\text{rows}-1)(\#\text{columns}-1)$, which would be $3\cdot1$ here. The probability that such a statistic exceeds 1.03, that is, the p-value of the test, is about .776, so we cannot reject independence of the two factors. On the basis of this condition, we therefore cannot reject independence of the successive difflogs either.

If indeed the logged price process is not a random walk, the investor would want to know in what specific way it fails, and whether analytical models are possible that would allow for prediction of future price values. In the field of statistical time series there are more sophisticated processes called ARIMA (auto-regressive, integrated, moving average) processes that actually allow for forecasting based on past observations. As a part of this topic there are tests, both graphical and analytical, for possible time correlations of observations that are similar to the lag 1 autocorrelation that we investigated above. We do not want to venture so far into this more advanced field here, however, but the reader should keep it in mind as a highly relevant subject to pursue for finance.

Exercises 14-15 ask you to apply independence testing to the Procter & Gamble data.

Exercises 5.4

1. For the Procter & Gamble and Target data, given the means and standard deviations stated in the text, how many observations are expected to be outside a 3-standard deviation interval around the mean?

2. (Technology required) For the Target difflog data, experiment with a few possible values of the degree of freedom parameter to find a non-central t-distribution that provides a good fit. Graph several choices, together with the data histogram as in the section. Comment on your best choice.

3. (Technology required) Compute chi-square goodness-of-fit statistics for the Target difflog data for both the normal distribution and a non-central t-distribution with 5 degrees of freedom. As in the section, use the deciles to separate the data into 10 equally likely categories. Which distribution seems to fit the data better?

4. Verify that the mean of the Laplace distribution with parameters μ and β is μ.

5. The **method of maximum likelihood** is another common way of producing estimates of the parameters of a probability distribution based on a random sample $X_1, X_2, ..., X_n$ from the distribution. This method finds values of the parameters that maximize the product of factors $f(x_i; \{\text{parameters}\})$, where f is the density of the distribution and x_i are the observed values. The meaning of this is that we wish to find parameter values that maximize the joint density (likelihood) of the data that we actually observe. These parameter estimates "best explain" the observed data.

Show that the maximum likelihood estimates of the Laplace parameters μ and β are:

$$\hat{\mu} = \text{median of the } x_i's; \quad \hat{\beta} = \frac{1}{n} \sum_{i=1}^{n} |x_i - \hat{\mu}|.$$

(Hint: the parameters that maximize the likelihood are also the ones that maximize its logarithm. So set up and simplify the log of the product $f(x_1; \mu, \beta) \cdot f(x_2; \mu, \beta) \cdots f(x_n; \mu, \beta)$, and differentiate with respect to both μ and β.)

6. (Technology required) For the Procter & Gamble data, compare the values of chi-square goodness-of-fit statistics for normal and Laplace distributions with appropriate parameter estimates. Use the deciles of the two distributions to separate the data into 10 equally likely categories. Which distribution seems to fit the data better?

7. Suppose that we are willing to abandon the assumption that the difflogs $\log(S_t) - \log(S_{t-1}) = \log\left(\frac{S_t}{S_{t-1}}\right)$ are normally distributed, while keeping the assumptions that they are independent and identically distributed. Now consider a put option expiring in 52 weeks with strike price 50 on an asset currently at a price of 51. Past data show that the sample weekly mean difflog has been .002 and the sample standard deviation has been .01. Show that if parameters in a $t(4)$ model and a Laplace model are calibrated to the normal model by the method of moments, then there is no difference among the models with regard to the probability that the put finishes in the money. (Hint:

The Central Limit Theorem would still imply that the total of many of these difflogs is approximately normally distributed.)

8. (Technology required) Continuing Exercise 7, suppose that the option only has 1 week left until expiration, and that the current price of the underlying asset is 50.05. Compare the probabilities that the put finishes in the money for (a) a normal model; (b) a $t(4)$ model; and (c) a Laplace model, estimating parameters by the method of moments. Does the chosen distributional model matter now?

9. (Technology required) Using the Target difflog data and the methodology of the section, do an ANOVA study to test whether the means are the same for the first, second, third, and fourth years.

10. (Technology required) Using the Target difflog data and the methodology of the section, test whether the variances are the same for the first, second, third, and fourth years. If you see differences, perform an F-ratio test for equality of two variances that you suspect are different.

11. Consider an investor with risk aversion 10 who is interested in splitting wealth between a risky asset and a riskless asset whose continuous rate of return is .01. The investor models the motion of the asset as in Black-Scholes, that is, a geometric Brownian motion with parameters $\mu = .04$ and $\sigma = .08$. The investor will wait two time periods to cash in the portfolio.

    (a) Find the mean and variance of the rate of return, as approximated by difference of log price, on the risky asset.

    (b) Find the investor's optimal portfolio and mean and variance of the rate of return on the portfolio.

    (c) Suppose that unbeknownst to the investor, the risky asset changes parameters at time 1 to $\mu = .02$ and $\sigma = .16$. Find the mean and variance of the rate of return on the investor's optimal portfolio from part (b). Should the investor be disappointed?

12. Consider a call option on a risky asset with strike price 100 and expiration time 4 weeks. Suppose that the underlying asset has current price 98 and satisfies the Black-Scholes model with $\mu = .005$ and $\sigma = .01$.

    (a) Find the probability that the option ends in the money.

    (b) Now suppose that after week 2, with probability .1 instead of its usual week 2-week 3 transition, the asset rate of return experiences a shock that produces a difflog of exactly .006. Recalculate the probability that the call option finishes in the money.

13. Reconsider the Black-Scholes asset and call option of the previous exercise. Suppose now that at the end of week 2, the volatility rate σ triples to .03 and remains there until time 4. Compute the probability that the option is in the money upon expiration.

14. (Technology required) For the Procter & Gamble difflog data, working along the lines of the text, produce a scatterplot of lagged data vs. data using a lag of one time period, compute a sample correlation between the two, and carry out a test of significance. Do the same using a lag of two time periods. Is there any evidence of time dependence?

15. (Technology required) For the Proctor & Gamble data, do an independence check of the two factors of week within month and sign of the difflog, analogously to the text computations for Target.

16. Suppose that in checking the independence assumption we decided to use as factors the sign of the difflog and the quarter of the data (earliest times, times 25%-50% through the data, etc.) that the difflog appeared in. Is this only a test of the independence assumption, or is there another factor at play that could confound the results?

A

Multivariate Normal Distribution

A.1 Review of Matrix Concepts

To work with the joint normal distribution that we have in mind, vector and matrix notation and arithmetic are useful. Here is a very quick review of basic facts from matrix algebra.

We can think of a joint state of a group of n random variables $X_1, X_2, ..., X_n$, and the list of means of these random variables, as column vectors:

$$\boldsymbol{x} = \begin{pmatrix} x_1 \\ x_2 \\ \vdots \\ x_n \end{pmatrix}, \; \boldsymbol{\mu} = \begin{pmatrix} \mu_1 \\ \mu_2 \\ \vdots \\ \mu_n \end{pmatrix}. \tag{A.1}$$

Vector and matrix addition and subtraction are done componentwise, when the dimensions are compatible. Hence the vector $\boldsymbol{x} - \boldsymbol{\mu}$ makes sense, and it is a column vector whose i^{th} entry is $x_i - \mu_i$.

We know that the covariance structure of the random variables in a random vector can be stored in a variance-covariance matrix, which can also be written in the form:

$$\sum = \begin{pmatrix} \sigma_1^2 & \rho_{12}\sigma_1\sigma_2 & \cdots & \rho_{1n}\sigma_1\sigma_n \\ \rho_{12}\sigma_1\sigma_2 & \sigma_2^2 & \cdots & \rho_{2n}\sigma_2\sigma_n \\ \vdots & \vdots & \ddots & \vdots \\ \rho_{1n}\sigma_1\sigma_n & \rho_{2n}\sigma_2\sigma_n & \cdots & \sigma_n^2 \end{pmatrix}, \tag{A.2}$$

where σ_i is the standard deviation of X_i, and ρ_{ij} is the correlation between X_i and X_j.

Multiplication of matrices and vectors is well-defined if the dimensions are compatible. Multiplication of an $m \times p$ matrix A by a $p \times n$ matrix B results in an $m \times n$ matrix $C = A \cdot B$ whose row i column j entry c_{ij} is the dot product of row i of A with column j of B. The formula for this dot product is:

$$c_{ij} = \sum_{k=1}^{p} a_{ik} b_{kj}. \tag{A.3}$$

As special cases, a $1 \times p$ row vector $\boldsymbol{a}$ can multiply a $p \times n$ matrix B in this way to produce a $1 \times n$ row vector $\boldsymbol{a} \cdot B$, and a $p \times n$ matrix A can multiply an $n \times 1$ column vector $\boldsymbol{b}$ to produce a $p \times 1$ column vector $A \cdot \boldsymbol{b}$.

Recall also that the **inverse** of a square matrix A, written A^{-1} when it exists, is a square matrix of the same size such that:

$$A \cdot A^{-1} = I = A^{-1} \cdot A, \tag{A.4}$$

where I is an identity matrix of the same size as A. By an identity matrix we mean a matrix with 1s on its diagonal and 0s elsewhere. The inverse of a diagonal matrix D is the diagonal matrix whose entries are the reciprocals of the diagonal entries of D.

The **transpose** of a matrix A, written as A^t or A', is the matrix containing the same entries as A, but for which the rows of A become the columns of A^t, and hence the columns of A are the rows of A^t. The transpose of a row vector is a column vector, and vice versa. The transpose of a **symmetric matrix** (that is, a matrix such that $a_{ij} = a_{ji}$ for all pairs i, j) is the same as the matrix itself.

A common construction in probability and statistics and other fields is the **quadratic form**, which is a three-way product of a row vector, a square matrix, and a column vector as follows:

$$\boldsymbol{x}^t \cdot Q \cdot \boldsymbol{x} = \begin{pmatrix} x_1 & x_2 & \cdots & x_n \end{pmatrix} \begin{pmatrix} q_{11} & q_{12} & \cdots & q_{1n} \\ q_{21} & q_{22} & \cdots & q_{2n} \\ \vdots & \vdots & \ddots & \vdots \\ q_{n1} & q_{n2} & \cdots & q_{nn} \end{pmatrix} \begin{pmatrix} x_1 \\ x_2 \\ \vdots \\ x_n \end{pmatrix}. \tag{A.5}$$

When multiplied out, $Q \cdot \boldsymbol{x}$ is an $n \times 1$ column vector, and so $\boldsymbol{x}^t \cdot (Q \cdot \boldsymbol{x})$ is a 1×1 matrix, equivalent to a scalar value. In applications of quadratic forms, Q is usually a symmetric matrix as well, that is, $q_{ij} = q_{ji}$. For example, in a two-dimensional case we might have:

$$
\begin{aligned}
\begin{pmatrix} x_1 & x_2 \end{pmatrix} \begin{pmatrix} 1 & 2 \\ 2 & 4 \end{pmatrix} \begin{pmatrix} x_1 \\ x_2 \end{pmatrix} &= \begin{pmatrix} x_1 & x_2 \end{pmatrix} \begin{pmatrix} x_1 + 2x_2 \\ 2x_1 + 4x_2 \end{pmatrix} \\
&= x_1^2 + 2x_1x_2 + 2x_1x_2 + 4x_2^2 \\
&= x_1^2 + 4x_1x_2 + 4x_2^2 \\
&= (x_1 + 2x_2)^2.
\end{aligned}
$$

The reason for the term "quadratic form" is apparent.

Finally, the **determinant** of a square matrix is a real number associated with the matrix, whose computation is defined recursively. The determinant of a 1×1 matrix is the single entry in the matrix. The determinant of a 2×2 matrix is the quantity:

$$\det(A) = \det \begin{pmatrix} a & b \\ c & d \end{pmatrix} = ad - bc. \tag{A.6}$$

Notice that this is the first top row element a times the determinant of the submatrix (d) with the row and column of a deleted, minus the second top row element b times the determinant of the submatrix (c) with again the row and column of b deleted. The determinant of a 3×3 matrix continues that pattern:

$$
\begin{aligned}
\det(A) &= \det \begin{pmatrix} a & b & c \\ d & e & f \\ g & h & i \end{pmatrix} \\
&= a \cdot \det \begin{pmatrix} e & f \\ h & i \end{pmatrix} - b \cdot \det \begin{pmatrix} d & f \\ g & i \end{pmatrix} + c \cdot \det \begin{pmatrix} d & e \\ g & h \end{pmatrix}.
\end{aligned}
$$
$$(A.7)$$

For higher dimensional matrices, the determinant is again a sum, with alternating signs, of top row elements times the determinants of submatrices in which the row and column of the top row element are deleted. It is easy to show that the determinant of a diagonal matrix is the product of its diagonal elements. We will rarely have to work with matrices larger than 3×3.

A.2   Multivariate Normal Distribution

Definition. A vector of random variables $X = (X_1, X_2, ..., X_n)$ is said to have the **n-dimensional normal distribution** with mean vector μ and covariance matrix $\sum$ if its joint density is:

$$
f(x) = \frac{1}{(2\pi)^{n/2}\sqrt{\det(\sum)}} e^{\frac{1}{2}(x-\mu)^t \sum^{-1}(x-\mu)}. \quad\blacksquare
$$
$$(A.8)$$

Compare this formula to the familiar 1-dimensional version, which can be written somewhat pedantically as:

$$
f(x) = \frac{1}{(2\pi)^{1/2}\sqrt{\sigma^2}} e^{-\frac{1}{2}(x-\mu)\frac{1}{\sigma^2}(x-\mu)}.
$$
$$(A.9)$$

Since in the 1-dimensional case, $\sum = (\sigma^2)$, $\det(\sum) = \sigma^2$ and $\sum^{-1} = (1/\sigma^2)$, formula (8) specializes to formula (9) when $n = 1$.

In the $n = 2$ case, or the **bivariate normal density**, the density function would look like:

$$
f(x_1, x_2) = \cfrac{1}{2\pi \sqrt{\det \begin{pmatrix} \sigma_1^2 & \rho_{12}\sigma_1\sigma_2 \\ \rho_{12}\sigma_1\sigma_2 & \sigma_2^2 \end{pmatrix}}}
$$

$$
\cdot \exp\left[-\tfrac{1}{2} \cdot \begin{pmatrix} x_1 - \mu_1 & x_2 - \mu_2 \end{pmatrix} \begin{pmatrix} \sigma_1^2 & \rho_{12}\sigma_1\sigma_2 \\ \rho_{12}\sigma_1\sigma_2 & \sigma_2^2 \end{pmatrix}^{-1} \right.
$$

$$
\left. \cdot \begin{pmatrix} x_1 - \mu_1 \\ x_2 - \mu_2 \end{pmatrix} \right]
$$

(A.10)

For ease of notation in the two-variable case, drop the subscripts in ρ_{12} and simply write ρ . In the leading constant factor, the determinant of the covariance matrix is:

$$
\det \begin{pmatrix} \sigma_1^2 & \rho\sigma_1\sigma_2 \\ \rho\sigma_1\sigma_2 & \sigma_2^2 \end{pmatrix} = \sigma_1^2\sigma_2^2 - \rho^2\sigma_1^2\sigma_2^2 = \left(1 - \rho^2\right)\sigma_1^2\sigma_2^2. \tag{A.11}
$$

You can check that in general the inverse of a 2×2 matrix A is:

$$
A^{-1} = \begin{pmatrix} a & b \\ c & d \end{pmatrix}^{-1} = \frac{1}{ad - bc} \begin{pmatrix} d & -b \\ -c & a \end{pmatrix} = \frac{1}{\det(A)} \begin{pmatrix} d & -b \\ -c & a \end{pmatrix},
$$

(A.12)

when the determinant in the denominator is non-zero. This means that the inverse matrix in the exponent of the density is:

$$
\begin{pmatrix} \sigma_1^2 & \rho\sigma_1\sigma_2 \\ \rho\sigma_1\sigma_2 & \sigma_2^2 \end{pmatrix}^{-1} = \frac{1}{(1 - \rho^2)\sigma_1^2\sigma_2^2} \begin{pmatrix} \sigma_2^2 & -\rho\sigma_1\sigma_2 \\ -\rho\sigma_1\sigma_2 & \sigma_1^2 \end{pmatrix}. \tag{A.13}
$$

Substituting these formulas into (10) and expanding the matrix product yields the following formula for the bivariate normal probability density function:

$$
f(x_1, x_2) = \frac{1}{2\pi\sigma_1\sigma_2\sqrt{1-\rho^2}}
$$

$$
\cdot \exp\left[\frac{-1}{2(1-\rho^2)} \left(\frac{(x_1-\mu_1)^2}{\sigma_1^2} - 2\rho \frac{(x_1-\mu_1)(x_2-\mu_2)}{\sigma_1\sigma_2} + \frac{(x_2-\mu_2)^2}{\sigma_2^2} \right) \right].
$$

(A.14)

Even in two variables, the multivariate normal density is bulky and hard to work with in its full form. In the general case with 3 or more variables, we prefer not to work with the full form of the density, although it is illuminating to look at one special case. Consider a three-variable normal distribution in which all correlations ρ_{12}, ρ_{13}, and ρ_{23} are zero. Under this assumption, the covariance matrix is diagonal. Thus, we have special results for Σ, and its determinant and inverse.

$$
\Sigma = \begin{pmatrix} \sigma_1^2 & \rho_{12}\sigma_1\sigma_2 & \rho_{13}\sigma_1\sigma_3 \\ \rho_{12}\sigma_1\sigma_2 & \sigma_2^2 & \rho_{23}\sigma_2\sigma_3 \\ \rho_{13}\sigma_1\sigma_3 & \rho_{23}\sigma_2\sigma_3 & \sigma_3^2 \end{pmatrix} = \begin{pmatrix} \sigma_1^2 & 0 & 0 \\ 0 & \sigma_2^2 & 0 \\ 0 & 0 & \sigma_3^2 \end{pmatrix};
$$

$$\det(\Sigma) = \sigma_1^2 \sigma_2^2 \sigma_3^2; \ \Sigma^{-1} = \begin{pmatrix} 1/\sigma_1^2 & 0 & 0 \\ 0 & 1/\sigma_2^2 & 0 \\ 0 & 0 & 1/\sigma_3^2 \end{pmatrix}.$$

Therefore, the constant coefficient in the formula for the trivariate normal density is:

$$\frac{1}{(2\pi)^{n/2}\sqrt{\det(\Sigma)}} = \frac{1}{\left(\sqrt{2\pi}\right)^3 \sigma_1 \sigma_2 \sigma_3}. \tag{A.15}$$

The product in the exponent of the density is:

$$-\tfrac{1}{2}(\boldsymbol{x} - \boldsymbol{\mu})^t \Sigma^{-1}(\boldsymbol{x} - \boldsymbol{\mu})$$

$$= -\tfrac{1}{2}\begin{pmatrix} x_1 - \mu_1 & x_2 - \mu_2 & x_3 - \mu_3 \end{pmatrix} \begin{pmatrix} 1/\sigma_1^2 & 0 & 0 \\ 0 & 1/\sigma_2^2 & 0 \\ 0 & 0 & 1/\sigma_3^2 \end{pmatrix} \begin{pmatrix} x_1 - \mu_1 \\ x_2 - \mu_2 \\ x_3 - \mu_3 \end{pmatrix}$$

$$= -\tfrac{1}{2}\begin{pmatrix} x_1 - \mu_1 & x_2 - \mu_2 & x_3 - \mu_3 \end{pmatrix} \begin{pmatrix} \frac{x_1 - \mu_1}{\sigma_1^2} \\ \frac{x_2 - \mu_2}{\sigma_2^2} \\ \frac{x_3 - \mu_3}{\sigma_3^2} \end{pmatrix}$$

$$= -\tfrac{1}{2}\left(\left(\tfrac{x_1 - \mu_1}{\sigma_1}\right)^2 + \left(\tfrac{x_2 - \mu_2}{\sigma_2}\right)^2 + \left(\tfrac{x_3 - \mu_3}{\sigma_3}\right)^2 \right).$$

$$\tag{A.16}$$

But then the constant factors, and the exponential of the sum is the product of the exponentials; hence the full density factors:

$$f(x_1, x_2, x_3) = \left(\frac{1}{\sqrt{2\pi}\sigma_1} e^{-\tfrac{1}{2}\left(\frac{x_1 - \mu_1}{\sigma_1}\right)^2} \right) \cdot \left(\frac{1}{\sqrt{2\pi}\sigma_1} e^{-\tfrac{1}{2}\left(\frac{x_2 - \mu_2}{\sigma_2}\right)^2} \right)$$

$$\cdot \left(\frac{1}{\sqrt{2\pi}\sigma_1} e^{-\tfrac{1}{2}\left(\frac{x_3 - \mu_3}{\sigma_3}\right)^2} \right). \tag{A.17}$$

This is easily recognizable as the product of three marginal normal densities. So we have determined that when the pairwise correlations are zero, the three random variables are independent and normally distributed, with their appropriate mean and variance parameters μ_i and σ_i^2. It is easy to generalize this result to more than three multivariate normal variables.

Turning to the parameters ρ_{ij}, it is true that $\rho_{ij} = \text{Corr}(X_i, X_j)$. We will derive this only in the special bivariate case with means $\mu_1 = \mu_2 = 0$, and variances $\sigma_1^2 = \sigma_2^2 = 1$, which shows the nature of the computation. It also happens to be true, and we use this hereafter without proof or comment, that if a random vector $\boldsymbol{X} = (X_1, X_2, \ldots, X_n)$ has the multinormal distribution, then any pair (X_i, X_j) of two of the random variables is bivariate normal and inherits their parameters, including their correlation, from the parameters of the full random vector.

The correlation and covariance are equal in the simple case with variances equal to 1, and also $\text{Cov}(X_1, X_2) = E[X_1 X_2]$, since the means are zero. If ρ is the parameter in the bivariate normal density, then the correlation $\rho_{12} = \text{Corr}(X_1, X_2)$ is:

$$\rho_{12} = \int_{-\infty}^{\infty} \int_{-\infty}^{\infty} x_1 x_2 \frac{1}{2\pi\sqrt{1-\rho^2}} \exp\left[\frac{-1}{2(1-\rho^2)}\left(x_1^2 - 2\rho x_1 x_2 + x_2^2\right)\right] dx_2 dx_1.$$

Completing the square on x_2 in the exponent gives us:

$$\begin{aligned} x_1^2 - 2\rho x_1 x_2 + x_2^2 &= \left(x_2^2 - 2\rho x_1 x_2 + \rho^2 x_1^2\right) - \rho^2 x_1^2 + x_1^2 \\ &= \left(x_2 - \rho x_1\right)^2 + \left(1 - \rho^2\right) x_1^2. \end{aligned}$$

Therefore:

$$\begin{aligned} \rho_{12} &= \int_{-\infty}^{\infty} x_1 \cdot \frac{1}{\sqrt{2\pi}} \exp\left(\frac{-\left(1-\rho^2\right)x_1^2}{2(1-\rho^2)}\right) \\ &\quad \cdot \int_{-\infty}^{\infty} x_2 \cdot \frac{1}{\sqrt{2\pi}\sqrt{1-\rho^2}} \cdot \exp\left(\frac{-(x_2-\rho x_1)^2}{2(1-\rho^2)}\right) dx_2 dx_1. \end{aligned}$$

The inner integral on x_2 is the expected value of a normal random variable with mean $\rho\, x_1$. This means that the remaining integral on x_1 becomes:

$$\begin{aligned} \rho_{12} &= \int_{-\infty}^{\infty} \rho x_1^2 \cdot \frac{1}{\sqrt{2\pi}} \exp\left(\frac{-\left(1-\rho^2\right)x_1^2}{2(1-\rho^2)}\right) dx_1 \\ &= \rho \cdot \int_{-\infty}^{\infty} x_1^2 \cdot \frac{1}{\sqrt{2\pi}} \exp\left(\frac{-x_1^2}{2}\right) dx_1. \end{aligned}$$

The integral is 1 because it is $E\left[X^2\right]$ for a $N(0,1)$ random variable; hence $\rho_{12} = \rho$.

Applying completing the square in a way similar to the last computation in the case of general means and variances of two bivariate normally distributed random variables X_1 and X_2, we get the identity:

$$\begin{aligned} &\frac{1}{2(1-\rho^2)}\left(\frac{(x_1-\mu_1)^2}{\sigma_1^2} - 2\rho\frac{(x_1-\mu_1)(x_2-\mu_2)}{\sigma_1\sigma_2} + \frac{(x_2-\mu_2)^2}{\sigma_2^2}\right) \\ &= \frac{(x_1-\mu_1)^2}{2\sigma_1^2} + \frac{\left(x_2-\left[\mu_2+\frac{\rho\sigma_2}{\sigma_1}(x_1-\mu_1)\right]\right)^2}{2\sigma_2^2(1-\rho^2)}. \end{aligned} \tag{A.18}$$

This makes the following important result very easy to prove. Note that we are not assuming independence.

Theorem 1. If X_1 and X_2 have the bivariate normal distribution with parameters $\mu_1, \mu_2, \sigma_1, \sigma_2$, and ρ, then:
  (a) The marginal density of X_1 is $N\left(\mu_1, \sigma_1^2\right)$;
  (b) The conditional density of X_2 given $X_1 = x_1$ is $N\left(\mu_{2|1}, \sigma_{2|1}^2\right)$, where:

$$\mu_{2|1} = \mu_2 + \frac{\rho\sigma_2}{\sigma_1}\left(x_1 - \mu_1\right), \quad \sigma_{2|1}^2 = \sigma_2^2\left(1-\rho^2\right). \tag{A.19}$$

  (c) The marginal density of X_2 is $N\left(\mu_2, \sigma_2^2\right)$;

(d) The conditional density of X_1 given $X_2 = x_2$ is $\mathcal{N}\left(\mu_{1|2}, \sigma_{1|2}^2\right)$, where:

$$\mu_{1|2} = \mu_1 + \frac{\rho\sigma_1}{\sigma_2}\left(x_2 - \mu_2\right), \quad \sigma_{1|2}^2 = \sigma_1^2\left(1 - \rho^2\right). \quad\quad \text{(A.20)}$$

Proof. By identity (18), we can write the joint density of X_1 and X_2 as:

$$f\left(x_1, x_2\right) = \frac{1}{\sqrt{2\pi}\sigma_1}e^{-(x_1-\mu_1)^2/2\sigma_1^2} \cdot \frac{1}{\sqrt{2\pi}\sigma_{2|1}}e^{-(x_2-\mu_{2|1})^2)/2\sigma_{2|1}^2}.$$

The marginal density of X_1 is found by integrating $f\left(x_1, x_2\right)$ from $-\infty$ to $+\infty$ with respect to x_2. The two leading factors come out of that integral, and the remaining two factors form a valid normal density function, which integrates to 1. The marginal density is therefore $\mathcal{N}\left(\mu_1, \sigma_1^2\right)$, as claimed in part (a). By rewriting the previous identity, we also have:

$$f\left(x_1, x_2\right) = f_1\left(x_1\right) \cdot \frac{1}{\sqrt{2\pi}\sigma_{2|1}}\exp\left(\frac{-\left(x_2-\mu_{2|1}\right)^2}{2\sigma_{2|1}^2}\right)$$

$$\implies f_{2|1}\left(x_2|x_1\right) = \frac{f(x_1,x_2)}{f_1(x_1)} = \frac{1}{\sqrt{2\pi}\sigma_{2|1}}\exp\left(\frac{-\left(x_2-\mu_{2|1}\right)^2}{2\sigma_{2|1}^2}\right).$$

This establishes part (b). Parts (c) and (d) are proved similarly, or by observing the symmetry of the bivariate normal distribution in the two variables. ∎

As an example, recall that Section 4.2 looked at two assets in the food industry, McDonald's and Yum! If we examine rates of return for a third restaurant business over the same time period, Darden Restaurants, which includes Red Lobster, Olive Garden and others, we see again the characteristic bell-shaped single-variable data histogram in Figure A.1. Figure A.2 shows the same dependence of the Darden rates of return on each of the other two rates. But also, the 3-dimensional plot of the triples in Figure A.3 is very interesting. The rates form a roughly ellipsoidal cloud in $\mathbb{R}^3$. In the *Mathematica* version of this text, you can grab the picture and rotate in order to see the shape more clearly than can be presented on a printed page.

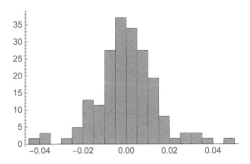

FIGURE A.1 Histogram of Darden rates, 2016.

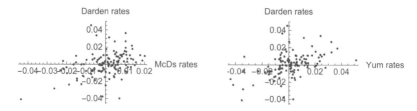

FIGURE A.2 Scatterplots of Darden rates vs. McDonald's and Yum!.

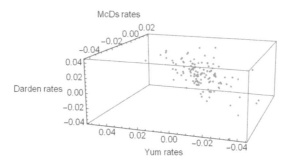

FIGURE A.3 3-D scatter plot of McDonald's, Yum!, and Darden rates.

The parameter estimates for the first two assets are repeated below, and the sample mean for the third asset, Darden, its sample standard deviation, and its correlations with the other two are reported.

$$\bar{X}_1 = .000240; \bar{X}_2 = .001234; \bar{X}_3 = .000169;$$
$$S_1 = .010303; S_2 = .015431; S_3 = .013764;$$
$$R_{12} = .584455; R_{13} = .406588; R_{23} = .494185.$$

Example. Reconsider the two real assets of Example 10 of Section 4.2 together with the Darden asset above, and suppose that the three rates of return are multivariate normal random variables with parameters as estimated by the sample statistics that were listed above. Compute:
  (a) the probability that Darden's weekly rate of return is at least .0009;
  (b) the probability that Darden's rate is at least .0009 given that McDonald's rate is .0006; and
  (c) the probability that Darden's rate is at least .0009 given that the Yum! rate of return is .0007.

Solution. (a) The marginal distribution of the Darden rate of return is normal, with mean estimated to be .000169 and standard deviation .013764. Then, using technology to find the appropriate c.d.f. value, we obtain:

$$P[X_3 \geq .0009] = 1 - P[X_3 \leq .0009] = 1 - F_3(.0009) \approx .478822.$$

(b) Together, McDonald's and Darden have the bivariate normal distribution, and the conditional distribution of the Darden rate given the McDonald's rate of .0006 is normal, with parameters:

$$
\begin{aligned}
\mu_{3|1} &= \mu_3 + \frac{\rho_{13}\sigma_3}{\sigma_1}(x_1 - \mu_1) \\
&\approx .000169 + \frac{(.406588)(.013764)}{.010303}(.0006 - .000240) \\
&= .000365, \\
\sigma_{3|1}^2 &= \sigma_3^2\left(1 - \rho_{13}^2\right) \approx (.013764)^2(1 - .406588)^2 = .000158.
\end{aligned}
$$

So, given a value of the McDonald's rate of .0006, which is greater than the mean rate of return on that asset, the conditional mean return for Darden has increased from .000169 to about .000365. Since the two assets have correlated rates of return, we would expect the conditional variance and standard deviation of the Darden rate of return to be smaller than their corresponding marginal quantities. And in fact, since $\sigma_{3|1} = \sqrt{\sigma_{3|1}^2} = \sqrt{.000158} = .012568$, the conditional standard deviation has decreased as compared to our estimated $\sigma_3 = .013764$. We can now compute the desired probability using the normal c.d.f. with these conditional parameters:

$$
\begin{aligned}
P[X_3 \geq .0009 | X_1 = .0006] &= 1 - P[X_3 \leq .0009 | X_1 = .0006] \\
&= 1 - F_{3|1}(.0009) \\
&\approx .483023.
\end{aligned}
$$

It may seem surprising to you that this is actually only slightly greater than the marginal probability of .478822 that was computed in part (a), despite the fact that the given McDonald's rate is higher than its mean. The reason that this happens is that the standard deviation of the Darden rate has gone down, which makes the .0009 cutoff just a bit harder for the rate of return to reach, in spite of the fact that the mean return has increased.

(c) We can proceed as in part (b). This time the conditional mean, variance, and standard deviation are:

$$
\begin{aligned}
\mu_{3|2} &= \mu_3 + \frac{\rho_{23}\sigma_3}{\sigma_2}(x_2 - \mu_2) \\
&\approx .000169 + \frac{(.494185)(.013764)}{.015431}(.0007 - .001234) \\
&= -.000066, \\
\sigma_{3|2}^2 &= \sigma_3^2\left(1 - \rho_{23}^2\right) \approx (.013764)^2(1 - .494185)^2 = .000143, \\
\sigma_{3|2} &= \sqrt{\sigma_{3|2}^2} = \sqrt{.000143} = .011958.
\end{aligned}
$$

So, as compared to the marginal parameters, the conditional mean return for Darden has decreased from .000169 to $-.000066$, and the standard deviation has reduced from .013764 to .011958, as we would expect, since the given Yum! rate of .0007 was substantially lower than its unconditional mean rate of .001234. The desired probability is:

$$\begin{aligned} P\left[X_3 \geq .0009 | X_2 = .0007\right] &= 1 - P\left[X_3 \leq .0009 | X_2 = .0007\right] \\ &= 1 - F_{3|2}(.0009) \\ &\approx .467807. \end{aligned}$$

The observed small rate of return on Yum! and the positive correlation have decreased the probability that Darden exceeds the threshold value of .0009. ∎

B

Answers to Selected Exercises

Section 1.1

1. (b)$1/6$; (c) \$4 .67.   2. .159.   3. .582.   5. $-\$2.59$.
6. $f(28.12) = .5^3 = .125$; $f(26.23) = 3(.5)^3 = .375$;
    $f(24.46) = 3(.5)^3 = .375$; $f(22.82) = .5^3 = .125$.
8. .079.
9. case $\sigma^2 = 10 : .013$; case $\sigma^2 = 20 : .057$; case $\sigma^2 = 30 : .098$.
10. .495.   11. In the second case, $q = (r - a)/(b - a)$.

Section 1.2

1. .12; .16.   2. $\mu = 3.75\sigma^2 + .02$.
3. The asset with mean return $\mu = .02$ and variance $\sigma^2 = .05$.
4. $a < 5$.   6. $a < 11.54$.   7. $10 < a < 20$.
9. $\mu_p = .033$.   10. .03.   12. $w = .125$.
13. $w = .833$.   14. $w = 1/3$; .004.

Section 1.3

1. 2; 5.   2. np.   3. $\frac{1}{\lambda}$; $\frac{2}{\lambda^2}$.
4. $e^{\mu(e^t - 1)}$.
7. (c) $P[X + Y \leq 1] = \frac{1}{2}$; (d) $E[X] = \frac{2}{3}$, $E[Y] = \frac{2}{3}$, and $E[X + Y] = \frac{4}{3}$.
8. .875.   9. 5.   10. $\frac{1}{3}$.
12. $c = \frac{1}{4}$; $E[X_1 X_2] = \frac{16}{9}$.   15. 0.
16. $\text{Cov}(X, Y) \approx .01778$; $\rho \approx .492$.
18. (a) .045; .00075. (b) .045; .000467.
20. .05667; .0003328.   21. .0425; .0009.

Section 1.4

1. .0575; .01375.   2. .046; .000054.
3. (a) $w_1 = .9875, w_2 = .0125$; (b) $w_1 = .9925, w_2 = .0075$; (c) $w_1 = .995, w_2 = .005$.
4. $w_2 = .125 - \frac{\epsilon}{.16}$.   5. $r = .005$; .06125.
8. (a) $w_A = -.035, w_B = 1.035$; (b) $w_A = .312, w_B = .688$.

9. (a) $w_1 = .6625, w_2 = .3375$; (b) $w_1 = .6646, w_2 = .3354$; (c) $w_1 = .6653, w_2 = .3347$.

10. $a \geq .0375$.   12. $w_1 = .623, w_2 = .377$.

13. $w_1 = .2$, $w_2 = .8$; mean: .046 variance: $\sigma_p^2 = 3.39 \times 10^{-21}$.

14. $w_1 = .146$, $w_2 = .854$; mean: .0756 variance: .00278.

15. $w = \frac{(.0132 - .0288\rho)}{2(.0312 - .0288\rho)}$.

Section 1.5

1. (a) \$200; (b) −\$50; (c) −\$100; expected profit: \$16 .67.

2. (a) \$375,000; (b) \$377,500; (c) \$382,500; (d) \$385,000; expected price \$381,250.

3. 100.

4. If he does not do a futures contract, variance = 4,296,880; standard deviation = \$2072 .89. If he does do a 60% futures contract, variance = 687,500; standard deviation = \$829 .16.

10. (a) expected payoff: \$3.50; (b) expected payoff: \$1.

11. \$2882 .94.   12. (a) 2.50; (b) − 1.20; (c) 0; (d) −1.20.

13. If the Dow index ends at \$16,000, the dealer loses all of his \$400. If he had not purchased the call options, he would have lost a total of \$10000. In the case that the final value of the Dow Jones index is \$14,000, the dealer profits by \$9,600 if he purchases the options, or \$10,000 if not.

14. Largest and smallest possible payoffs: 2 and 0.

16. At least 97 options are sufficient in the worst possible case.

17. The final value is always E.

19. (a) \$1.50; (b) \$1.50.   20. (a) $q = \frac{r-a}{b-a}$.

Section 1.6

1. $\Delta = \frac{4}{5}$, $c = -\$37.4757$. Price: \$2.5243.

2. $\Delta = -\frac{2}{9}$, $c = \$22.8758$. Price: \$.6536.

3. $\Delta = \frac{5}{8}$, $c = -\$73.3173$. Price: \$1.6827.

4. \$0.198.   6. \$1.307.   7. \$2.451.   8. \$5.00.

9. (a) up case: profit = \$175.97; down case: profit = \$1425.97. (b) up case: profit = −\$1074; down case: profit = \$2675 .97.

11. (a) up case: profit = \$1010; down case: profit = −\$490. (b) up case: profit = \$10; down case: profit = \$10. (c) up case: profit = \$510; down case: profit = −\$240.

12. (a) up case: profit = \$514.85; down case: profit = \$514.85. (b) up case: profit = \$886.14; down case: profit = −\$113.86.

13. $E = \$47.31$.   14. .6c.

15. \$62.3762Δ.   16. \$2.9703.

Section 2.1

1. The solution is $w_2 = .0285$, $w_3 = .0259$, hence $w_1 = .9456$. Mean rate of return: $.0127$; Variance: $.000334$.
2. The weights are $w_2 = .2589$, $w_3 = .2296$, $w_4 = .2087$, and so $w_1 = .3028$.
3. $w_2 = .5$, $w_3 = .5102$, $w_4 = .4630$; therefore $w_1 = -.4732$.
5. 18.5185.   6. $\mu_2 \geq .067$.   7. $r \geq .0201$.
8. $w_1 = .2012$, $w_2 = .4288$, $w_3 = .3700$.
10. $w_2 = .2604$, $w_3 = .2315$, and so $w_1 = .5081$.
11. $w_1 = .6023$, $w_2 = .2193$, $w_3 = .1784$.
13. (a) $w_2^* = .3394$, $w_3^* = .3463$, $w_4^* = .3143$; (b) $.0625$.
14. (a) $w_2^* = .4675$, $w_3^* = .2597$, $w_4^* = .2727$; (b) $.0263$.
15. For the investor with risk aversion 2: $w_2 = .25$, $w_3 = .1389$, $w_4 = .1458$; for the investor with risk aversion 18: $w_2 = .0278$, $w_3 = .0154$, $w_4 = .0162$.
16. The market portfolio weights are: $w_2^* = .708$, $w_3^* = .292$; for the investor with risk aversion 10: $w_2 = .958$, $w_3 = .396$, and so the non-risky weight is $w_1 = -.354$.
17. $w_2^* = .229$, $w_3^* = .587$, $w_4^* = .183$.

Section 2.2

1. Optimal portfolio: $(.8013, .6232, .3616)$;
   market portfolio: $(.448628, .348933, .202439)$.
2. Market portfolio: $(.5271, .1757, .1098, .1054, .0820)$; mean $= .0604$; variance $= .007097$. Optimal portfolio for investor (risky assets) $(.1, .03333, .02083, .02, .01556)$,
   risk-free asset: $.81028$.
3. Market portfolio: $(.5185, .4815)$; mean $= .0696$; standard deviation $= .04645$.
6. Market portfolio: $(1.1616, .8656, -1.0272)$.
7. (b) $(.04068, .07483)$; (c) $(.75, .25)$; $\sigma_p^2 = .001656$; $\sigma_p = .0407$.
10. $r < .011$.   11. $\mu_4 \geq .0584$.
14. $p =$ proportion of wealth in asset $1 = 1/3$; $\sigma_p^2 = .1244$.
15. Min variance portfolio: $(.3292, .2633, .2194, .1881)$, mean: $.0553$, variance: $.02633$, standard deviation: $.1623$. For mean return $.06$, portfolio: $(.25, .25, .25, .25)$, standard deviation: $.1658$.
16. (a) $(.0493, .034)$; (b) $(.0497, .035)$; (c) $(.5625, .25, .1875)$; (d) Roughly $p = .57$ for the market and $1 - p = .43$ for the risk-free asset.
17. (a) $(.03707, .0524)$; (b) $(.03753, .05)$; (c) $(.3394, .3463, .3143)$; (d) Proportion $p = .253$ to the market and $1 - p = .747$ to the risk-free asset.

Section 2.3

1. Proportion $p = .8230$ in the market portfolio $\boldsymbol{w}^* = (.448628, .348933, .202439)$; $\sigma_p = .0339$.

2. Proportion $p = .4061$ in the market portfolio $w^* = (.09317, .22360, .31056,)$; $.37267\sigma_p^2 = .006054$.

3. $a = 1.652$.

4. Proportion $p = 1.221$ in the market portfolio $w^* = (\ 635\quad .365\ )$; $\sigma_p = .0789$; for $p = 1.3$, $\mu_p = .0732$ and $\sigma_p = .0840$.

6. Market portfolio $w^* = (\ .60886\quad .17435\quad .21679\ )$; proportion in the market $p = 1.74825$; variance $\sigma_p^2 = .0182924$.

7. $a = \frac{s}{2p}$, where $s = \sum_{j=2}^{4} \frac{\mu_j - r}{\sigma_j^2}$.

9. (a) .054; (b) .5.   11. $\beta_2 = .5831$; $\beta_3 = 1.165$; $\beta_4 = 2.039$.

12. $\beta_2 = .5127$; $\beta_3 = 1.0254$.   13. $\beta_4 = 1.0152$.

16. Asset 3: Market risk $= .0112$, Specific risk $= .0788$; Asset 4: Market risk $= .0159$, Specific risk $= .2341$; Asset 5: Market risk $= .0227$, Specific risk $= .4673$.

17. Asset 2: Market risk $= .001514$, Specific risk $= .078486$; Asset 3: Market risk $= .013620$, Specific risk $= .086380$; Asset 4: Market risk $= .037835$, Specific risk $= .082165$; Asset 5: Market risk $= .074137$, Specific risk $= .065863$.

19. No.   20. Underpriced.

21. (a) asset 2 : $S_2 = .2$, asset 3 : $S_3 = .083$, asset 4 : $S_4 = .06$, asset 5 : $S_5 = .05$; (b) .142.

Section 2.4

1. \$5940 with probability .4; \$6020 with probability .2; \$6100 with probability .2; and \$6220 with probability .2.

2. $h = .554$.   3. Y_h such that $h = \frac{1}{3}$.

4. $x = \$324$.   7. Beneath.

8. $m = 2$; $M = 2.8585$; CE $(Y_{.5}) = 2.028$; $E[Y_{.5}] = 2.429$.

9. (a) CE $(Y_{.2}) = 5.431$; CE $(Y_{.8}) = 8.193$; (b) $M = 11.389$; CE $(Y_{.3}) = 5.822$; CE $(Y_{.5}) = 6.718$.

10. $E[Y_{.5}] = .2705$; CE $(Y_{.5}) = .243$.

11. $c = m = 5.3132$; $b = .5667$; $M = 10.8079$.

12. 1.759.   13. 1.214.   14. CE$(X) = \$183.77$.

15. $\gamma \approx .5972$.   19. $A(x) = \frac{1}{x}$; $R(x) = 1$.

20. $w_1 = -30.8$, $w_2 = 1 - w_1 = 31.8$.

Section 2.5

2. The solution does not change.

3. $w_0 = w_1 = \dfrac{\left((p(b-r))^{3/2} - ((1-p)(r-a))^{3/2}\right)(1+r)}{((1-p)(r-a))^{3/2}(b-r) + (p(b-r))^{3/2}(r-a)}$.

4. $E\left[\dfrac{Z_1 - R}{R + w_1 \cdot (Z_1 - R)}\right] = 0$.

5. $\delta = \frac{2}{9}$; $\delta = .0509$.

7. $w_2 = w_1 = w_0 = .714$.

9. $c_1 = \frac{W_1}{1+(1.01)^{-1}} = .502488W_1$; $c_0 = \frac{W_0}{1+(1.01)^{-1}+(1.01)^{-2}} = .336656W_0$, $w_1 = w_0 = .271542$.

10. $c_1 = .502488W_1$; $c_0 = .336656W_0$ $w_1 = w_0 = -3.34566$.

12. $w_3 = -9.10909 + \frac{44.5632}{W_3}$, $w_2 = -9.10909 + \frac{43.6894}{W_2}$, $w_1 = -9.10909 + \frac{42.8328}{W_1}$, $w_0 = -9.10909 + \frac{41.9929}{W_0}$.

14. $w_1 = -8.74576 + \frac{84.7458}{W_1}$; $w_0 = -8.66$.

16. $w^* = 0$.

Section 3.1

1. (a) 51.9841; (b) 56.2754; (c) 51.4233

2. $E_p[S_3] = 50.7538$; $E_q[S_3] = 51.5151$.

5. $\Delta_{21} = .86518$, $c_{21} = -17.789$; $\Delta_{11} = .556812$, $c_{11} = -11.0083$; $\Delta_{01} = .358344$, $c_{01} = -6.81209$.

6. \$ .144133

7. (a) .353082; (b) .389273 (c) at node $(1,1)$, $\Delta = .306122; c = -15.4412$; at node $(1,2)$, $\Delta = 0; c = 0$; at node $(0,1)$, $\Delta = .180072; c = -8.65052$.

9. $v_{3,2} = 0$, $v_{3,3} = .3250$.   11. 4.90793.   12. 8.351.

14. .0334.

17. $q_{01} = .625$, $q_{11} = .6375$, $q_{12} = .6175$; option value 3.89.

18. (a) Between 61.01 and 61.02; (b) between 60.18 and 60.19.

Section 3.2

3. 128.   4. No; yes.   5. $\{0,1,2,3\}, \{4\}, \{5,6,7,8\}$.

9. No. $\mathcal{A}(Z) = \{\emptyset, [0,4], [0,2), [2,4]\}$.   10. $\{0\}, \{-1,1\}, \{-2,2\}, \{-3,3\}$.

11. $\mathcal{A}(X) = \{\emptyset, \{1,4\}, \{2,5\}, \{3,6\}, \{1,2,4,5\}, \{1,3,4,6\}, \{2,3,5,6\}, \{1,2,3,4,5,6\}\}$.

Section 3.3

1.

$$E[X|\mathcal{A}](\omega) = \begin{cases} 1 & \text{if } \omega \in A_1; \\ 2/3 & \text{if } \omega \in A_2; \\ 2 & \text{if } \omega \in A_3. \end{cases}$$

3.

$$E[X|\mathcal{A}](\omega) = \begin{cases} 3 & \text{if } \omega \in B_1; \\ 9 & \text{if } \omega \in B_2; \\ 16 & \text{if } \omega \in B_3. \end{cases}$$

11. (a) $(p(1+b) + (1-p)(1+a))^3 S_n$; (b) $(p(1+b) + (1-p)(1+a))^k S_n$.

12. $2X + X \cdot E[Y|\mathcal{B}] - X \cdot Z$; $2E[X] + E[X] \cdot E[Y] - E[XZ]$.   16. False.

Section 3.4

2. $\theta_1(\omega) = (0, 20)$, $\theta_2(\omega) = (0, 20)$, for all outcomes, and

$$\theta_3(\omega) = \begin{cases} (7983.79, 0) & \text{if } \omega \in \{\text{UUU}, \text{UUD}\}; \\ (0, 20) & \text{otherwise.} \end{cases}$$

3. The successive values are: $1000, $1010, $795.10, $803.05, $581,08, and $586.89.

4. Both sides of the equation equal 2163.20.

6. Let outcomes ω_1, ... , ω_8 correspond, respectively, to paths (U, U, U), (U, U, D), (U, D, U), (U, D, D), (D, U, U), (D, U, D), (D, D, U), and (D, D, D). Then:

$V_2(\omega) = 230.448, \theta_3^0 = 48.405, \theta_3^1 = .777767$ for $\omega = \omega_1, \omega_2$;
$V_2(\omega) = 212.436, \theta_3^0 = 89.5669, \theta_3^1 = .551023$ for $\omega = \omega_3, \omega_4$;
$V_2(\omega) = 211.064, \theta_3^0 = 44.3289, \theta_3^1 = .77779$ for $\omega = \omega_5, \omega_6$;
$V_2(\omega) = 195.849, \theta_3^0 = 124.302, \theta_3^1 = .319681$ for $\omega = \omega_7, \omega_8$;
$V_1(\omega) = 215.811, \theta_2^0 = 15.0752, \theta_2^1 = .935202$ for $\omega = \omega_1, \omega_2, \omega_3, \omega_4$;
$V_1(\omega) = 198.07, \theta_2^0 = 27.8981, \theta_2^1 = .862528$ for $\omega = \omega_5, \omega_6, \omega_7, \omega_8$;
$V_0 = 201.824, \theta_1^0 = 4.70214, \theta_1^1 = .985611$.

8. Let outcomes ω_1, ... , ω_8 correspond, respectively, to paths (U, U, U), (U, U, D), (U, D, U), (U, D, D), (D, U, U), (D, U, D), (D, D, U), and (D, D, D). Then:

$V_2(\omega) = 19.802, \theta_3^0 = 19.4118, \theta_3^1 = 0$ for $\omega = \omega_1, \omega_2$;
$V_2(\omega) = 19.4738, \theta_3^0 = 9.30277, \theta_3^1 = .5$ for $\omega = \omega_3, \omega_4$;
$V_2(\omega) = 18.9982, \theta_3^0 = 18.248, \theta_3^1 = .0192$ for $\omega = \omega_5, \omega_6$;
$V_2(\omega) = 17.9758, \theta_3^0 = 8.5872, \theta_3^1 = .5$ for $\omega = \omega_7, \omega_8$;
$V_1(\omega) = 19.4834, \theta_2^0 = 15.2293, \theta_2^1 = .1972$ for $\omega = \omega_1, \omega_2, \omega_3, \omega_4$;
$V_1(\omega) = 18.4301, \theta_2^0 = 5.5946, \theta_2^1 = .6656$ for $\omega = \omega_5, \omega_6, \omega_7, \omega_8$;
$V_0 = 18.8992, \theta_1^0 = 5.7332, \theta_1^1 = .6583$

9. The outcomes and their probabilities can be defined as:

$\omega_1 = ((U, U), (U, U))$; $\omega_2 = ((U, U), (U, D))$; $\omega_3 = ((U, U), (D, U))$;
$\omega_4 = ((U, U), (D, D))$; each with probability $= 9/100$;
$\omega_5 = ((U, D), (U, U))$; $\omega_6 = ((U, D), (U, D))$; $\omega_7 = ((U, D), (D, U))$;
$\omega_8 = ((U, D), (D, D))$; $\omega_9 = ((D, U), (U, U))$; $\omega_{10} = ((D, U), (U, D))$;
$\omega_{11} = ((D, U), (D, U))$; $\omega_{12} = ((D, U), (D, D))$; each with probability $= 6/100$;
$\omega_{13} = ((D, D), (U, U))$; $\omega_{14} = ((D, D), (U, D))$; $\omega_{15} = ((D, D), (D, U))$;
$\omega_{16} = ((D, D), (D, D))$; each with probability $= (1/25)$.

10. Option value: $V_0 = 1.24502$; path probabilities

$Q(U, U, U) = .130044$; $Q(U, U, D) = .123289$; $Q(U, D, U) = .124155$;
$Q(U, D, D) = .122511$; $Q(D, U, U) = .124994$; $Q(D, U, D) = .123339$;
$Q(D, D, U) = .124155$; $Q(D, D, D) = .127511$.

11. At levels 0 and 2, the up probabilities are $q = 1/2$. There are infinitely many possibilities for the probability triples q_{11}, q_{12}, q_{13} in the upper and lower parts of the graph at level 1, including .45, .3, and .25, such that $q_{12} = 3 - 6q_{11}$ and $q_{13} = 5q_{11} - 2$.
12. 1.5. 13. $V_0 = 3.604$. 14. 50.

Section 3.5

1. Initial value: 3.64528. 2. Initial value: .317. 3. Initial value: 2.068.
5. Initial value: .4425. 7. Initial value: .7683. 8. Initial value: 1.33607.
10. Initial value: .0204. 11. Initial value: .0752. 13. Initial value: 6.21063.
14. Initial value: 1.47698. 15. Initial value: 27.7666. 16. Initial value: 29.

Section 3.6

(Simulation estimates will vary.)

2. The width would be multiplied by about .58.
6. One particular run with 418 replications resulted in an average V_0 of .0371, with an error estimate of .0259.
7. No, in fact only the discounted values 0, .422, and 8.306 are possible, or undiscounted values 0, .435, and 8.5575.
8. A run with 6000 replications gave an estimate of about 3.897 with a precision of about .05.
10. About 3000 replications are needed. One simulation run gave a V_0 estimate of .4306 with a precision of .0199.
12. One particular experimental run gave an estimated value of about 7.56, with a precision of about .25.

Section 3.7

2. 0 or ∞. 6. .1316. 7. .00003.

Section 4.1

1. $c = 3$; 7/8. 2. mean: 3/4; variance: 3/80.
3. $c = 4$; .593994. 4. 1. 8. $\frac{(b-a)^2}{12}$.
9. $P[X > .5] = .46875$; mean: 0 ; variance: 2.
11. (a) .864334; (b) 2.249; (c) .454808.
12. $\mu = .025$. 13. .4726.

Section 4.2

1. 1/3; $\left(1 - e^{-1}\right)$; 1/4. 2. $1/12 \cdot x_1 \left(1 - e^{-x_2}\right) \cdot (x_3 - 1)$.

3. $\frac{7}{24}$.   5. $c = 1/(\log(2))^2$.

6. (a) $\frac{1}{8}(2x+2), \frac{1}{8}(2y+2)$; (b) $\frac{3}{8}, \frac{5}{8}$; (c) $\frac{1}{4}$; (d) $\frac{5}{8}$.

7. $F_1(x_1) = \frac{1}{3}x_1^2 + \frac{2}{3}x_1$; $F_2(x_2) = \frac{2}{3}x_2^2 + \frac{1}{3}x_2$.

12. (a) $\frac{\frac{1}{2}x_1}{1-\frac{1}{4}x_2^2}, x_2 \le x_1 \le 2$; (b) $\frac{2x_2}{x_1^2}, 0 \le x_2 \le x_1$; (c) $\frac{1}{6} \cdot \frac{(8-x_2^3)}{1-\frac{1}{4}x_2^2}$; (d) $\frac{2}{3}x_1$.

16. (b) $p(x) = 3x^2$, for $x \in [0,1]$, $q(y) = 2y$, for $y \in [0,1]$; (c) $3/5$.

17. $c = 2$; $3/4$.

18. (a) $f_1(x_1) = 1 + x_1$, if $x_1 \in [-1,0]$ and $f_1(x_1) = 1 - x_1$, if $x_1 \in (0,1)$;
$f_2(x_2) = 2 - 2x_2, x_2 \in [0,1]$;

    (b) $\frac{1}{1+x_1}$, if $x_1 \in [-1,0]$ and $x_2 \in [0, 1+x_1]$;

    $\frac{1}{1-x_1}$, if $x_1 \in (0,1)$ and $x_2 \in [0, 1 - x_1]$;

    (c) $\frac{(1+x_1)^2}{3}$ for $x_1 \in [-1,0]$ and $\frac{(1-x_1)^2}{3}$ for $x_1 \in (0,1)$;

    (d) 0.

20. (a) $.4733$; (b) $\mu_{1|2} = -.0001635$, $\sigma^2_{1|2} = .00006989$; (c) $.5126$.

21. $\rho \ge .6$.

Section 4.3

9. $E[X \cdot I_{A_2}] = E[Y \cdot I_{A_2}] = \frac{1}{3}$;
$E[X \cdot I_{A_3}] = E[Y \cdot I_{A_3}] = \frac{5}{9}$;
$E[X] = E[Y] = 1$.

10. $E[Z|\mathcal{A}] = \frac{1}{12}I_{[0,1/2)}(\omega) + \frac{19}{48}I_{[1/2,3/4)}(\omega) + \frac{37}{48}I_{[3/4,1]}(\omega)$.

14. $X^+(\omega) = \begin{cases} X(\omega) & \text{if } X(\omega) \ge 0; \\ 0 & \text{otherwise,} \end{cases}$

    $X^-(\omega) = \begin{cases} 0 & \text{if } X(\omega) \ge 0; \\ -X(\omega) & \text{if } X(\omega) < 0. \end{cases}$

15. $E[X](3E[Y|\sigma(Z)] - Z)$.

Section 4.4

1. $E[X_m] = (1/2)m$; $\text{Var}(X_m) = (1/12)m$.

2. (a) $.281851$; (b) $.281851$.   3. (a) $.4931$; (b) $.4634$.

4. (a) $.9214$; (b) $.9214$; (c) $.7603$.   6. (a) $.7734$; (b) $.7022$.

8. $\sigma^2 s$.   9. $.1437$.   10. $.4822$.

11. (a) $.41136$; (b) $.03972$.   12. $x_0^n e^{(n\mu + (n^2/2)\sigma^2)t}$.

Section 4.5

1. $.4580$.

2. Non-standard Brownian motion with initial state 0 and parameters $\mu = .05$ and $\sigma = .1$.

3. $dY_t = \mu dt + \sigma dB_t$.   5. Mean: 0; variance: $\int_a^b x_t^2 dt$.

7. $dY_t = d(Z_t^2) = dt + 2Z_t dZ_t$.

Section 5.1

3. .759761. 4. 5.93381. 6. .005.

Section 5.2

2. Each T in the formula is replaced by the remaining time $T - t$.

10. $p = E - \frac{A_0^-}{B_0^-}$. 11. $V_0 = k \cdot B_0^- - A_0^-$.

12. $V_0 = A_0^+(E) - E \cdot B_0^+(E) - A_0^+(E + c) + (E + c) \cdot B_0^+(E - c)$.

13. $V_0 = \left(A_0^+(k_2) - k_2 \cdot B_0^+(k_2)\right) - \left(k_1 \cdot B_0^-(k_1) - A_0^-(k_1)\right)$.

14. $$\left(A_0^+(E - c) - (E - c) \cdot B_0^+(E - c)\right) + \left(A_0^+(E + c) - (E + c) \cdot B_0^+(E - c)\right) \\ - 2\left(A_0^+(E) - E \cdot B_0^+(E)\right).$$

Section 5.3

3. (a) $V_T = C \cdot I_{\{E - X_T \geq 0\}}$; (b) $V_T = X_T \cdot I_{\{X_T - E \geq 0\}}$; (c) $V_T = X_T - E$;
(d) $V_T = c_1 \cdot I_{\{S_T - E_1 \geq 0\}} + c_2 \cdot I_{\{E_2 - S_T \geq 0\}}$

Section 5.4

1. P & G: .564; Target: .561.

3. For the $t(5)$- distribution the statistic is about 2.96, while for normal it is about 17.9.

6. The chi-square statistic for the normal model is 8.75, and for the Laplace model is 14.59.

8. (a) .3821; (b) .3472; (c) .3277.

11. (a) $\mu_2 = .0736; \sigma_2^2 = .0128$; (b) risky weight: .2086; $\mu_p = .0313; \sigma_p^2 = .000557$; (c) $\mu_p = .0171; \sigma_p^2 = .00139$.

12. (a) .4920; (b) .4943. 13. .4893. 14. lag 1: $-.078$, lag 2: .0368.

Bibliography

[1] Baxter, Martin and Andrew Rennie. *Financial Calculus*. Cambridge University Press, Cambridge, United Kingdom, 1996.

[2] Black, Fischer and Myron Scholes. "The Pricing of Options and Corporate Liabilities," *Journal of Political Economy*. 81 (3): 637-654, 1970.

[3] Buchen, Peter. *An Introduction to Exotic Option Pricing*. Chapman & Hall/CRC Press, Boca Raton, Florida, 2012.

[4] Capiński, Marek and Tomasz Zastawniak. *Mathematics for Finance: An Introduction to Financial Engineering*. Springer-Verlag, London, 2003.

[5] Chung, Kai Lai. *A Course in Probability Theory, 3rd Ed.* Academic Press, San Diego, 2001.

[6] Çinlar, Erhan. *Introduction to Stochastic Processes*. Prentice-Hall, Inc., Englewood Cliffs, New Jersey, 1975.

[7] Cvitanić, Jakša and Fernando Zapatero. *Introduction to the Economics and Mathematics of Financial Markets*. MIT Press, Cambridge, Massachusetts, 2004.

[8] Friedman, Avner. *Stochastic Differential Equations and Applications, Vol. 1*. Academic Press, London, 1975.

[9] Hastings, Kevin J. *Introduction to the Mathematics of Operations Research with Mathematica, 2nd ed.* Chapman & Hall/CRC Press, Boca Raton, Florida, 2006.

[10] Hastings, Kevin J. *Introduction to Probability with Mathematica, 2nd ed.* CRC Press, Boca Raton, Florida, 2010.

[11] Hastings, Kevin J. *Introduction to Financial Mathematics*. CRC Press, Boca Raton, Florida, 2010.

[12] Hogg, Robert V, and Elliot A. Tanis. *Probability and Statistical Inference, 8th ed.* Prentice-Hall, Inc., Upper Saddle River, New Jersey, 2010.

[13] Longstaff, Francis A. and Eduardo S. Schwartz, "Valuing American Options by Simulation: A Simple Least Squares Approach", *The Review of Financial Studies*, Spring 2001, Vol. 14, No. 1, pp. 113-147.

[14] McDonald, Robert L. *Derivatives Markets, 3rd ed.* Pearson, Upper Saddle River, New Jersey, 2013.

[15] Oksendal, Bernt. *Stochastic Differential Equations: An Introduction with Applications, 5th ed.* Springer, Berlin, 2000.

[16] Pliska, Stanley R. *Introduction to Mathematical Finance.* Blackwell Publishers, Malden, Massachusetts, 1997.

[17] Roman, Steven. *Introduction to the Mathematics of Finance.* Springer, New York, 2004.

[18] Ross, Sheldon M. *An Elementary Introduction to Mathematical Finance, 2nd ed.* Cambridge University Press, Cambridge, United Kingdom, 2003.

[19] Ross, Sheldon M. *Simulation, 4th ed.* Elsevier/Academic Press, Burlington, Massachusetts, 2006.

[20] Ross, Sheldon M. *Stochastic Processes, 2nd ed.* John Wiley & Sons, Inc., New York, 1996.

[21] Sharpe, William F. *Portfolio Theory and Capital Markets.* McGraw-Hill, New York, 1970.

[22] Stampfli, Joseph and Victor Goodman. *The Mathematics of Finance.* Brooks/Cole, Pacific Grove, California, 2001.

[23] Wang, Hui. *Monte Carlo Simulation with Applications to Finance.* Chapman & Hall/CRC Press, Boca Raton, Florida, 2012.

[24] Williams, R.J. *Introduction to the Mathematics of Finance.* American Mathematical Society, Providence, Rhode Island, 2006.

Index